Zhang-Gradient Control

Yunong Zhang • Binbin Qiu • Xiaodong Li

Zhang-Gradient Control

Springer

Yunong Zhang
School of Data and Computer Science
Sun Yat-sen University
Guangzhou, Guangdong, China

Binbin Qiu
School of Data and Computer Science
Sun Yat-sen University
Guangzhou, Guangdong, China

Xiaodong Li
School of Intelligent Systems Engineering
Sun Yat-sen University
Guangzhou, Guangdong, China

ISBN 978-981-15-8259-2 ISBN 978-981-15-8257-8 (eBook)
https://doi.org/10.1007/978-981-15-8257-8

This Springer imprint is published by the registered company Springer Nature Singapore Pte Ltd.
The registered company address is: 152 Beach Road, #21-01/04 Gateway East, Singapore 189721, Singapore

To our parents and ancestors, as always

Preface

The tracking-control problems of nonlinear systems have been widely encountered in various applications, such as flight control, pendulum control, and robot control. For the purpose of tracking control, we need to design a controller in terms of control input for nonlinear systems such that the actual output can track the desired output. For solving the tracking-control problems of nonlinear systems, a number of methods have been presented and investigated, such as the input–output linearization (IOL) method, the optimal control method, and the backstepping method. However, most of the conventional control methods are relatively complex for their design procedures of controllers and practical implementations. Therefore, it is necessary and significant for practitioners to propose, develop, and investigate a simple and effective control method for the design of controllers.

From the viewpoint of time-varying (or say, dynamic) problem solving, the tracking control of nonlinear systems can be investigated in a unique manner. In recent years, a special class of neural dynamics has been exploited for the online solution of time-varying problems. As this neural-dynamic method is proposed by Zhang et al. and zeroes out each element of error function, it is called Zhang dynamics (also known as zeroing dynamics, ZD). Specifically, ZD is designed on the basis of an indefinite matrix-/vector-/scalar-valued error function (termed Zhang function, ZF) and takes full advantage of the time-derivative information of time-varying parameters. The ZD method is an error-based dynamic method, of which the core is the ZD design formula that forces each element of ZF to converge to zero exponentially. Such an idea can actually be found in the control field, i.e., forcing the error between the actual output and the desired output to be zero (or near zero in practice). Differing from the ZD, the conventional gradient dynamics (GD) is designed on the basis of a scalar-valued nonnegative error function (termed energy function, EF). The GD method is an energy-based minimization method, of which the core is the GD design formula such that the minimum point of the EF can be reached along the negative gradient direction. Besides, the GD method designed intrinsically for time-invariant (or say, static, constant) problem solving has been extended to solve time-varying problems. It is worth pointing out that such two methods both aim at forcing the error functions to be zero, which is essentially consistent with the objective of tracking control. However, in the previous studies, the ZD method and the GD method are generally exploited for problem solving individually and comparatively, and other researchers rarely consider combining them to utilize the advantage of each method as well as the superiority of their combination.

In this book, by effectively combining the ZD and GD methods together, a simple and effective controller-design method is developed and presented, which is termed Zhang-gradient (ZG) method. Accordingly, based on the ZG method, a special kind of controllers termed ZG controllers are designed, developed, and investigated for tracking control of various nonlinear systems (including linear systems as a special case), i.e., chaotic systems, integrator systems, pendulum systems, affine-form nonlinear (AFN) systems, as well as time-varying linear and nonlinear systems. In general, under the framework of the ZG method, a ZG controller obtained by adopting the ZD method m times and the GD method n times is called a $zmgn$ controller. Specifically, the $zmg0$ controllers are designed by adopting the ZD method m times and without using the GD method, which can be viewed as a special case of ZG controllers and thus often termed ZD controllers directly for comparisons with the ZG controllers using the GD method; besides, the $zmg1$ controllers are designed by adopting the ZD method m times and the GD method 1 time. It is worth pointing out that, in most cases, the ZG controllers refer to the $zmg1$ controllers, which can elegantly conquer the knotty division-by-zero (DBZ) problem. In traditional investigations, the DBZ problem is rarely considered and studied since it is a knotty problem for conventional controller design. In the conventional controller design, the divisor of a controller is simply assumed to be nonzero at any time instant, which often leads to contradictions between theoretical investigations and practical applications. Note that the DBZ problem has existed for thirteen centuries. However, past efforts have been spent on studying the problem under a time-invariant premise, i.e., studying the division operation with fixed operands at a certain time instant. By contrast, this book mainly focuses on investigating the DBZ problem from the perspective of temporal evolution instead of under a time-invariant premise. The simple and effective ZG method presented in this book is capable of designing the ZG controllers in a division-free manner. That is, the ZG controllers get rid of the potential possibility of encountering the DBZ problem and thus remain valid at the DBZ points encountered during the tracking-control process of nonlinear systems. Through the related theoretical analyses, the ZD and ZG controllers (more specifically, the $zmg0$ and $zmg1$ controllers under the framework of the ZG method) both possess the global and exponential convergence performance, which theoretically guarantee the efficacy of controllers. Computer simulations with various illustrative examples are further performed to substantiate the feasibility and efficacy of the presented ZD and ZG controllers (as well as the ZG method) for tracking control of various nonlinear systems. More importantly, the superiority of ZG controllers in conquering the DBZ problem is also illustrated by comparative simulation results. In brief, the main highlights of this book can be listed as follows.

(1) This book is the first book on the ZG method for controller design in connection with nonlinear/linear, time-varying/time-invariant, and multi-class or various systems.
(2) This book overcomes the challenges of control singularity and system collapse posed by the DBZ problem.
(3) This book provides detailed theoretical analyses, as well as abundant and comparative simulation results.

The idea for this book on neural dynamics and control was conceived during the classroom teaching as well as the research discussion in the laboratory and at international academic meetings. Most of the materials of this book are derived from the authors' papers published in journals and proceedings of the international conferences. In fact, since the early 1980s, the field of neural dynamics has undergone the phases of exponential growth, generating many new theoretical concepts and tools (including the authors' ones). At the same time, these theoretical results have been successfully applied to the solution of many practical problems. Our first priority is thus to cover each central topic in enough details to make the material clear and coherent; in other words, each part (and even each chapter) is written in a relatively self-contained manner.

In this book, Chap. 1 presents the introduction, concepts, and preliminaries, and the remainder contains 16 chapters that are classified into the following 5 parts:

- Part I: Chaotic Systems Using ZG Control (Chaps. 2–4);
- Part II: Integrator Systems Using ZG Control (Chaps. 5–7);
- Part III: Pendulum Systems Using ZG Control (Chaps. 8–10);
- Part IV: AFN Systems Using ZG Control (Chaps. 11–14);
- Part V: Time-Varying Systems Using ZG Control (Chaps. 15–17).

Chapter 2—In this chapter, we investigate the tracking-control problems of Lorenz, Chen, and Lu (also written as Lü) chaotic systems. By combining the ZD and GD methods together, a simple and effective controller-design method, termed ZG method, is presented for tracking control of the three chaotic systems. Both theoretical analyses and simulative verifications substantiate that the presented ZG controllers can achieve satisfactory tracking accuracy and successfully conquer the DBZ problem encountered during the tracking-control process.

Chapter 3—In this chapter, the ZG method is investigated for chaos synchronization with multiple inputs (i.e., three or two inputs). Based on the ZG method, the traditional three-input chaos synchronization problem can be successfully solved with desirable convergence rate and satisfactory accuracy. Besides, an important extension of the ZG method is investigated to solve the thorny two-input chaos synchronization problem. Simulation results illustrate that the controller groups designed by the ZG method not only achieve satisfactory synchronization accuracy and exponential convergence rate on the three-input chaos synchronization problem but also successfully solve the chaos synchronization problem with only two inputs.

Chapter 4—In this chapter, the ZG method is studied for solving the tracking-control problem of the modified Lorenz nonlinear system via additive input or mixed inputs (i.e., the mixture of additive and multiplicative inputs). Both theoretical analyses and simulative verifications validate that the ZG controllers with additive input or mixed inputs not only achieve satisfactory tracking accuracy but also successfully conquer the DBZ problem encountered during the tracking-control process.

Chapter 5—In this chapter, we apply the ZG method to the tracking control of Brockett integrator. Based on the ZG method, different types of controller groups are designed for Brockett integrator. Both theoretical analyses and simulative verifications indicate that the tracking errors are bounded and exponentially convergent. More importantly, comparative simulation results illustrate that the ZG controller group is superior to the ZD controller group in conquering the DBZ problem encountered during the tracking-control process.

Chapter 6—In this chapter, the ZG controllers for explicit and implicit tracking control of a double-integrator (DI) system are designed and presented. In addition, we conduct the corresponding computer simulations with different values of the design parameter λ used to illustrate the efficacy of ZG controllers. However, different settings of simulation options in MATLAB ordinary differential equation (ODE) solvers may lead to different simulation results (e.g., failure and success). The successful and failed simulation results are both presented to remind us to pay more attention to MATLAB defaults and options during conducting such simulations.

Chapter 7—In this chapter, the tracking-control problems of multiple-integrator (MI) systems are investigated by using the ZG method. Several types of ZD and ZG controllers are presented for tracking control of MI systems, e.g., triple-integrator (TI) systems. As an example, the design procedures of ZD and ZG controllers for TI systems with a linear output function (LOF) and a nonlinear output function (NOF) are presented. Corresponding theoretical analyses are given to guarantee the convergence performance of ZD and ZG controllers for TI systems. Computer simulations concerning the tracking control of MI systems with different types of output functions are further performed to substantiate the feasibility and efficacy of ZD and ZG controllers for tracking-control problem solving. Moreover, comparative simulation results for the tracking control of MI systems with NOFs substantiate that the ZG controllers can effectively conquer the DBZ problem.

Chapter 8—In this chapter, we firstly design ZD controllers for the explicit and implicit tracking control of a simple pendulum system. For achieving the DBZ-containing implicit tracking control, ZG controllers are further designed for conquering the DBZ problem. Computer simulations with an explicit tracking example and two implicit tracking examples are conducted. Comparative simulation results have substantiated the superiority of the ZG controllers for the DBZ-containing implicit tracking control of simple pendulum system.

Chapter 9—In this chapter, the cart path tracking control of an inverted-pendulum-on-a-cart (IPC) system is considered and investigated. Based on the ZG method, several types of ZD and ZG controllers are developed to achieve the tracking-control purpose. Besides, theoretical analyses are presented to guarantee the global and exponential convergence performance of both ZD and ZG controllers.

Computer simulations are further performed to illustrate the feasibility and efficacy of both ZD and ZG controllers. More importantly, comparative simulation results indicate that ZG controllers can effectively conquer the DBZ problem.

Chapter 10—In this chapter, two tracking controllers based on the ZG method are designed for the IPC system. Importantly, the presented ZG controller not only realizes the simultaneous control of pendulum swinging up and pendulum angle tracking but also conquers the DBZ problem elegantly without using any switching strategy. Besides, corresponding theoretical analyses on the convergence performance of both ZD and ZG controllers are provided. Computer simulations with three illustrative examples are further conducted to show the efficacy of both ZD and ZG controllers for the pendulum tracking control of the IPC system. In particular, comparative simulation results substantiate the superiority of the z2g1 controller for the control of pendulum tracking (including swinging up) of the IPC system in conquering the DBZ problem.

Chapter 11—In this chapter, we incorporate the GD into IOL, which leads to the GD-aided IOL method for conquering the DBZ problem encountered in the AFN system, with the proposition of the loose condition on relative degree. Corresponding theoretical analyses on tracking-error bound and convergence performance of the GD-aided IOL controller are provided. Moreover, comparative simulation results further substantiate that the GD-aided IOL controller is capable of fulfilling the tracking-control task with the DBZ problem conquered.

Chapter 12—In this chapter, a classic nonlinear system of Van der Pol oscillator in the affine-control form is investigated. By applying the ZG method, a ZG controller is designed for trajectory generation of the aforementioned nonlinear oscillator. Simulation results illustrate the feasibility and efficacy of the ZG controller with the DBZ problem conquered. In addition, the effects of ZD and GD design parameters on the performance of ZG controller are further studied.

Chapter 13—In this chapter, by following the ZG method, a ZD controller and a ZG controller are presented for tracking control of AFN system, which may encounter the DBZ problem. For comparison, the conventional IOL controller is also presented. The ZD, ZG, and IOL controllers are compared in different relative-degree cases, i.e., the standard relative-degree case, the pseudo-DBZ (PDBZ) relative-degree case, and the true-DBZ (TDBZ) relative-degree case. In addition, the theoretical analyses on ZD and ZG controllers are provided. Corresponding computer simulations are further performed to illustrate the tracking performance of the ZD, ZG, and IOL controllers, as well as to show the superiority of the ZG controller in conquering the TDBZ problem for tracking control of AFN system.

Chapter 14—In this chapter, according to the impact of DBZ points on the state variables of the controlled nonlinear system, the concepts of the PDBZ problem and the TDBZ problem are presented. Besides, the two classes of DBZ problems are solved under the framework of the ZG method. Specific examples are investigated to illustrate such two concepts and the efficacy of the ZG controllers in conquering PDBZ and TDBZ problems. The practical application to a two-wheeled mobile robot further substantiates the efficacy of the ZG method for tracking control of nonlinear system with physical meaning while conquering the TDBZ problem.

Chapter 15—In this chapter, the output tracking of time-varying linear (TVL) system is investigated. For solving such an output-tracking problem, three different types of controllers are presented, i.e., the conventional controller, ZD controller, and ZG controller. Simulation results with two illustrative examples show that such three types of controllers are feasible and effective for output-tracking problem solving. Especially, the presented ZG controller is capable of conquering the DBZ problem of TVL system.

Chapter 16—In this chapter, the stabilization of TVL system is investigated with PDBZ phenomenon shown. Based on the ZG method, a ZD stabilization controller and a ZG stabilization controller are designed. Simulation results indicate that the ZD stabilization controller is able to realize the stabilization of the TVL system in spite of the controller itself containing DBZ points, and that the ZG stabilization controller not only realizes the stabilization of the TVL system but also solves the PDBZ problem contained in the ZD stabilization controller.

Chapter 17—In this chapter, the ZG method is utilized to design ZD and ZG controllers for the output tracking of TVL and time-varying nonlinear (TVN) systems. Particularly, the investigated TVL and TVN systems may both have PDBZ phenomena. From the simulation results, although the presented ZD and ZG controllers fulfill well the output tracking of TVL and TVN systems, the infinite value of the former and the finite value of the latter at DBZ time instants indicate that the ZG controller is more effective in dealing with the PDBZ problem.

In summary, this book presents a simple and effective ZG method for solving the tracking-control problems of various nonlinear systems in the control field and further applies such a method to the tracking control of practical systems, e.g., IPC system and two-wheeled mobile robot (showing its application prospect). This book is written for undergraduate and postgraduate students as well as academic and industrial researchers studying in the developing fields of neural dynamics/neural networks, nonlinear control, computer mathematics, time-varying problem solving, modeling and simulation, analog hardware, and robotics. It provides a comprehensive view of the combined research of these fields, in addition to its accomplishments, potentials, and perspectives. We do hope that this book will generate curiosity and also happiness to its readers for learning more in the fields and the research, and that it will provide new challenges to seek new theoretical tools and practical applications.

At the end of this preface, it is worth pointing out that, in this book, a new and inspiring direction on the control method is provided for the design of controllers, together with the notorious DBZ problem conquered effectively, which has existed and has been investigated for more than 1300 years in academia and has stood in the tracking-control area of nonlinear systems for several decades (specifically, since the work of Alberto Isidori in 1985). This completely opens the door to the theoretical researches, simulative verifications, and practical/industrial applications of the DBZ-conquering ZG controllers designed by the ZG method, as the knotty DBZ problem has now been solved truly, systematically, and methodologically. It may promise to become a major inspiration for studies and researches in neural dynamics/neural networks, nonlinear control, computer mathematics, time-varying problem solving, modeling and simulation, analog hardware, and robotics. Without doubt, this book can be extended. Any comments or suggestions are welcome. The authors can be contacted via e-mails: zhynong@mail.sysu.edu.cn, qiubb6@mail.sysu.edu.cn, and lixd@mail.sysu.edu.cn. The web page of Yunong Zhang is available at http://sdcs.sysu.edu.cn/content/2477.

Guangzhou, China — Yunong Zhang
Guangzhou, China — Binbin Qiu
Guangzhou, China — Xiaodong Li
July 2020

Acknowledgements

This book is basically composed of many original research papers of the authors' research group, which have done a lot of meticulous and creative research work. Therefore, we are very grateful to our contributors for their high-quality work. During the process of preparing this book, we have the opportunity to discuss its various aspects and the results with many contributors and students. We highly appreciate their contributions, especially the great improvements in the presentation and quality of this book. We are very grateful for the valuable help and suggestions provided by Jinjin Guo, Min Yang, Jian Li, Yang Shi, Chaowei Hu, Dechao Chen, Huanchang Huang, Mengling Xiao, Huihui Gong, Zhiyuan Qi, Zhongxian Xue, Liu He, Shuo Yang, and so on.

The continuous aid by the National Natural Science Foundation of China (with number 61976230), the Project Supported by Guangdong Province Universities and Colleges Pearl River Scholar Funded Scheme (with number 2018), the China Postdoctoral Science Foundation (with number 2018M643306), the Guangdong Basic and Applied Basic Research Foundation (with number 2019A1515012128), the Key-Area Research and Development Program of Guangzhou (with number 202007030004), the Shenzhen Science and Technology Plan Project (with number JCYJ20170818154936083), and also the Fundamental Research Funds for the Central Universities (with number 19lgpy227) is gratefully acknowledged here.

Moreover, we would like to thank the editors (especially Editor Jasmine Dou) sincerely for their very important and constructive comments and suggestions provided, in addition to their time and effort spent in handling this book.

We are always very grateful to the nice people (especially the staff in Springer) for their strong support during the preparation and publishing of this book.

Contents

About the Authors

Yunong Zhang received his B.S. degree from Huazhong University of Science and Technology, Wuhan, China, in 1996, M.S. degree from South China University of Technology, Guangzhou, China, in 1999, and Ph.D. degree from Chinese University of Hong Kong, Shatin, Hong Kong, China, in 2003. He is currently a Professor at School of Data and Computer Science, Sun Yat-sen University, Guangzhou, China. Before joining Sun Yat-sen University in 2006, he had been with National University of Singapore, University of Strathclyde, and National University of Ireland at Maynooth, since 2003. His main research interests include system control, neural dynamics/neural networks, robotics, computation, and optimization. He has been working on the researches and applications of neural dynamics/neural networks for 20 years. He has now published totally 562 scientific works of various types with the number of SCI citations being 2486 and the number of Google citations being 5560. These include 13 monographs/books, 153 SCI papers (with 72 SCI papers published in recent 5 years), 47 IEEE Transactions/Magazine papers, and 10 single-authored works, crosswise. He was supported by the Program for New Century Excellent Talents in Universities in 2007, was presented the Best Paper Award of ISSCAA in 2008 and the Best Paper Award of ICAL in 2011, and was among the Highly Cited Scholars of China selected and published by Elsevier from 2014 to 2019.

Binbin Qiu received his B.S. degree from Jiangxi University of Science and Technology, Ganzhou, China, in 2013, and Ph.D. degree from Sun Yat-sen University, Guangzhou, China, in 2018. He is currently a Postdoctoral Fellow at School of Data and Computer Science, Sun Yat-sen University, Guangzhou, China. His main research interests include nonlinear systems, neural dynamics/neural networks, robotics, numerical computation, and optimization. He has authored/co-authored more than 60 scientific papers, including 23 SCI papers and 8 IEEE Transactions/Magazine papers, crosswise.

Xiaodong Li received his B.S. degree from Shaanxi Normal University, Xi'an, China, in 1987, M.S. degree from Nanjing University of Science and Technology, Nanjing, China, in 1990, and Ph.D. degree from City University of Hong Kong, Hong Kong, China, in 2007. He is currently a Professor at School of Intelligent Systems Engineering, Sun Yat-sen University, Guangzhou, China. His main research interests include intelligent control, 2D system theory, and artificial intelligence.

Acronyms

AFN	Affine-form nonlinear
DBZ	Division-by-zero
DI	Double-integrator
EF	Energy function
GD	Gradient dynamics
IOL	Input–output linearization
IPC	Inverted-pendulum-on-a-cart
LOF	Linear output function
MI	Multiple-integrator
MIMO	Multiple-input multiple-output
NOF	Nonlinear output function
ODE	Ordinary differential equation
PDBZ	Pseudo-DBZ
TDBZ	True-DBZ
TI	Triple-integrator
TVL	Time-varying linear
TVN	Time-varying nonlinear
UBIBS	Uniformly bounded-input bounded-state
ZD	Zhang dynamics
ZF	Zhang function
ZG	Zhang-gradient
ZNN	Zhang neural network

List of Figures

List of Tables

Chapter 1
Introduction, Concepts and Preliminaries

Abstract In this chapter, we firstly discuss the importance for the research of the division-by-zero (DBZ) problem, and briefly introduce the Zhang-gradient (ZG) method investigated in this book. Then, we explain the essential concepts of several terms frequently used in this book, together with an illustrative example provided. Moreover, we introduce the basic principle of the ZG method for control as well as the general design procedure of a DBZ-conquering ZG controller systematically and methodologically. Finally, for comparative purposes, we present the differences between the ZG method and other types of control methods in detail.

1.1 Introduction

For decades, the tracking-control problems of nonlinear systems have been widely investigated since it is frequently encountered in practical applications [1–3]. For solving such problems, a number of methods have been presented and studied. For example, the input-output linearization (IOL) method is an effective method for solving the tracking-control problems of nonlinear systems [1, 4]. However, due to the existence of the division-by-zero (DBZ) problem, for which the nonlinear systems cannot have a well-defined relative degree, the IOL method fails to solve the corresponding tracking-control problems. It is worth pointing out that other conventional methods also often bring in the division operation, which leads to the potential possibility of generating DBZ points. If the nonlinear systems investigated encounter the DBZ problem, the controllers designed by the conventional methods are generally difficult to be implemented. Evidently, the DBZ problem is a challenging subject in the control field [2, 3, 5, 6].

Differing from the conventional controller-design methods, the simple and effective Zhang-gradient (ZG) method is presented, analyzed and simulated in this book, which combines the Zhang dynamics (ZD) method and the gradient dynamics (GD) method as a whole, and utilizes the advantage of each method as well as the superiority of their combination for the tracking-control problem solving. Based on the ZG method, the controllers can be designed in a division-free manner, termed

Y. Zhang et al., *Zhang-Gradient Control*,
https://doi.org/10.1007/978-981-15-8257-8_1

DBZ-conquering ZG controllers, which can effectively solve the tracking-control problems of various nonlinear systems and successfully conquer the knotty DBZ problem encountered during the tracking-control process.

1.2 Concepts

In this section, for readers' convenience, the essential concepts of several terms frequently used in this book are given firstly. Then, via an illustrative example, the procedures of using the ZD method and the GD method for time-varying problem solving are presented separately and comparatively.

1.2.1 Concept of ZF

The Zhang function (or termed, zeroing function, ZF), which is the design basis of ZD, differs from the usual error functions defined in the conventional methods. Specifically, compared with the energy function (EF), the ZF (1) is indefinite (i.e., can be positive, zero, or negative, in addition to being bounded, unbounded, or even lower unbounded), (2) can be matrix-valued, vector-valued or scalar-valued (when solving a time-varying matrix-valued, vector-valued or scalar-valued problem), and (3) can be real-valued or complex-valued (corresponding to a real-valued or complex-valued time-varying problem solving), so as to fully monitor and control the process of time-varying problem solving [7–9].

1.2.2 Concept of EF

The energy function (EF), which is usually associated with GD, is a norm-based or square-based scalar-valued nonnegative or at least lower-bounded function (i.e., a form of error function) [7, 10].

1.2.3 Concept of ZD

For better understanding, it is necessary to firstly introduce the concept of Zhang neural network (or termed, zeroing neural network, ZNN). The ZNN, originating and extending from the research of Hopfield neural network [11], is a special class of recurrent neural networks proposed by Zhang et al. since March 2001 [12]. (1) It is proposed as a systematic approach for real-time solution of various time-varying problems; (2) it is based on the elimination of every element of a

ZF; (3) it is depicted generally in implicit dynamics (with some exceptions in explicit dynamics), coinciding better with the systems in practice and in nature; (4) it belongs to a predictive approach that utilizes the time-derivative information and can thus accurately and efficiently solve various time-varying problems [13].

The Zhang dynamics (or termed, zeroing dynamics, ZD) has been generalized from the ZNN since December 2008 [14, 15], of which the state dimension can be multiple or one. It is viewed as a systematic approach to solving various time-varying problems with the scalar situation included. It differs from the conventional GD in terms of the problem to be solved, error function, design formula, dynamic equation, and the utilization of time-derivative information [8, 9]. Besides, the simplest ZD design formula can be formulated as $\dot{z}(t) = -\lambda z(t)$, with the ZD design parameter $\lambda > 0$, where $z(t)$ can be a scalar-valued, vector-valued or matrix-valued indefinite error function (i.e., ZF) [8].

1.2.4 Concept of GD

The gradient dynamics (GD), which is also referred to as gradient-based dynamics, has been proposed and known for decades. It corresponds to gradient neural network [16, 17] and Hopfield neural network with the state dimension being multiple or one. It is viewed as a conventional method for solving time-invariant (or termed, static, constant) problems. For a specific problem solving, it starts with the definition of a norm-based or square-based scalar-valued nonnegative or at least lower-bounded error function (i.e., EF) ϵ. Then, the GD design formula can be formulated as $\dot{x}(t) = -\gamma \partial\epsilon/\partial x$, with the GD design parameter $\gamma > 0$, where the unknown variable $x(t)$ to be obtained can be scalar-valued, vector-valued or matrix-valued [10, 17, 18].

1.2.5 Concept of ZG Control

The Zhang-gradient (or termed, zeroing-gradient, ZG) control proposed by Zhang et al. since December 2012 [19–21] is the combination of ZD and GD, which is a simple and effective controller-design method for tracking control of various nonlinear systems, together with the notorious DBZ problem conquered effectively. Specifically, the ZG control only needs to repeatedly construct a series of ZFs and apply the simple ZD design formula until the expansion of the ZD design formula includes the explicit expression of the control input $u(t)$, and then a ZD controller in the form of $u(t)$ is obtained without using GD; to obtain a DBZ-conquering ZG controller, we need to further define an EF and employ the GD design formula to construct a ZG controller in the form of $\dot{u}(t)$ [21].

1.2.6 Illustrative Example

To illustrate the ZD and GD methods for time-varying problem solving, this subsection presents the procedures of using the ZD method and the GD method separately and comparatively to solve the time-varying linear equation system.

Specifically, the following problem formulation of time-varying linear equation system is considered:

$$A(t)\mathbf{x}(t) = \mathbf{b}(t), \quad \text{with } t \in [0, +\infty), \tag{1.1}$$

where coefficient matrix $A(t) \in \mathbb{R}^{n\times n}$ and vector $\mathbf{b}(t) \in \mathbb{R}^n$ are smoothly time-varying, and $\mathbf{x}(t) \in \mathbb{R}^n$ is the unknown vector to be obtained in real time t.

The design procedure of the ZD model for solving the time-varying linear equation system (1.1) is firstly presented as follows.

Step 1 To monitor and control the solving process, the following ZF is defined:

$$\mathbf{z}(t) = A(t)\mathbf{x}(t) - \mathbf{b}(t). \tag{1.2}$$

Note that the vector-valued ZF (1.2) can be positive, zero, negative, bounded or unbounded (even including lower unbounded).

Step 2 The time derivative of $\mathbf{z}(t)$, i.e., $\dot{\mathbf{z}}(t)$, can be constructed such that every element of ZF (1.2) converges to zero. Here, we employ the following ZD design formula:

$$\dot{\mathbf{z}}(t) = \frac{\mathrm{d}\mathbf{z}(t)}{\mathrm{d}t} = -\lambda\mathbf{z}(t), \tag{1.3}$$

where the design parameter $\lambda \in \mathbb{R}^+$ is used to scale the convergence rate of the ZD solution.

Step 3 Expanding the ZD design formula (1.3) yields the following ZD model for solving (1.1):

$$A(t)\dot{\mathbf{x}}(t) = -(\dot{A}(t) + \lambda A(t))\mathbf{x}(t) + (\dot{\mathbf{b}}(t) + \lambda\mathbf{b}(t)), \tag{1.4}$$

where $\mathbf{x}(t)$, starting with a randomly generated initial state $\mathbf{x}(0) \in \mathbb{R}^n$, is the state vector of the ZD model corresponding to time-varying theoretical solution $\mathbf{x}^*(t)$ of (1.1). For ZD model (1.4), we have detailed theoretical results on its global and exponential convergence [8, 13].

Then, the design procedure of the GD model for solving the time-varying linear equation system (1.1) is presented as follows.

Step 1 To monitor and control the solving process, the following EF is defined:

$$\epsilon = \frac{1}{2}\|A(t)\mathbf{x}(t) - \mathbf{b}(t)\|_2^2, \tag{1.5}$$

where $\|\cdot\|_2$ denotes the two-norm of a vector. It is worth pointing out that the minimum point of EF (1.5) is achieved with $\epsilon = 0$, if and only if $\mathbf{x}(t)$ is the time-varying theoretical solution $\mathbf{x}^*(t)$ of (1.1).

Step 2 A computational scheme can be designed to evolve along the negative gradient direction of EF (1.5). Here, we adopt the following GD design formula:

$$\dot{\mathbf{x}}(t) = -\gamma\frac{\partial\epsilon}{\partial\mathbf{x}}, \tag{1.6}$$

where the design parameter $\gamma \in \mathbb{R}^+$ is used to scale the convergence rate of the GD solution.

Step 3 Expanding the GD design formula (1.6) yields the following GD model for solving (1.1):

$$\dot{\mathbf{x}}(t) = -\gamma A^{\mathrm{T}}(t)(A(t)\mathbf{x}(t) - \mathbf{b}(t)),$$

where the superscript “T” denotes the transpose operator, and $\mathbf{x}(t)$ is the state vector of the GD model corresponding to the time-varying theoretical solution $\mathbf{x}^*(t)$ of (1.1). For such a GD model or similar ones, we have theoretical results on convergence and error bound [10, 16, 22].

1.3 Preliminaries

In this section, we firstly introduce the basic principle of the ZG method for control as well as the general design procedure of a DBZ-conquering ZG controller systematically and methodologically. Afterwards, for readers' convenience and also for comparative purposes, the differences between the ZG method and other types of control methods, i.e., the optimal control method, the backstepping method and the IOL method, are presented in detail.

1.3.1 Principle of ZG Method for Control

As presented above, the previous studies generally exploit the ZD method and the GD method individually and comparatively, and other researchers rarely consider combining them for problem solving or discover the superiority of their combination. In this book, by combining the ZD method and the GD method together, the ZG method is developed and presented to solve the tracking-control problems of various nonlinear systems in a division-free manner, which can effectively conquer the DBZ problem. Note that the ZG method reduces to the ZD method when the GD method is not utilized.

For illustrating the basic principle of the ZG method for control as well as the general design procedure of a DBZ-conquering ZG controller, the multiple-input multiple-output (MIMO) nonlinear system is taken as an illustrative example. Let us consider the following MIMO nonlinear system [21, 23]:

$$\begin{cases} \dot{\mathbf{x}}(t) = \mathbf{f}(\mathbf{x}(t), \mathbf{u}(t)), \\ \mathbf{y}(t) = \mathbf{y}(\mathbf{x}(t), \mathbf{u}(t)), \end{cases}$$

where state vector $\mathbf{x}(t) = [x_1(t), x_2(t), \cdots, x_n(t)]^\mathrm{T} \in \mathbb{R}^n$, control input vector $\mathbf{u}(t) = [u_1(t), u_2(t), \cdots, u_m(t)]^\mathrm{T} \in \mathbb{R}^m$, and output vector $\mathbf{y}(t) = [y_1(t), y_2(t), \cdots, y_m(t)]^\mathrm{T} \in \mathbb{R}^m$. Besides, $\mathbf{f}(\mathbf{x}(t), \mathbf{u}(t)) = [f_1(\mathbf{x}(t), \mathbf{u}(t)), f_2(\mathbf{x}(t), \mathbf{u}(t)), \cdots, f_n(\mathbf{x}(t), \mathbf{u}(t))]^\mathrm{T} \in \mathbb{R}^n$ and $\mathbf{y}(\mathbf{x}(t), \mathbf{u}(t)) = [y_1(\mathbf{x}(t), \mathbf{u}(t)), y_2(\mathbf{x}(t), \mathbf{u}(t)), \cdots, y_m(\mathbf{x}(t), \mathbf{u}(t))]^\mathrm{T} \in \mathbb{R}^m$ are the continuous and smooth function vectors, and $\mathbf{y}_\mathrm{d}(t) = [y_{\mathrm{d}1}(t), y_{\mathrm{d}2}(t), \cdots, y_{\mathrm{d}m}(t)]^\mathrm{T} \in \mathbb{R}^m$ is the given smooth and bounded desired output vector. The objective is to design a ZG controller for the above system such that the output vector $\mathbf{y}(t)$ can track the desired output vector $\mathbf{y}_\mathrm{d}(t)$ and the corresponding tracking error vector $\mathbf{e}(t) = \mathbf{y}(t) - \mathbf{y}_\mathrm{d}(t)$ asymptotically converges to zero (or near zero in practice). To design the ZG controller for the tracking control of the MIMO nonlinear system, the general framework of the ZG method for the controller design is presented as follows.

Step 1 We repeatedly construct a series of ZFs with regard to $y_1(t)$ and $y_{\mathrm{d}1}(t)$, and apply the ZD design formula, until the expansion of ZF includes the explicit expression of $\mathbf{u}(t)$ (specifically, including a component of $\mathbf{u}(t)$), such as the κ_1th ZF $z_{1\kappa_1}(\mathbf{x}(t), \mathbf{u}(t))$ with κ_1 being a positive integer. By this time, the first step of the ZG method for the controller design is finished. Therefore, in the first step, we can define the following ZF vector (which is actually a scalar in this case):

$$\mathbf{z}_1(\mathbf{x}(t), \mathbf{u}(t)) = z_{1\kappa_1}(\mathbf{x}(t), \mathbf{u}(t)) \in \mathbb{R}.$$

Note that, if the explicit expression of $\mathbf{u}(t)$ is still not available, no matter how many ZFs are constructed (specifically, n ZFs), such a ZG controller cannot be designed for the MIMO nonlinear system to track the desired output $y_{\mathrm{d}1}(t)$. Then, proceed to the second step below.

Step 2 We repeatedly construct a series of ZFs with regard to $y_2(t)$ and $y_{\mathrm{d}2}(t)$, and apply the ZD design formula, until the expansion of ZF includes the explicit expression of $\mathbf{u}(t)$, such as the κ_2th ZF

$z_{2\kappa_2}(\mathbf{x}(t), \mathbf{u}(t))$ with κ_2 being a positive integer. Therefore, in the second step, we can define the following ZF vector:

$$\mathbf{z}_2(\mathbf{x}(t), \mathbf{u}(t)) = [z_{1\kappa_1}(\mathbf{x}(t), \mathbf{u}(t)), z_{2\kappa_2}(\mathbf{x}(t), \mathbf{u}(t))]^{\mathrm{T}} \in \mathbb{R}^2.$$

If the explicit expression of $\mathbf{u}(t)$ is still not available, no matter how many ZFs are constructed (specifically, n ZFs), such a ZG controller cannot be designed for the MIMO nonlinear system to track the desired output $y_{\mathrm{d}2}(t)$. Then, proceed to the third step below.

Step 3 Similar to the above two steps, we repeatedly construct a series of ZFs with regard to $y_3(t)$ and $y_{\mathrm{d}3}(t)$, and apply the ZD design formula, until the expansion of ZF includes the explicit expression of $\mathbf{u}(t)$, such as the κ_3th ZF $z_{3\kappa_3}(\mathbf{x}(t), \mathbf{u}(t))$ with κ_3 being a positive integer. Therefore, in the third step, we can define the following ZF vector:

$$\mathbf{z}_3(\mathbf{x}(t), \mathbf{u}(t)) = [z_{1\kappa_1}(\mathbf{x}(t), \mathbf{u}(t)), z_{2\kappa_2}(\mathbf{x}(t), \mathbf{u}(t)), z_{3\kappa_3}(\mathbf{x}(t), \mathbf{u}(t))]^{\mathrm{T}} \in \mathbb{R}^3.$$

Similarly, if the explicit expression of $\mathbf{u}(t)$ is still not available, no matter how many ZFs are constructed (specifically, n ZFs), such a ZG controller cannot be designed for the MIMO nonlinear system to track the desired output $y_{\mathrm{d}3}(t)$ ...

Step m Similar to the above design procedures, the κ_mth ZF $z_{m\kappa_m}(\mathbf{x}(t), \mathbf{u}(t))$ with regard to $y_m(t)$ and $y_{\mathrm{d}m}(t)$ can be obtained, and then the following ZF vector in the mth step can be defined:

$$\begin{aligned}\mathbf{z}_m(\mathbf{x}(t), \mathbf{u}(t)) =& [z_{1\kappa_1}(\mathbf{x}(t), \mathbf{u}(t)), z_{2\kappa_2}(\mathbf{x}(t), \mathbf{u}(t)), z_{3\kappa_3}(\mathbf{x}(t), \mathbf{u}(t)), \cdots, \\ & z_{m\kappa_m}(\mathbf{x}(t), \mathbf{u}(t))]^{\mathrm{T}} \in \mathbb{R}^m.\end{aligned}$$

Likewise, if the explicit expression of $\mathbf{u}(t)$ is still not available, no matter how many ZFs are constructed (specifically, n ZFs), such a ZG controller cannot be designed for the MIMO nonlinear system to track the desired output $y_{\mathrm{d}m}(t)$. Then, proceed to the last step below.

Step m + 1 We firstly define a scalar-valued EF as $\epsilon = \|\mathbf{z}_m(\mathbf{x}(t), \mathbf{u}(t))\|_2^2/2$. Then, the GD design formula is adopted as $\dot{\mathbf{u}}(t) = -\gamma \partial\epsilon/\partial\mathbf{u}$. Finally, by substituting the EF ϵ into the GD design formula, a ZG controller in the form of $\dot{\mathbf{u}}(t)$ can be obtained for the tracking control of the MIMO nonlinear system.

It is worth particularly pointing out that a ZD controller in the form of $\mathbf{u}(t)$ can be obtained without using the last step of the ZG method, and that the above ZG controller in the form of $\dot{\mathbf{u}}(t)$ can effectively conquer the knotty DBZ problem.

For better illustration and understanding, as shown in Fig. 1.1, we also present the flowchart of controller design using the ZG method for the tracking control of the MIMO nonlinear system.

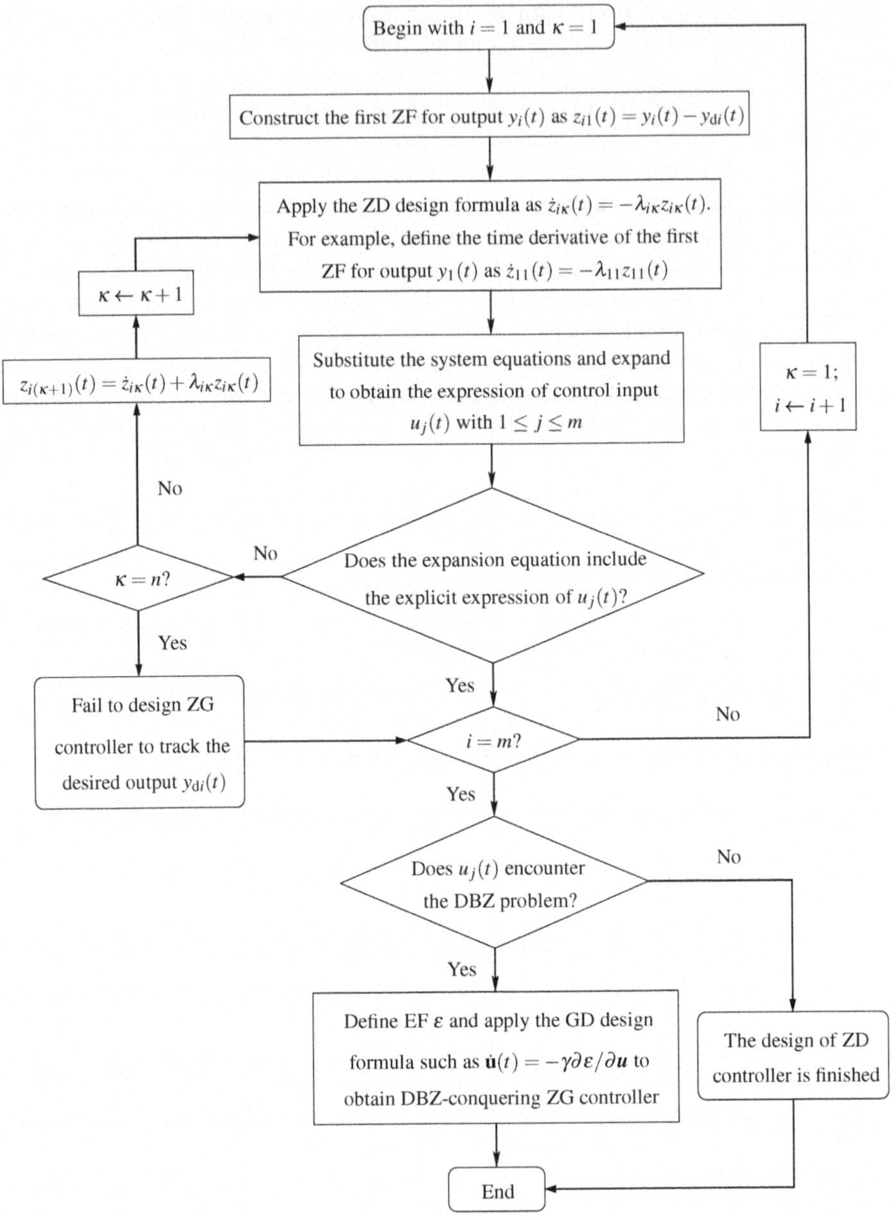

Fig. 1.1 Flowchart of controller design using ZG method for tracking control of MIMO nonlinear system

1.3.2 Comparison with Other Methods

For better understanding and comparative purposes, the differences between the ZG method and the backstepping method as well as the IOL method are presented as below [1, 4, 10, 20, 24, 25].

(1) To lay a basis for further comparison with the ZG method, the main design steps in the backstepping method are outlined as follows. To design a backstepping controller, the whole system is firstly divided into several subsystems. Secondly, an error or regulatory variable is defined, and a Lyapunov function is designed accordingly. Thirdly, a virtual control law is chosen to make the derivative of the Lyapunov function negative definite. A stabilizing function, which equals the virtual control, is then found to stabilize the subsystem. The above design procedure is repeated till the last subsystem. Besides, the actual control law can be designed in the same manner. Evidently, the design of control law is a recursive process in backstepping control [25].
(2) In the design of linear systems, it can be found that the ZG method and the backstepping method have a certain degree of connection. However, in the design of nonlinear systems, the distinctions between such two methods are more than their connections. For instance, in the design of the ZG method, the ZG controller is constructed with only one negligible restriction (i.e., with design parameters larger than zero). However, the Lyapunov function has to be introduced in each step in the design of the backstepping method, and the derivation process can be much more complex than that in the ZG method. Besides, there are many systematic/parametric requirements and limitations in the backstepping control.
(3) The usage of the ZG method is simple and effective, which is reflected in the following two aspects: (1) it does not need to introduce any Lyapunov function during the design process; (2) it does not need to define any virtual control. In contrast, with the increase of design steps, the backstepping method becomes more and more complicated, due to the Lyapunov functions introduced in each design step, their time-derivative derivation, and the concept of virtual control.
(4) A strict definition of relative degree is unnecessary for the ZG method, and even systems without the definition of relative degree can be addressed by it. Nevertheless, in the backstepping method or the IOL method, the system of interest relates to a certain relative degree well defined, and then the system can be handled by these methods.
(5) In the design of control systems, the ZG method can conquer the DBZ problem effectively. In contrast, the backstepping method and the IOL method may introduce but cannot address the DBZ problem.

For readers' convenience and also for comparison, the differences between the ZG method and the optimal control method are presented as below [21, 26–28].

(1) The optimal control method is related to finding a control strategy that drives a dynamic system to a desired solution in an optimal manner; e.g., finding an

optimal controller such that the actual output can track the desired output, and that a predefined performance index (or cost function) with an integral form on the whole time interval is minimized. In contrast, the ZG method is the combination of the ZD and GD methods. Specifically, the ZD method is an error-based dynamic method, of which the core is the ZD design formula that makes the instantaneous ZF converge to zero exponentially; the GD method is an energy-based minimization method, of which the core is the GD design formula such that the minimum point of the EF can be reached along the negative gradient direction.

(2) The design of the optimal control method, which is though mathematically elegant, is usually obtained offline and requires the complete knowledge of system dynamics to be known on the whole time interval, e.g., $[0, +\infty)$. That is, the computation involved in the optimal control method includes not only the present and previous data information but also the future one. However, it is usually impossible that the future data information is known at present time instant in reality. As a consequence, the system may be hard to work accurately in an optimal manner at present time instant. Besides, in view of the uncertainties in system dynamics, the unknown system parameters need to be updated/estimated online by using the tracking error, thereby making the application of optimal control to adaptation potentially less satisfactory. In contrast, the design of the ZG method is just based on the present (or previous) data information, which may thus be more practical and applicable in the control field.

(3) To design an optimal controller, the performance functional and constraint conditions are firstly selected and determined. Secondly, the Hamilton function of system is constructed. Thirdly, according to the necessary conditions of obtaining the functional extremum, we can obtain several equations, e.g., the governing equation and the canonical equation. Fourthly, solve these equations and then determine the integral constants by using the boundary conditions. Finally, calculate the optimal control and its optimal trajectory. In contrast, to design a ZG controller, the ZG method only needs to repeatedly construct a series of ZFs and apply the simple ZD design formula until the expansion of the ZD design formula includes the explicit expression of the control input $u(t)$. Then, a ZD controller in the form of $u(t)$ can be obtained without using the GD method. To obtain a DBZ-conquering ZG controller, we can further define an EF and employ the GD design formula to construct a ZG controller in the form of $\dot{u}(t)$. Evidently, the design procedure of an optimal controller is generally more complex than that of a ZG controller.

(4) As a special case of the optimal control, the bang-bang control has been studied extensively, of which the main characteristic is that each component of the control input vector is selected as one of the boundary values in control domain and then switched between these values. However, the bang-bang control might possibly cause an undesired oscillation problem, as the optimal control is frequently switched from a boundary value to another one (though it may have better effectiveness). Evidently, the optimal control input generated is not

smooth. In contrast, with the time-derivative information fully exploited, the control result generated by the ZG method is usually smooth. Thus, the ZG method can effectively avoid the above problem.

(5) In the linear quadratic optimal control system, the design procedure of the optimal control method involves the Riccati matrix differential equation solving, where the computationally-expensive matrix inversion is required and may result in the control singularity problem. Moreover, the Riccati matrix differential equation is a type of nonlinear differential equation, which is generally difficult to obtain an analytical solution or is relatively complex to obtain a numerical solution, especially for complicated high-dimensional systems. Though the Riccati equation for the linear quadratic optimal control could be solved before the control performing, its solution is obtained offline, which may be less desirable for real-time control systems in practice. In contrast, the above problems do not exist in the design and implementation of the ZG method.

1.4 Chapter Summary

In this chapter, the importance for the research of the DBZ problem has been discussed, and the ZG method investigated in this book has been briefly introduced. Besides, the essential concepts of several terms frequently used in this book have been explained. For better understanding, the basic principle of the ZG method as well as the general design procedure of a DBZ-conquering ZG controller has been introduced systematically and methodologically. At last, for comparative purposes, the differences between the ZG method and other types of control methods have been presented detailedly.

References

1. Isidori A (1985) Nonlinear control systems: an introduction. Springer, New York
2. Hauser J, Sastry S, Kokotovic P (1992) Nonlinear control via approximate input-output linearization: the ball and beam example. IEEE Trans Autom Control 37(3):392–398
3. Guardabassi GO, Savaresi SM (2001) Approximate linearization via feedback—an overview. Automatica 37(1):1–15
4. Slotine JE, Li W (1991) Applied nonlinear control. Prentice Hall, New Jersey
5. Tomlin CJ, Sastry SS (1997) Switching through singularities. Syst Control Lett 35(3):145–154
6. Zhang Y, Yu X, Yin Y, Xiao L, Fan Z (2013) Using GD to conquer the singularity problem of conventional controller for output tracking of nonlinear system of a class. Phys Lett A 377(25–27):1611–1614
7. Zhang Y, Li Z, Guo D, Li W, Chen P (2013) Z-type and G-type models for time-varying inverse square root (TVISR) solving. Soft Comput 17(11):2021–2032
8. Zhang Y, Guo D (2015) Zhang functions and various models. Springer, Berlin

9. Zhang Y, Li M, Yin Y, Jin L, Yu X (2013) Controller design of nonlinear system for fully trackable and partially trackable paths by combining ZD and GD. In: Proceedings of the 25th Chinese control and decision conference, pp 209–214
10. Zhang Y, Xiao L, Xiao Z, Mao M (2015) Zeroing dynamics, gradient dynamics, and Newton iterations. CRC Press, Florida
11. Hopfield JJ (1982) Neural networks and physical systems with emergent collective computational abilities. Proc Natl Acad Sci USA 79(8):2554–2558
12. Zhang Y, Jiang D, Wang J (2002) A recurrent neural network for solving Sylvester equation with time-varying coefficients. IEEE Trans Neural Netw 13(5):1053–1063
13. Zhang Y, Yi C (2011) Zhang neural networks and neural-dynamic method. Nova Science Publishers, New York
14. Chen K, Zhang L, Zhang Y (2008) Cyclic motion generation of multi-link planar robot performing square end-effector trajectory analyzed via gradient-descent and Zhang et al.'s neural-dynamic methods. In: Proceedings of the 2nd international symposium on systems and control in aerospace and astronautics, pp 1–6
15. Zhang Y, Yi C, Ma W (2008) Comparison on gradient-based neural dynamics and Zhang neural dynamics for online solution of nonlinear equations. In: Proceedings of the 3rd international symposium on intelligence computation and applications, pp 269–279
16. Zhang Y, Chen K, Tan H-Z (2009) Performance analysis of gradient neural network exploited for online time-varying matrix inversion. IEEE Trans Autom Control 54(8):1940–1945
17. Xiao L, Zhang Y (2011) Zhang neural network versus gradient neural network for solving time-varying linear inequalities. IEEE Trans Neural Netw 22(10):1676–1684
18. Jin L, Zhang Y (2016) Continuous and discrete Zhang dynamics for real-time varying nonlinear optimization. Numer Algorithms 73(1):115–140
19. Zhang Y, Liu J, Yin Y, Guo D, Luo F (2012) Zhang-gradient tracking controllers of Z1G0 and Z1G1 types for time-invariant linear systems, In: Proceedings of the 2nd international conference on computer science and network technology, pp 146–150
20. Jin L, Zhang Y, Qiao T, Tan M, Zhang Y (2016) Tracking control of modified Lorenz nonlinear system using ZG neural dynamics with additive input or mixed inputs. Neurocomputing 196:82–94
21. Zhang Y, Qiu B, Liao B, Yang Z (2017) Control of pendulum tracking (including swinging up) of IPC system using zeroing-gradient method. Nonlinear Dyn 89(1):1–25
22. Zhang Y, Yang Y, Ruan G (2011) Performance analysis of gradient neural network exploited for online time-varying quadratic minimization and equality-constrained quadratic programming. Neurocomputing 74(10):1710–1719
23. Ren J, Qiu J, Wang Y, Yin Y, Zhang Y (2014) Framework and outlook of ZG control method for multiple-input and multiple-output nonlinear systems. J Dalian Univ 35(6):27–29
24. Liu X, Lin Z (2012) On the backstepping design procedure for multiple input nonlinear systems. Int J Robust Nonlinear Control 22(8):918–932
25. Adhikary N, Mahanta C (2013) Integral backstepping sliding mode control for underactuated systems: swing-up and stabilization of the cart-pendulum system. ISA Trans 52(6):870–880
26. Dusek F, Honc D, Sharma KR, Havlicek L (2016) Inverted pendulum optimal control based on first principle model. Adv Intell Sys Comput 466:63–74
27. Zhang Y, Chen D, Jin L, Zhang Y, Yin Y (2016) GD-aided IOL (input-output linearisation) controller for handling affine-form nonlinear system with loose condition on relative degree. Int J Control 89(4):757–769
28. Lewis FL, Vrabie DL, Syrmos VL (2012) Optimal control, 3rd edn. Wiley, New Jersey

Part I
Chaotic Systems Using ZG Control

Chapter 2
ZG Tracking Control of a Class of Chaotic Systems

Abstract This chapter investigates the tracking-control problems of Lorenz, Chen and Lu chaotic systems. Note that the IOL method cannot solve these tracking-control problems because of the existence of DBZ points, at which such chaotic systems fail to have a well-defined relative degree. By combining the ZD and GD methods together, a simple and effective controller-design method, termed ZG method, is presented for tracking control of the three chaotic systems. This ZG method, with DBZ points conquered, is capable of solving the tracking-control problems of the chaotic systems. Both theoretical analyses and simulative verifications substantiate that the tracking controllers based on the ZG method can achieve satisfactory tracking accuracy and successfully conquer the DBZ problem encountered during the tracking-control process.

2.1 Introduction

The study of chaos has attracted a lot of attention over the past decades [1–4]. In 1963, the first chaotic attractor was found by Lorenz [1]:

$$\begin{cases} \dot{x}_1(t) = f_1(x_1(t), x_2(t)) = a(x_2(t) - x_1(t)), \\ \dot{x}_2(t) = f_2(x_1(t), x_2(t), x_3(t)) = cx_1(t) - x_1(t)x_3(t) - x_2(t), \\ \dot{x}_3(t) = f_3(x_1(t), x_2(t), x_3(t)) = x_1(t)x_2(t) - bx_3(t), \end{cases}$$

where $x_1(t)$, $x_2(t)$ and $x_3(t)$ are the system states with constants $a = 10$, $b = 8/3$ and $c = 28$. Notably, in the rest of this book, the argument t is frequently omitted for presentation convenience, e.g., by writing $x_1(t)$ as x_1. The Lorenz chaotic system is classified into a generalized Lorenz system family [5]. Then, the Chen chaotic

Y. Zhang et al., *Zhang-Gradient Control*,
https://doi.org/10.1007/978-981-15-8257-8_2

system was discovered in 1999 [2] as

$$\begin{cases} \dot{x}_1 = f_1(x_1, x_2) = a(x_2 - x_1), \\ \dot{x}_2 = f_2(x_1, x_2, x_3) = (c - a)x_1 - x_1x_3 + cx_2, \\ \dot{x}_3 = f_3(x_1, x_2, x_3) = x_1x_2 - bx_3, \end{cases}$$

where $a = 35$, $b = 3$ and $c = 28$. This chaotic system is classified into another canonical family of chaotic systems [6].

Although the Lorenz and Chen chaotic systems are classified into different families of chaotic systems, in 2002, literature [3] investigated the connection between them. Specifically, the Lu chaotic system was found in [3], which bridges the gap between the Lorenz system and Chen system in both mathematical form and dynamic behavior. Mathematically, the Lu chaotic system can be described by the following equations:

$$\begin{cases} \dot{x}_1 = f_1(x_1, x_2) = a(x_2 - x_1), \\ \dot{x}_2 = f_2(x_1, x_2, x_3) = -x_1x_3 + cx_2, \\ \dot{x}_3 = f_3(x_1, x_2, x_3) = x_1x_2 - bx_3, \end{cases} \tag{2.1}$$

which displays a chaotic attractor when $a = 36$, $b = 3$ and $c = 20$ [3, 4]. Moreover, when $a = 36$ and $b = 3$ are fixed while c varies, the dynamic behavior of the Lu chaotic system is similar to that of the Lorenz system with $12.7 < c < 17.0$ and similar to that of the Chen system with $23.0 < c < 28.5$ [3].

As for the investigations of chaotic systems, two main directions are frequently considered. The first direction is to utilize and generate chaotic behaviors. On the one hand, the chaotic behaviors are utilized for practical applications, such as true random number generators [7] as well as encryption and secure communication [8]. On the other hand, chaotic systems with more chaotic behaviors are generated, such as hyperchaotic systems [9]. The other direction is to get rid of chaotic behaviors existing in physical, biological and other systems. For example, in electrical circuits [10] and coronary artery system [11], the systems may perform unexpected behaviors under the influence of chaos. The investigations of the above two directions involve the control of chaos especially its tracking control, which attracts lots of attention [10, 11]. This chapter [12] focuses on the tracking control of a class of chaotic systems containing the Lorenz, Chen and Lu chaotic systems. Note that many other chaotic systems are modified or evolved from these three typical types of chaotic systems [13, 14].

In nonlinear science, as an important topic, chaos control has been investigated widely and many methods have been developed since the first consideration by Ott et al. [15]. It has been shown that IOL is an effective method for solving nonlinear tracking-control problems [16]. However, when a nonlinear system fails to have a well-defined relative degree for some division-by-zero (DBZ) points, the IOL method is no longer valid. For example, the IOL method fails to solve the tracking-control problems of Lorenz, Chen and Lu chaotic systems because of the existence of DBZ points. To handle this DBZ problem, the technique of approximate IOL presented in [17] neglects some nonlinear terms that may generate DBZ points. As pointed out in [18], that method may not work well when the system is away from the DBZ points, because of the approximation error generated by the dropped terms. In [19], a switching scheme was provided with an idea of switching between approximate and exact IOL, so as to avoid the DBZ points and enhance the tracking performance. As one of the main techniques, that switching-type control scheme has been developed to solve this problem [20]. Nevertheless, the switching-type control scheme may cost much in terms of the implementation since it requires two or more controllers for solving a single problem.

It is also worth pointing out that the aforementioned techniques generally bring in the operation of division, which may generate DBZ points. In this chapter [12], by combining the ZD and GD methods together, a simple and effective controller-design method termed ZG method is presented. Based on the ZG method, tracking controllers can be designed in a division-free manner, which truly solve the DBZ problems of Lorenz, Chen and Lu chaotic systems, as well as many other chaotic systems modified or evolved from these three chaotic systems [13, 14].

Specifically, as a dynamic method for time-varying problem solving, ZD has been developed to solve time-varying linear matrix equation, time-varying quadratic minimization, time-varying nonlinear inequality, and so on [21, 22]. On the other hand, as another type of dynamic method, GD is intrinsically feasible and efficient to solve time-invariant (or say, static) problems and has been generalized to solve time-varying problems, such as time-varying matrix inversion, time-varying quadratic minimization and equality-constrained quadratic programming [23, 24]. The previous studies generally exploited ZD and GD methods independently and comparatively, and other researchers rarely consider combining them for solving a problem or discover the superiority of their combination. It is worth pointing out that the main contribution of this chapter [12] lies on the superiority of the ZG method for conquering the DBZ points during the tracking control of chaotic systems.

2.2 Systems and Controllers

In this section, the tracking-control problems of Lorenz, Chen and Lu chaotic systems are first presented. For achieving the purpose of the tracking control of these systems, this section takes the Lu chaotic system as an illustrative example of a class of chaotic systems, and two types of controllers, compared with the conventional IOL controller, are then designed by the ZG method.

2.2.1 Chaotic Systems with Single Input

The tracking-control problem of the Lu chaotic system equipped with a single control input u is first presented as an illustrative example of a class of chaotic systems. The DBZ problem encountered in the tracking control of Lorenz, Chen and Lu chaotic systems is then illustrated. Specifically, let us consider the following Lu chaotic system equipped with a single control input u:

$$\begin{cases} \dot{x}_1 = f_1(x_1, x_2) = a(x_2 - x_1), \\ \dot{x}_2 = f_2(x_1, x_2, x_3, u) = -x_1x_3 + cx_2 + u, \\ \dot{x}_3 = f_3(x_1, x_2, x_3) = x_1x_2 - bx_3. \end{cases} \tag{2.2}$$

Besides, $y = x_3$ denotes the output of system (2.2). The objective of tracking control is to design a controller such that y tracks the bounded and smoothly time-varying desired trajectory y_d, with the tracking error $e = y - y_\mathrm{d}$ asymptotically approaching zero (or say, near zero in practice). Notably, to better focus on the DBZ problem, in the controller design, all the states are considered to be available and all the system parameters are considered to be known.

According to the IOL method [16], when $x_1 \neq 0$, the following conventional IOL controller can be obtained to solve the above tracking-control problem:

$$u = \frac{1}{x_1}\left(x_1^2x_3 + (a + b - c - k_1)x_1x_2 - ax_2^2 + \alpha_1\right), \tag{2.3}$$

where $\alpha_1 = (k_1b - b^2 - k_0)x_3 + \ddot{y}_\mathrm{d} + k_1\dot{y}_\mathrm{d} + k_0y_\mathrm{d}$, with k_0 and k_1 satisfying the condition that all roots of characteristic polynomial $P(s) = s^2 + k_1s + k_0$ are in the open left-half complex plane. Evidently, the conventional IOL controller (2.3) may not be implementable in reality, since it is possible to encounter the DBZ problem. Specifically, when the value of x_1 is close or even equal to zero, the magnitude of IOL controller (2.3) becomes extremely large or even infinite. This DBZ situation may lead to the crash of the system when equipped with the conventional IOL controller (2.3) during the tracking-control process. For example, with $y_\mathrm{d} = \sin(t) + 1.01$, Lu chaotic system (2.2) equipped with IOL controller (2.3) crashes when x_1 approaches zero, as shown in Fig. 2.1.

For completeness, the conventional controllers designed by the IOL method and the corresponding DBZ problems for the tracking control of Lorenz, Chen and Lu chaotic systems with a single control input are presented in Table 2.1. For the tracking control of these systems, this chapter [12] presents and investigates a simple and effective method to conquer the DBZ problem.

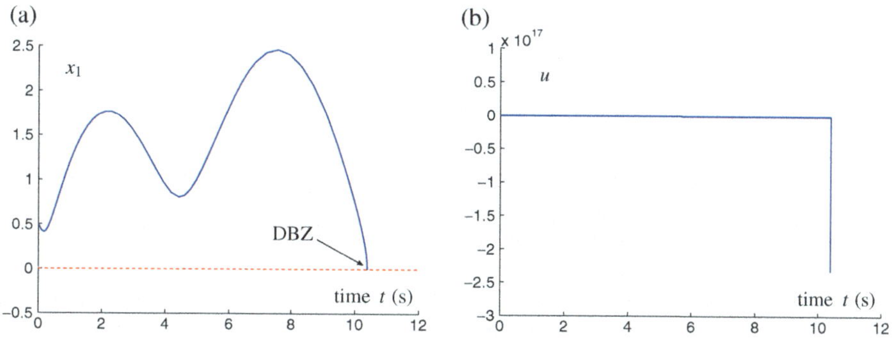

Fig. 2.1 Crash of Lu chaotic system (2.2) equipped with conventional IOL controller (2.3) for desired trajectory $y_d = \sin(t) + 1.01$ when x_1 approaches zero. (**a**) Trajectory of x_1. (**b**) Control input

Table 2.1 Conventional IOL controllers and DBZ problems of Lorenz, Chen and Lu chaotic systems with single control input

System	Conventional IOL controller	DBZ plane
Lorenz	$u = \left(x_1^2 x_3 + (a+1+b-k_1)x_1x_2 - cx_1^2 - ax_2^2 + \alpha_1\right)/x_1$	$x_1 = 0$
Chen	$u = \left(x_1^2 x_3 + (a-c+b-k_1)x_1x_2 + (a-c)x_1^2 - ax_2^2 + \alpha_1\right)/x_1$	$x_1 = 0$
Lu	$u = \left(x_1^2 x_3 + (a+b-c-k_1)x_1x_2 - ax_2^2 + \alpha_1\right)/x_1$	$x_1 = 0$

*Note that $\alpha_1 = (k_1 b - b^2 - k_0)x_3 + \ddot{y}_d + k_1\dot{y}_d + k_0 y_d$

2.2.2 *Design of ZD and ZG Controllers*

This subsection shows the design procedure of controllers for solving the DBZ problem in the tracking control. Without loss of generality, this subsection takes the Lu chaotic system as an illustrative example of a class of chaotic systems.

To monitor and control the tracking process, the first ZF is constructed as

$$z_1 = y - y_d = x_3 - y_d.$$

Based on the ZD method [21, 22], the following ZD design formula is adopted:

$$\dot{z}_1 = \frac{dz_1}{dt} = -\lambda\varphi(z_1), \tag{2.4}$$

where the design parameter $\lambda \in \mathbb{R}^+$ is used to scale the convergence rate of the ZD solution, and $\varphi(\cdot) : \mathbb{R} \to \mathbb{R}$ denotes a general activation-function mapping. It is worth pointing out that different values of λ and different choices of activation functions can affect the convergence performance of the tracking controllers. Generally speaking, any odd monotonically-increasing activation function can be used for the ZD construction [21, 22]. For simplicity, the linear activation function

$\varphi(z_1) = z_1$ is chosen in this chapter [12]. Thus, the ZD design formula (2.4) reduces to

$$\dot{z}_1 = -\lambda z_1. \tag{2.5}$$

Expanding (2.5), we can obtain $\dot{x}_3 - \dot{y}_d = -\lambda(x_3 - y_d)$. It follows from the third equation of system (2.2) that

$$x_1x_2 + (\lambda - b)x_3 - \dot{y}_d - \lambda y_d = 0. \tag{2.6}$$

Since there does not exist an explicit expression of u in (2.6), the second ZF is further constructed as $z_2 = x_1x_2 + (\lambda - b)x_3 - \dot{y}_d - \lambda y_d$. Applying the ZD method again, we have the ZD design formula $\dot{z}_2 = -\lambda z_2$, and then $\dot{x}_1x_2 + x_1\dot{x}_2 + (\lambda - b)\dot{x}_3 - \ddot{y}_d - \lambda\dot{y}_d = -\lambda(x_1x_2 + (\lambda - b)x_3 - \dot{y}_d - \lambda y_d)$. It follows from system (2.2) that

$$x_1^2x_3 + (a + b - c - 2\lambda)x_1x_2 - ax_2^2 - x_1u + \alpha_2 = 0, \tag{2.7}$$

where $\alpha_2 = (2\lambda b - b^2 - \lambda^2)x_3 + \ddot{y}_d + 2\lambda\dot{y}_d + \lambda^2 y_d$. Thus, a tracking controller in the form of u for Lu chaotic system (2.2) is directly obtained from (2.7), i.e.,

$$u = \frac{1}{x_1}\left(x_1^2x_3 + (a + b - c - 2\lambda)x_1x_2 - ax_2^2 + \alpha_2\right). \tag{2.8}$$

Since ZD is exploited twice during the design of controller (2.8) (while GD is not exploited), such a controller is termed the z2g0 controller. Note that, for z2g0 controller (2.8), there also exists a DBZ plane of $x_1 = 0$.

Evidently, the possibility of DBZ points mostly lies on the division operation, such as those in controllers (2.3) and (2.8). To conquer the DBZ point $x_1 = 0$, the basic idea is to transform the direct control (2.8) into the solution of a time-varying minimization problem. Therefore, the GD method is further exploited for the design of a controller in the form of $\dot{u}$ (termed z2g1 controller) on the basis of (2.8).

From (2.8), we define $\phi = x_1^2x_3 + (a+b-c-2\lambda)x_1x_2 - ax_2^2 - x_1u + \alpha_2$. Based on the GD method [22, 23], the square-based EF is defined as

$$\epsilon = \frac{1}{2}\phi^2. \tag{2.9}$$

Then, we adopt the following GD design formula:

$$\dot{u} = -\gamma\frac{\partial\epsilon}{\partial u}, \tag{2.10}$$

where the design parameter $\gamma \in \mathbb{R}^+$ is used to scale the convergence rate of the GD solution. From (2.9) and (2.10), a tracking controller in the form of $\dot{u}$ (i.e., the z2g1

Table 2.2 Tracking controllers designed by ZG method for Lorenz, Chen and Lu chaotic systems with single control input

System	z2g0 controller	z2g1 controller
Lorenz	$u=\big(x_1^2x_3+(a+b+1-2\lambda)x_1x_2-ax_2^2-cx_1^2+\alpha_2\big)/x_1$	$\dot{u}=\gamma x_1\big(x_1^2x_3+(a+b+1-2\lambda)x_1x_2-ax_2^2-cx_1^2-x_1u+\alpha_2\big)$
Chen	$u=\big(x_1^2x_3+(a+b-c-2\lambda)x_1x_2+(a-c)x_1^2-ax_2^2+\alpha_2\big)/x_1$	$\dot{u}=\gamma x_1\big(x_1^2x_3+(a+b-c-2\lambda)x_1x_2+(a-c)x_1^2-ax_2^2-x_1u+\alpha_2\big)$
Lu	$u=\big(x_1^2x_3+(a+b-c-2\lambda)x_1x_2-ax_2^2+\alpha_2\big)/x_1$	$\dot{u}=\gamma x_1\big(x_1^2x_3+(a+b-c-2\lambda)x_1x_2-ax_2^2-x_1u+\alpha_2\big)$

*Note that $\alpha_2=(2\lambda b-b^2-\lambda^2)x_3+\ddot{y}_\text{d}+2\lambda\dot{y}_\text{d}+\lambda^2y_\text{d}$

controller) for Lu chaotic system (2.2) is finally obtained as

$$\dot{u}=\gamma x_1\phi=\gamma x_1(x_1^2x_3+(a+b-c-2\lambda)x_1x_2-ax_2^2-x_1u+\alpha_2). \tag{2.11}$$

Notably, compared with the conventional IOL controller (2.3) and z2g0 controller (2.8), z2g1 controller (2.11) is free of division operation.

For completeness, the z2g0 and z2g1 controllers for the tracking control of Lorenz, Chen and Lu chaotic systems are presented in Table 2.2. To sum up, the ZG method, which is based on the combination of ZD and GD, contains two procedures. First, a controller in the form of u is designed by constructing a series of ZFs. Then, based on such a standard controller, GD is used to obtain a division-free $\dot{u}$-form controller.

2.3 Convergence Performance Analyses

This section presents the convergence performance analyses on z2g0 controller (2.8) and z2g1 controller (2.11) for the tracking control of Lu chaotic system (2.2).

2.3.1 Analysis on ZD Controller

As an illustrative example, this subsection presents the theoretical result and analysis of Lu chaotic system (2.2) equipped with z2g0 controller (2.8).

Theorem 2.1 *Consider Lu chaotic system (2.2) equipped with z2g0 controller (2.8) for smooth and bounded desired trajectory* y_d. *Starting with bounded initial state* $\mathbf{x}(0)=[x_1(0),x_2(0),x_3(0)]^\text{T}\in\mathbb{R}^3$, *the tracking error of the system exponentially converges to zero on a large scale, provided that* $x_1\neq 0$, $\forall t\in[0,+\infty)$.

Proof According to the ZD design formulas about z_1 and z_2, we readily obtain the following equation by substituting $z_2 = \dot{z}_1 + \lambda z_1$ into $\dot{z}_2 = -\lambda z_2$:

$$\ddot{z}_1 + 2\lambda\dot{z}_1 + \lambda^2 z_1 = 0.$$

Then, solving the above equation, we have

$$z_1 = c_1 \exp(-\lambda t) + c_2 t \exp(-\lambda t), \tag{2.12}$$

where c_1 and c_2 are constants and $t \in [0, +\infty)$. In view of $\lambda > 0$, there exist $\bar{c} > 0$ and $\bar{\lambda} > 0$ such that $z_1 \leq \bar{c} \exp(-\bar{\lambda} t)$ [25]. Thus, $z_1 = e$ exponentially converges to zero. Therefore, the tracking error e of the system exponentially converges to zero on a large scale. The proof is thus completed. □

Note that, similar to the above theorem and proof for the Lu chaotic system, the convergence analyses on the tracking errors for the Lorenz and Chen chaotic systems equipped with the corresponding z2g0 controllers shown in Table 2.2 can also be obtained.

2.3.2 *Analyses on ZG Controller*

This subsection investigates the theoretical analyses of z2g1 controller (2.11) for the tracking control of Lu chaotic system (2.2) with the DBZ point $x_1 = 0$ conquered. Note that the performance analyses of z2g1 controllers shown in Table 2.2 for the Lorenz and Chen chaotic systems can correspondingly be obtained by following these analyses of z2g1 controller (2.11) for the Lu chaotic system.

2.3.2.1 Preliminary

As aforementioned, in order to conquer DBZ points, the fundamental idea of z2g1 controller (2.11) is to transform direct control (2.8) into the solution of a time-varying minimization problem. That is, the direct computational manner of controller (2.8) is transformed to the manner of minimizing the time-varying EF (2.9). For further discussion, controller (2.8) is formulated as $u = \alpha_3/x_1$, where $\alpha_3 = x_1^2 x_3 + (a + b - c - 2\lambda)x_1 x_2 - a x_2^2 + \alpha_2$. Then, accordingly, $\phi = \alpha_3 - x_1 u$. Thus, the time-varying solution of minimizing (2.9) by using GD can be rewritten as

$$\dot{u} = \gamma x_1 \phi = \gamma x_1(\alpha_3 - x_1 u). \tag{2.13}$$

Besides, let $u^* = \alpha_3/x_1$ denote the desired time-varying solution of this minimization problem.

2.3.2.2 Tracking-Error Bound

The convergence performance analyses on the tracking error of Lu chaotic system (2.2) equipped with z2g1 controller (2.11) are presented below.

Theorem 2.2 *Consider Lu chaotic system (2.2) equipped with z2g1 controller (2.11) for smooth and bounded desired trajectory y_{d}. Starting with bounded initial state $\mathbf{x}(0) \in \mathbb{R}^3$ and control input $u(0) \in \mathbb{R}$, the following results are achieved on a large scale for the tracking control of the system.*

- *For the case of $x_1 \neq 0$ (i.e., the non-DBZ case), the tracking error of the system converges toward or stays within the error bound $\chi\varrho/(\lambda^2\gamma\varsigma)$, provided that (i) $\sqrt{\varsigma} \leq |x_1| \leq \chi$, $\exists 0 < \sqrt{\varsigma} \leq \chi < +\infty$, and (ii) $|\dot{u}^*| = |\mathrm{d}u^*/\mathrm{d}t| \leq \varrho$, $\exists 0 \leq \varrho < +\infty$.*
- *For the case of $x_1 = 0$ (i.e., the DBZ case), the tracking error of the system is bounded.*

Proof We analyze the following two cases.

(i) For the case of $x_1 \neq 0$ (i.e., the non-DBZ case).

Let us define a solution error of z2g1 controller (2.11) or (2.13) for solving the time-varying minimization problem of ϵ as $\psi = u - u^*$ with $u^* = \alpha_3/x_1$ denoting the desired time-varying solution. Then, we can prove that the solution error $|\psi|$ of z2g1 controller (2.11) for solving the time-varying minimization problem of ϵ is upper bounded by $\varrho/(\gamma\varsigma)$ when the solving process enters steady state. That is, in mathematics, the steady-state solution error

$$|\psi| = |u - u^*| \leq \frac{\varrho}{\gamma\varsigma}, \tag{2.14}$$

where $t \geq t_{\mathrm{e}}$ with t_{e} being large enough. Note that the proof of (2.14) is given in Appendix 1.

From the derivation process of z2g1 controller (2.11), we have $\phi = -\dot{z}_2 - \lambda z_2 = -\ddot{z}_1 - 2\lambda\dot{z}_1 - \lambda^2 z_1 = -\ddot{e} - 2\lambda\dot{e} - \lambda^2 e = \alpha_3 - x_1 u = x_1(u^* - u) = -x_1\psi$. Together with (2.14), when $t \geq t_{\mathrm{e}}$, we have

$$\begin{cases} -\chi\varrho/(\gamma\varsigma) \leq -|x_1||\psi| \leq \ddot{e} + 2\lambda\dot{e} + \lambda^2 e, \\ \ddot{e} + 2\lambda\dot{e} + \lambda^2 e \leq |x_1||\psi| \leq \chi\varrho/(\gamma\varsigma). \end{cases} \tag{2.15}$$

Then, based on Gronwall inequality [26], the two-sided inequality

$$-\tilde{c}_1 \exp(-\lambda t) - \tilde{c}_2 t \exp(-\lambda t) - \frac{\chi\varrho}{\lambda^2\gamma\varsigma} \leq e \leq \bar{c}_1 \exp(-\lambda t) + \bar{c}_2 t \exp(-\lambda t) + \frac{\chi\varrho}{\lambda^2\gamma\varsigma} \tag{2.16}$$

holds true when $t \geq t_\mathrm{e}$, where $\tilde{c}_1$, $\tilde{c}_2$, $\bar{c}_1$ and $\bar{c}_2$ are constants. Note that the proof of (2.16) is presented in Appendix 2. Furthermore, we have

$$|e| \leq |c_1| \exp(-\lambda t) + |c_2| t \exp(-\lambda t) + \frac{\chi \varrho}{\lambda^2 \gamma \varsigma}, \tag{2.17}$$

where $|c_1| = \max\{|\tilde{c}_1|, |\bar{c}_1|\}$ and $|c_2| = \max\{|\tilde{c}_2|, |\bar{c}_2|\}$, and $t \geq t_\mathrm{e}$. Thus, similar to the analysis about Eq. (2.12), the right-hand side of (2.17) is exponentially convergent to $\chi\varrho/(\lambda^2\gamma\varsigma)$. That is, we finally have

$$\limsup_{t \to +\infty} |e| \leq \frac{\chi \varrho}{\lambda^2 \gamma \varsigma}.$$

Thus, the tracking error of Lu chaotic system (2.2) equipped with z2g1 controller (2.11) converges toward or stays within the error bound $\chi\varrho/(\lambda^2\gamma\varsigma)$ in this case.

(ii) For the case of $x_1 = 0$ (i.e., the DBZ case).

For the system equipped with z2g1 controller (2.11) to track a time-varying desired trajectory y_d, it can be readily derived that $\lim_{t \to t_\mathrm{s}} \dot{u} = \lim_{x_1 \to 0} \dot{u} = 0$ in view of $\dot{u} = \gamma x_1(\alpha_3 - x_1 u)$, where t_s denotes the time instant of the DBZ point, with the subscript "s" corresponding to the word "singularity" (being a synonym of DBZ in this book). Therefore, the control input at t_s is the same as that at the previous time instant $t_{\mathrm{s}-}$, which implies that $u(t_\mathrm{s}) = u(t_{\mathrm{s}-})$. Similarly, at $t_{\mathrm{s}+}$ (which is the time instant after the DBZ point), $u(t_\mathrm{s}) = u(t_{\mathrm{s}+})$. Then, we have the result that $u(t_{\mathrm{s}-}) = u(t_\mathrm{s}) = u(t_{\mathrm{s}+})$ and they are bounded. For a bounded input, the output of Lu chaotic system (2.2) is bounded. Since the desired trajectory y_d is bounded, the tracking error is bounded at the time instants $t_{\mathrm{s}-}$, t_s and $t_{\mathrm{s}+}$. After getting through the DBZ point (i.e., for the time instants after $t_{\mathrm{s}+}$), the tracking error converges toward an error bound again, which implies that Lu chaotic system (2.2) equipped with z2g1 controller (2.11) finally conquers the DBZ points in a bounded and convergent manner.

By the above analyses, the proof is thus completed. □

2.3.2.3 Exponential Convergence Rate

In Theorem 2.2, the tracking error is just asymptotic convergence, which may be less desirable in practical applications. In this subsection, we obtain the following results on the exponential convergence of z2g1 controller (2.11) to a relatively loose tracking-error bound.

Theorem 2.3 *Consider Lu chaotic system (2.2) equipped with z2g1 controller (2.11) for smooth and bounded desired trajectory* y_d. *Starting with bounded initial state* $\mathbf{x}(0) \in \mathbb{R}^3$ *and control input* $u(0) \in \mathbb{R}$, *the following results are achieved on a large scale for the tracking control of the system.*

- *For the case of* $x_1 \neq 0$ *(i.e., the non-DBZ case), the tracking error of the system exponentially converges toward or stays within the error bound* $\chi\varrho/(\omega\lambda^2\gamma\varsigma)$ *with loosening parameter* $\omega \in (0, 1)$*, provided that (i)* $\sqrt{\varsigma} \leq |x_1| \leq \chi$, $\exists 0 < \sqrt{\varsigma} \leq \chi < +\infty$*, and (ii)* $|\dot{u}^*| \leq \varrho$, $\exists 0 \leq \varrho < +\infty$.
- *For the case of* $x_1 = 0$ *(i.e., the DBZ case), the tracking error of the system is bounded.*

Proof For the case of $x_1 \neq 0$ (i.e., the non-DBZ case), according to [23, 24], we know that the solution error $|\psi|$ exponentially converges toward or stays within a relatively loose error bound of $\varrho/(\omega\gamma\varsigma)$. Besides, the convergence time is $t_c = \ln(\omega\gamma\varsigma|\psi(0)|/\varrho)/((1-\omega)\gamma\varsigma)$. That is, in mathematics, the solution error $|\psi|$ satisfies the following inequality:

$$|\psi| = |u - u^*| \leq \frac{\varrho}{\omega\gamma\varsigma}, \ \forall t \geq t_c. \tag{2.18}$$

Note that the proof of (2.18) is presented in Appendix 3.

Similar to the derivation of (2.15), when $t \geq t_c$, we have

$$-\frac{\chi\varrho}{\omega\gamma\varsigma} \leq \ddot{e} + 2\lambda\dot{e} + \lambda^2 e \leq \frac{\chi\varrho}{\omega\gamma\varsigma}.$$

Then, similar to the derivation of (2.17), we obtain

$$|e| \leq |\hat{c}_1|\exp(-\lambda t) + |\hat{c}_2| t \exp(-\lambda t) + \frac{\chi\varrho}{\omega\lambda^2\gamma\varsigma} \tag{2.19}$$

holds true when $t \geq t_c$, where $\hat{c}_1$ and $\hat{c}_2$ are constants. Then, similar to the analysis about Eq. (2.12), the right-hand side of (2.19) exponentially converges toward a new error bound $\chi\varrho/(\omega\lambda^2\gamma\varsigma)$. Thus, the error $|e|$ of Lu chaotic system (2.2) equipped with z2g1 controller (2.11) exponentially converges toward or stays within error bound $\chi\varrho/(\omega\lambda^2\gamma\varsigma)$ in this case.

For the case of $x_1 = 0$ (i.e., the DBZ case), according to the proof in Theorem 2.2, we have the same conclusion that the tracking error of the system is bounded.

By the above analyses, the proof is thus completed. □

Remark 2.1 About the DBZ plane of $x_1 = 0$, we further have the following analyses and results for the system equipped with z2g1 controller (2.11). That is, (1) if $x_1(0) = x_2(0) = u(0) = 0$, then the system is trapped in the DBZ plane of $x_1 = 0$; and, (2) if at least one of $x_1(0)$, $x_2(0)$ and $u(0)$ is nonzero, then the system is guaranteed to escape from the DBZ plane of $x_1 = 0$.

- For the first case of $x_1(0) = x_2(0) = u(0) = 0$, we have $\dot{x}_1(0) = \dot{x}_2(0) = \dot{u}(0) = 0$ according to (2.2). At the next time instant, $x_1(0_+) = x_2(0_+) = u(0_+) = 0$ and $\dot{x}_1(0_+) = \dot{x}_2(0_+) = \dot{u}(0_+) = 0$. Then, it can be recursively derived that $x_1 = x_2 = u = 0$, $\forall t \geq 0$; i.e., the system is trapped in the DBZ plane of $x_1 = 0$.

- Let us analyze the second case via the contradiction technique. At first, assume that the system enters the DBZ plane of $x_1 = 0$ at t_s and is trapped, i.e., $x_1 = 0, \forall t \geq t_s$. According to (2.2), $x_2 = \dot{x}_2 = \dot{x}_1 = u = 0, \forall t \geq t_s$. Then, $\dot{u} = \dot{x}_2 = \dot{x}_1 = u = x_2 = x_1 = 0, \forall t \geq t_s$. According to the time-derivative definition as well as continuity, we have $\dot{u} = \dot{x}_2 = \dot{x}_1 = u = x_2 = x_1 = 0, \forall t \geq t_{s^-}$ and $\forall t \geq 0$ (derived recursively), which is finally in contradiction with the condition that at least one of $x_1(0)$, $x_2(0)$ and $u(0)$ is nonzero. Therefore, the system is guaranteed to escape from the DBZ plane of $x_1 = 0$ in this case.

2.4 Simulation, Verification and Comparison

In this section, the simulations are performed to substantiate the efficacy and superiority of the presented ZG method for the tracking control. The results are shown in Figs. 2.2, 2.3, 2.4, 2.5, 2.6, 2.7 and 2.8. Without loss of generality, the design parameters are set as $\lambda = 1$ and $\gamma = 10^5$ unless they are specified otherwise.

Starting with a randomly generated initial state $\mathbf{x}(0)$, Fig. 2.2 illustrates the results of Lu chaotic system (2.2) equipped with z2g0 controller (2.8) for desired trajectory $y_d = \sin(t) + 5$. From Fig. 2.2a, the system output y rapidly tracks the desired trajectory y_d. In addition, as illustrated in Fig. 2.2b, the absolute tracking error $|e|$ (i.e., the absolute value of the tracking error) exponentially converges to zero (specifically, being below the order of 10^{-10}). These results coincide well with the analyses on the z2g0 controller in Theorem 2.1, and illustrate the efficacy of the z2g0 controller for the tracking control of the system in the non-DBZ situation.

The simulation results of Lu chaotic system (2.2) equipped with z2g1 controller (2.11) for desired trajectory $y_d = \sin(t) + 5$ are illustrated in Figs. 2.3 and 2.4. As seen from Fig. 2.3a, with randomly generated initial state $\mathbf{x}(0)$ and

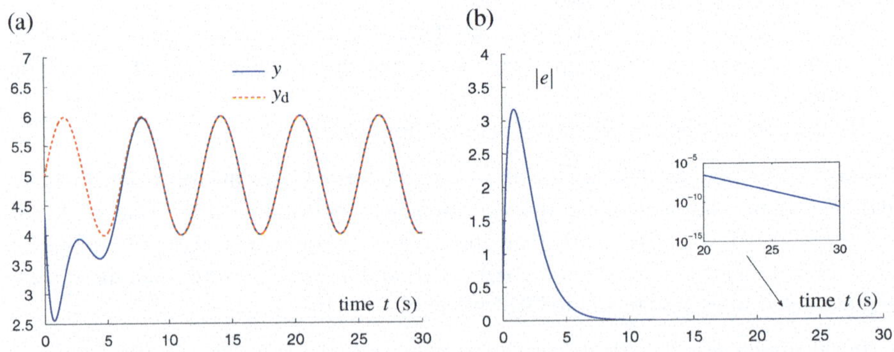

Fig. 2.2 Tracking performance of Lu chaotic system (2.2) equipped with z2g0 controller (2.8) for desired trajectory $y_d = \sin(t) + 5$. (**a**) Output trajectory and desired trajectory. (**b**) Absolute tracking error

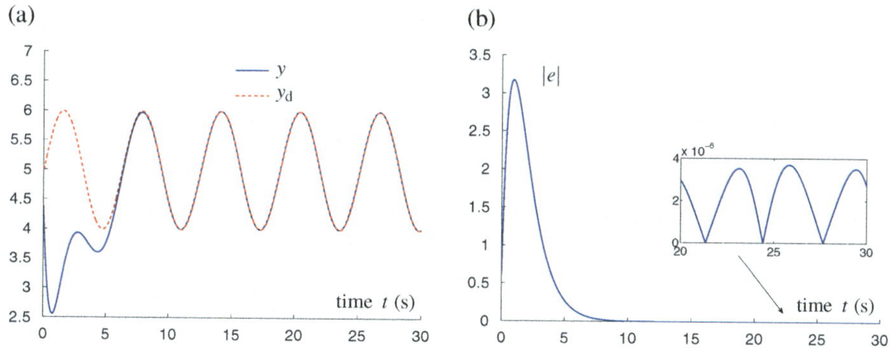

Fig. 2.3 Tracking performance of Lu chaotic system (2.2) equipped with z2g1 controller (2.11) for desired trajectory $y_d = \sin(t) + 5$. (**a**) Output trajectory and desired trajectory. (**b**) Absolute tracking error

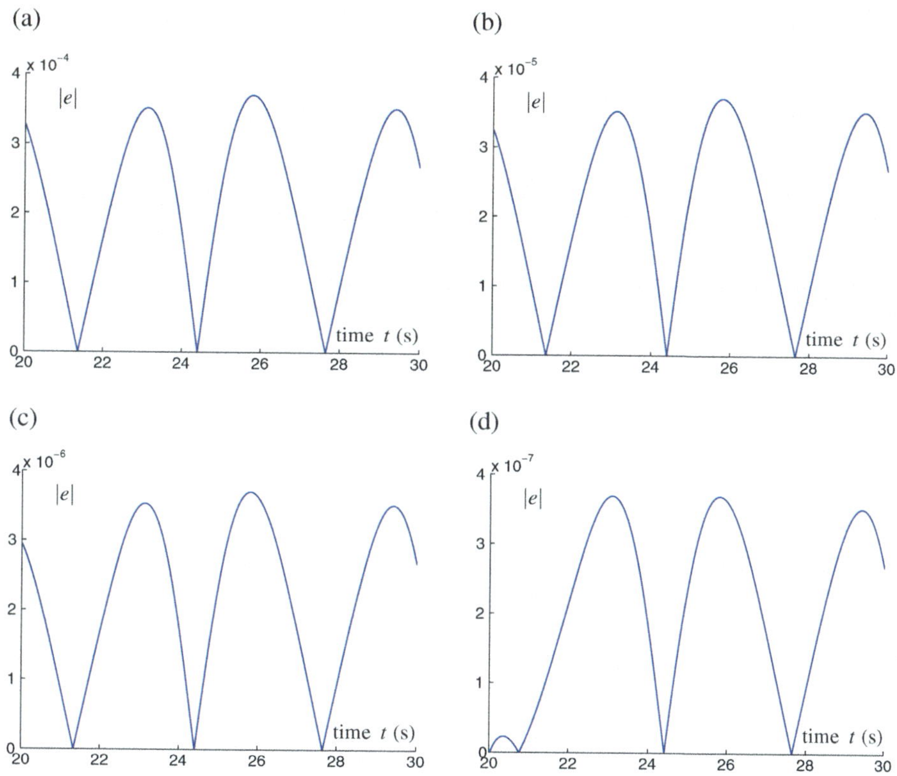

Fig. 2.4 Effect of parameter γ on convergence error bound of absolute tracking error $|e|$ for Lu chaotic system (2.2) equipped with z2g1 controller (2.11) to track desired trajectory $y_d = \sin(t)+5$. (**a**) $|e|$ in steady state with $\gamma = 10^3$. (**b**) $|e|$ in steady state with $\gamma = 10^4$. (**c**) $|e|$ in steady state with $\gamma = 10^5$. (**d**) $|e|$ in steady state with $\gamma = 10^6$

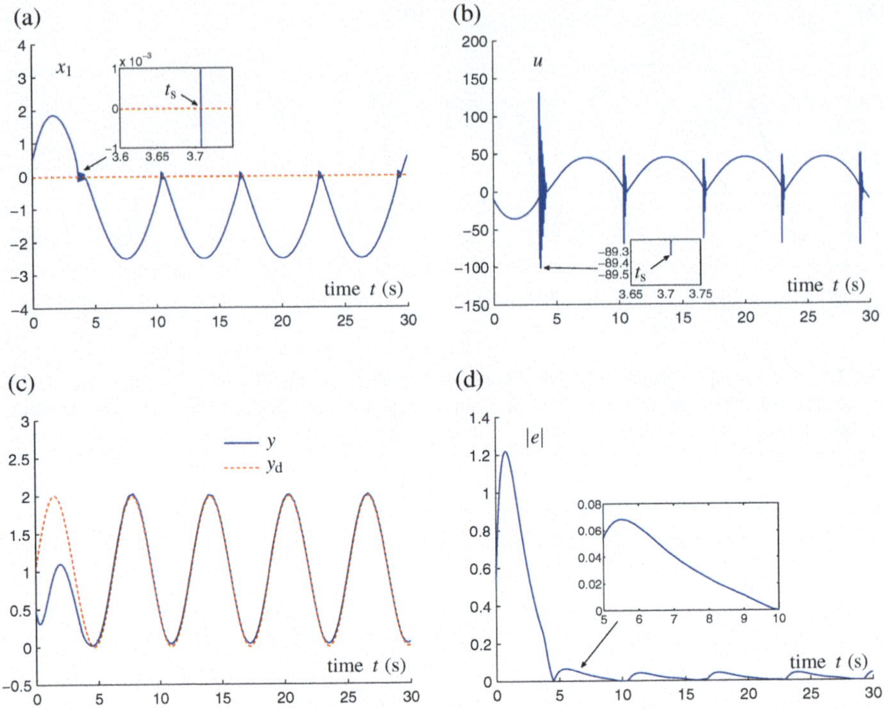

Fig. 2.5 Tracking performance of Lu chaotic system (2.2) equipped with z2g1 controller (2.11) for desired trajectory $y_d = \sin(t)+1.01$ encountering DBZ points. (**a**) Trajectory of x_1. (**b**) Control input. (**c**) Output trajectory and desired trajectory. (**d**) Absolute tracking error

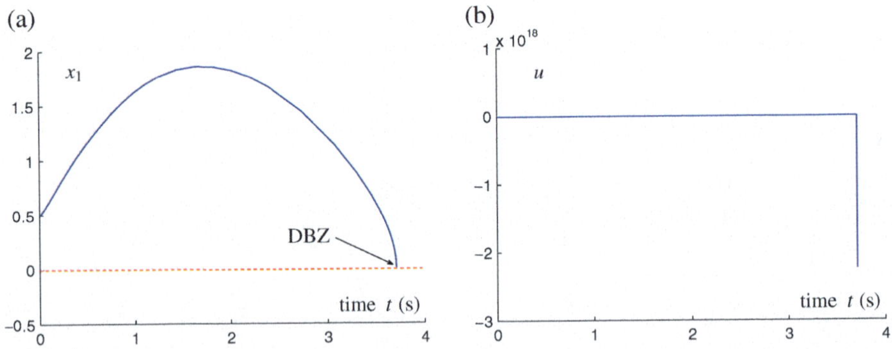

Fig. 2.6 Tracking performance of Lu chaotic system (2.2) equipped with z2g0 controller (2.8) for desired trajectory $y_d = \sin(t) + 1.01$ encountering DBZ point. (**a**) Trajectory of x_1. (**b**) Control input

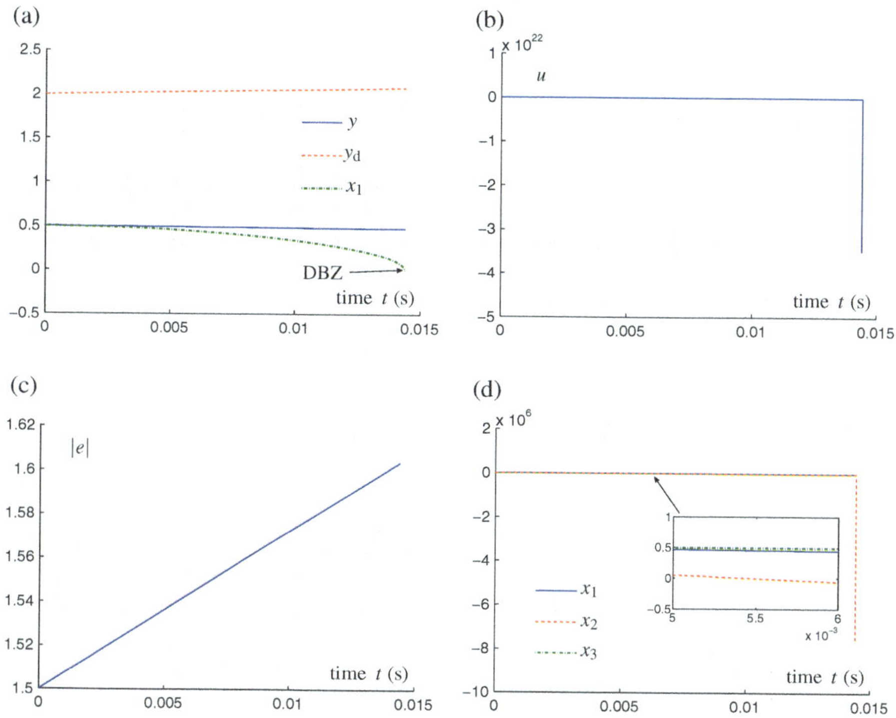

Fig. 2.7 Tracking performance of Lu chaotic system (2.2) equipped with conventional IOL controller (2.3) for desired trajectory $y_d = 2\cos(5t) + 3\sin(2t)$ encountering DBZ point. (**a**) Trajectories of y, y_d and x_1. (**b**) Control input. (**c**) Absolute tracking error. (**d**) System states

control input $u(0)$, the Lu chaotic system completes the tracking-control task well. From Fig. 2.3b, the tracking error exponentially converges toward an error bound, which is of order 10^{-6}. These results coincide well with the analyses on the z2g1 controller in Theorems 2.2 and 2.3 in the case of $x_1 \neq 0$. Furthermore, as shown in Fig. 2.4, the upper bounds of $|e|$ in the steady state become 10^{-4}, 10^{-5}, 10^{-6} and 10^{-7} for $\gamma = 10^3$, $\gamma = 10^4$, $\gamma = 10^5$ and $\gamma = 10^6$, respectively. These results show that the error bound of $|e|$ can be decreased effectively by increasing the value of γ according to practical needs.

More importantly, the DBZ-conquering superiority of z2g1 controller (2.11) for the tracking control of Lu chaotic system (2.2) is illustrated in Fig. 2.5. Specifically, as seen from subfigures (a) through (c) of Fig. 2.5, the Lu chaotic system equipped with z2g1 controller (2.11) encounters many DBZ points of $x_1 = 0$ during the tracking process but still works well, and the control input is acceptable. Besides, from Fig. 2.5d, the error $|e|$ slightly increases in the neighborhood of the DBZ points, and rapidly decreases to a level of tiny values when the Lu chaotic system is away from the DBZ points. These results coincide well with the DBZ-

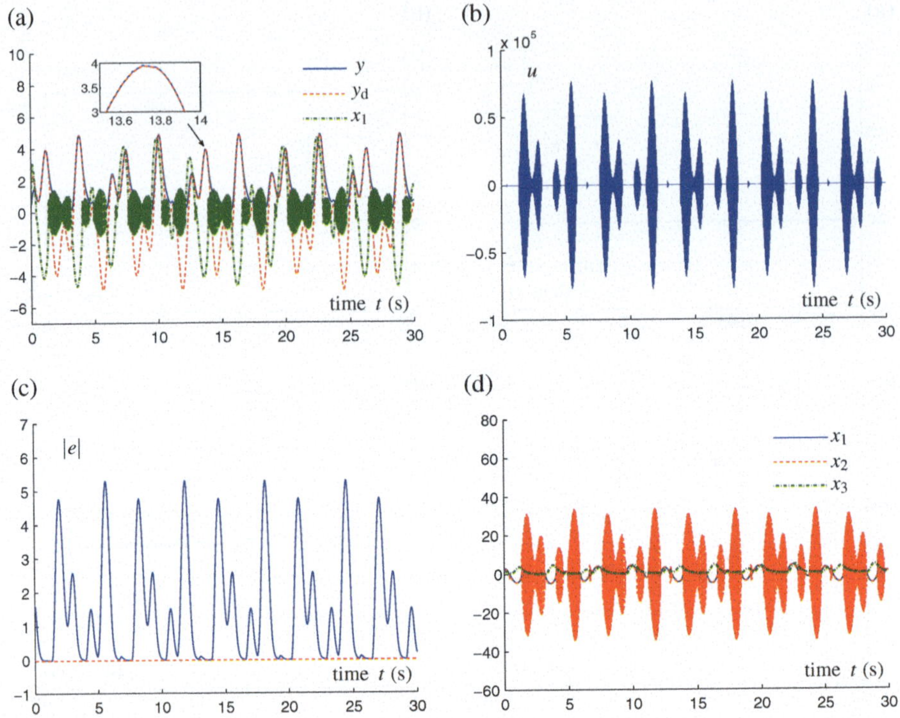

Fig. 2.8 Tracking performance of Lu chaotic system (2.2) equipped with z2g1 controller (2.11) for desired trajectory $y_d = 2\cos(5t) + 3\sin(2t)$ encountering many DBZ points. (**a**) Trajectories of y, y_d and x_1. (**b**) Control input. (**c**) Absolute tracking error. (**d**) System states

conquering analyses in Theorems 2.2 and 2.3, and substantiate the superiority of z2g1 controller (2.11) based on the ZG method.

For comparison, Fig. 2.6 presents the results of Lu chaotic system (2.2) equipped with z2g0 controller (2.8) for the desired trajectory $y_d = \sin(t) + 1.01$. As seen from the figure, when the value of x_1 approaches zero, the control input u tends to infinity. This DBZ problem, which the z2g0 controller cannot tackle well, leads to the system crash. These results are similar to those shown in Fig. 2.1. Comparing these results in Figs. 2.1, 2.5 and 2.6, we know that the tracking process of the Lu chaotic system equipped with z2g1 controller (2.11) succeeds even when it encounters many DBZ points. In contrast, the Lu chaotic system equipped with the conventional IOL controller (2.3) or z2g0 controller (2.8) crashes at the first DBZ time instant and cannot complete the tracking-control task. Specifically, as seen from Fig. 2.1 or 2.6, when the value of x_1 is close or even equal to zero (denoted by DBZ), the magnitude of control input u of conventional IOL controller (2.3) or z2g0 controller (2.8) becomes extremely large or even infinite. It leads to the crash of the Lu chaotic system equipped with the conventional IOL controller or the z2g0 controller. Thus, from the above comparative results, it can be confirmed that z2g1 controller (2.11) has the superiority of conquering the DBZ problem.

Moreover, compared with the conventional IOL controller (2.3), in order to further illustrate the efficacy and superiority of z2g1 controller (2.11) for conquering DBZ points (especially, many DBZ points), the desired trajectory is changed to $y_d = 2\cos(5t) + 3\sin(2t)$. Note that the design parameters are set as $\lambda = 10$ and $\gamma = 10^5$ in the simulation of z2g1 controller (2.11). The corresponding results are shown in Figs. 2.7 and 2.8. Specifically, as seen from Fig. 2.7a, the tracking process fails when the conventional IOL controller (2.3) encounters the first DBZ point (denoted by DBZ) at time instant $t = 0.0144$ s. More results are presented in subfigures (b) and (c) of Fig. 2.7. For comparison, Fig. 2.8 shows the results of z2g1 controller (2.11) for tracking the desired trajectory. As seen from Fig. 2.8a, z2g1 controller (2.11) still works well even when encountering many DBZ points (i.e., $x_1 = 0$). In addition, Fig. 2.8b, d show that the control input u and all states of the system are bounded. Besides, Fig. 2.8c shows that the error $|e|$ increases in the neighborhood of the DBZ points and then rapidly decreases to a level of tiny values when the Lu chaotic system leaves the DBZ point $x_1 = 0$. These results coincide again and well with the DBZ-conquering analyses in Theorems 2.2 and 2.3. Summarizing the above comparative results (i.e., Figs. 2.7 and 2.8), we can draw the conclusions that the Lu chaotic system equipped with the conventional IOL controller (2.3) crashes when it encounters the first DBZ point; by contrast, the system equipped with z2g1 controller (2.11) still works well even though it encounters many DBZ points. These observations further substantiate the efficacy and superiority of the presented z2g1 controller based on ZG method for the tracking control of the Lu chaotic system.

In summary, the efficacy of z2g0 controller (2.8) designed by the pure ZD method is verified. More importantly, via comparative simulations, the capability of z2g1 controller (2.11) for conquering the DBZ problem is illustrated. Note that the corresponding simulation results of Lorenz and Chen chaotic systems equipped with z2g0 or z2g1 controllers are similar to those of the Lu chaotic system, and thus omitted here. These simulation results substantiate well the feasibility and efficacy of the ZG method for the tracking control of Lorenz, Chen and Lu chaotic systems.

2.5 Chapter Summary

This chapter has combined the ZD method and the GD method together (termed ZG method) with theoretical analyses for the DBZ-conquering tracking control of a class of chaotic systems containing the Lorenz, Chen and Lu chaotic systems. The corresponding simulation results have substantiated that the controllers designed by the ZG method can achieve the satisfactory tracking performance and successfully conquer the DBZ problem encountered during the tracking-control process.

Appendix 1: Proof of Inequality (2.14)

Let us define a solution error of z2g1 controller (2.11) or (2.13) for solving the time-varying minimization problem of ϵ as $\psi = u - u^*$ with $u^* = \alpha_3/x_1$. Then, we have $u = \psi + u^*$ and its time derivative as

$$\dot{u} = \dot{\psi} + \dot{u}^*. \tag{2.20}$$

Substituting solution (2.13) into (2.20) yields $\gamma x_1(\alpha_3 - x_1 u) = \dot{\psi} + \dot{u}^*$ and further $\dot{\psi} = -\gamma x_1^2 \psi - \dot{u}^*$. For further analyses, let us define a Lyapunov function candidate as $L = \psi^2/2$. Evidently, L is positive definite in view of $L = \psi^2/2 > 0$ for $\psi \neq 0$ and $L = 0$ for $\psi = 0$ only. Then, taking its time derivative yields

$$\dot{L} = \psi\dot{\psi} = \psi\left(-\gamma x_1^2\psi - \dot{u}^*\right) = -\gamma x_1^2\psi^2 - \psi\dot{u}^*. \tag{2.21}$$

There are two terms in the time derivative $\dot{L}$, i.e., $-\gamma x_1^2\psi^2$ and $-\psi\dot{u}^*$. Let us handle these two terms individually. For the first term $-\gamma x_1^2\psi^2$, we have

$$-\gamma x_1^2\psi^2 \leq -\gamma\varsigma\psi^2, \tag{2.22}$$

where $\varsigma > 0$ is defined previously by $x_1^2 \geq \varsigma$. On the other hand, for the second term $-\psi\dot{u}^*$, we have the following result based on Cauchy inequality [27]:

$$-\psi\dot{u}^* \leq |\psi||\dot{u}^*| \leq \varrho|\psi|, \tag{2.23}$$

where ϱ is defined previously by $|\dot{u}^*| \leq \varrho$. Then, substituting (2.22) and (2.23) into (2.21) yields

$$\dot{L} = -\gamma x_1^2\psi^2 - \psi\dot{u}^* \leq -\gamma\varsigma\psi^2 + \varrho|\psi| = -|\psi|(\gamma\varsigma|\psi| - \varrho). \tag{2.24}$$

During the time evolution of solution error ψ, (2.24) falls into one of the following three situations: (1) $\gamma\varsigma|\psi| - \varrho > 0$; (2) $\gamma\varsigma|\psi| - \varrho = 0$; (3) $\gamma\varsigma|\psi| - \varrho < 0$. These three situations are analyzed in detail as follows.

- In the first situation (i.e., $|\psi| > \varrho/(\gamma\varsigma)$), $\dot{L} < 0$, which implies that ψ approaches zero (i.e., u approaches u^*) as time evolves.
- In the second situation (i.e., $|\psi| = \varrho/(\gamma\varsigma)$, a so-called ball surface), $\dot{L} \leq 0$, which implies that ψ approaches zero (i.e., u approaches u^*) or ψ stays on the ball surface with $|\psi| = \varrho/(\gamma\varsigma)$ (i.e., $|u - u^*| = \varrho/(\gamma\varsigma)$), in view of $\dot{L} \leq 0$ containing sub-situations $\dot{L} < 0$ and $\dot{L} = 0$, respectively. That is, ψ will not go outside the ball of $\varrho/(\gamma\varsigma)$ in this situation.
- In the third situation (i.e., $|\psi| < \varrho/(\gamma\varsigma)$, inside the ball), it follows from (2.24) that $\dot{L}$ is less than a positive scalar (containing sub-situations $\dot{L} \leq 0$ and $\dot{L} > 0$),

and thus ψ may not decrease again. Let us analyze the worst case, i.e., sub-situation $\dot{L} > 0$. In this case, with $\gamma\varsigma|\psi| - \varrho < 0$, L and $|\psi|$ would increase, which implies that $\gamma\varsigma|\psi| - \varrho$ increases as well. Evidently, there exists a certain time instant such that $\gamma\varsigma|\psi| - \varrho = 0$, which returns to the second situation, i.e., $\dot{L} \leq 0$.

Summarizing the above three situations, we can conclude that the solution error of z2g1 controller (2.11) or (2.13) for solving the time-varying minimization problem of ϵ is upper bounded by $\varrho/(\gamma\varsigma)$ when the solving process enters steady state. That is, in mathematics, the steady-state solution error $|\psi| = |u - u^*| \leq \varrho/(\gamma\varsigma)$, where $t \geq t_{\mathrm{e}}$ with t_{e} being large enough. The proof of inequality (2.14) is thus completed. □

Appendix 2: Proof of Two-Sided Inequality (2.16)

For completeness, we present the proof of two-sided inequality (2.16) as below.

From (2.15) and the derivation process of (2.11), we have

$$\ddot{e} + 2\lambda\dot{e} + \lambda^2 e = \dot{z}_2 + \lambda z_2 \leq \frac{\chi\varrho}{\gamma\varsigma},$$

where $t \geq t_{\mathrm{e}}$. Making use of Gronwall inequality [26], we have

$$z_2 \leq z_2(t_{\mathrm{e}})\exp(-\lambda t) + \int_{t_{\mathrm{e}}}^{t} \exp(-\lambda(t-\iota))\frac{\chi\varrho}{\gamma\varsigma}\mathrm{d}\iota \leq \hat{c}\exp(-\lambda t) + \frac{\chi\varrho}{\lambda\gamma\varsigma},$$

where $t \geq t_{\mathrm{e}}$ and $\hat{c}$ is a constant. Since $z_2 = \dot{e} + \lambda e$, we have

$$\dot{e} + \lambda e \leq \Omega,$$

where $\Omega = \hat{c}\exp(-\lambda t) + \chi\varrho/(\lambda\gamma\varsigma)$. Then, making use of Gronwall inequality again, we have

$$\begin{aligned} e &\leq e(t_{\mathrm{e}})\exp(-\lambda t) + \int_{t_{\mathrm{e}}}^{t} \exp(-\lambda(t-\iota))\Omega(\iota)\mathrm{d}\iota \\ &\leq \bar{c}_1\exp(-\lambda t) + \bar{c}_2 t\exp(-\lambda t) + \frac{\chi\varrho}{\lambda^2\gamma\varsigma}, \end{aligned} \tag{2.25}$$

where $\bar{c}_1$ and $\bar{c}_2$ are constants, and $t \geq t_{\mathrm{e}}$.

From (2.15), we also have

$$-\ddot{e} - 2\lambda\dot{e} - \lambda^2 e = -\dot{z}_2 - \lambda z_2 \leq \frac{\chi\varrho}{\gamma\varsigma}.$$

Similar to the derivation process of (2.25), we have

$$e \geq -\tilde{c}_1 \exp(-\lambda t) - \tilde{c}_2 t \exp(-\lambda t) - \frac{\chi\varrho}{\lambda^2\gamma\varsigma}, \tag{2.26}$$

where $\tilde{c}_1$ and $\tilde{c}_2$ are constants, and $t \geq t_e$. Combining (2.25) and (2.26) yields (2.16). The proof of (2.16) is thus completed. □

Appendix 3: Proof of Inequality (2.18)

Theorem 2.2 proves that the solution error $|\psi|$ of z2g1 controller (2.11) asymptotically converges toward the ball of $\varrho/(\gamma\varsigma)$. In the following, we show that $|\psi|$ exponentially converges toward a relatively loose error bound of $\varrho/(\omega\gamma\varsigma)$ with $0 < \omega < 1$. From inequality (2.24), we have $\dot{L} \leq -\gamma\varsigma\psi^2 + \varrho|\psi|$ and then

$$\dot{L} \leq -(1-\omega)\gamma\varsigma\psi^2 + (-\omega\gamma\varsigma\psi^2 + \varrho|\psi|), \tag{2.27}$$

where the loosening parameter $\omega \in (0, 1)$. On the right-hand side of (2.27), the first term $-(1-\omega)\gamma\varsigma\psi^2 \leq 0$. Then, the analysis of solution error ψ contains the following two situations.

- For the solution error ψ satisfying $-\omega\gamma\varsigma\psi^2 + \varrho|\psi| \leq 0$, (i.e., $|\psi| \geq \varrho/(\omega\gamma\varsigma)$, outside or on the surface of new ball $\varrho/(\omega\gamma\varsigma)$), it follows from (2.27) that

  $$\dot{L} \leq -(1-\omega)\gamma\varsigma\psi^2 + (-\omega\gamma\varsigma\psi^2 + \varrho|\psi|) \leq -(1-\omega)\gamma\varsigma\psi^2 = -2(1-\omega)\gamma\varsigma L.$$

  Then, we have $L \leq \exp(-2(1-\omega)\gamma\varsigma t)L(0)$ and

  $$|\psi| \leq |\psi(0)| \exp(-(1-\omega)\gamma\varsigma t), \forall t \in [0, t_c],$$

  where the exponential convergence rate is $(1-\omega)\gamma\varsigma$, and the convergence time $t_c = \ln(\omega\gamma\varsigma|\psi(0)|/\varrho)/((1-\omega)\gamma\varsigma)$ in view of $|\psi(0)| \exp(-(1-\omega)\gamma\varsigma t_c) = \varrho/(\omega\gamma\varsigma)$.
- For the solution error ψ satisfying $-\omega\gamma\varsigma\psi^2 + \varrho|\psi| > 0$, (i.e., $|\psi| < \varrho/(\omega\gamma\varsigma)$, inside the new ball $\varrho/(\omega\gamma\varsigma)$), such ψ can never leave the new ball. The analysis is similar to that of the third situation of Eq. (2.24).

Thus, with the loosening parameter $\omega \in (0, 1)$ selected, we have the following facts.

- If $|\psi(0)| \geq \varrho/(\omega\gamma\varsigma)$, then

$$|\psi| \begin{cases} \leq |\psi(0)| \exp(-(1-\omega)\gamma\varsigma t), \forall t \in [0, t_{\mathrm{c}}], \\ \leq \varrho/(\omega\gamma\varsigma), \forall t \in [t_{\mathrm{c}}, +\infty). \end{cases}$$

- If $|\psi(0)| \leq \varrho/(\omega\gamma\varsigma)$, then $|\psi| \leq \varrho/(\omega\gamma\varsigma), \forall t \in [0, +\infty)$.

In summary, the steady-state solution error $|\psi| = |u - u^*| \leq \varrho/(\omega\gamma\varsigma),\ \forall t \geq t_{\mathrm{c}}$. The proof is thus completed. □

References

1. Lorenz EN (1963) Deterministic nonperiodic flows. J Atmos Sci 20(2):130–141
2. Chen G, Ueta T (1999) Yet another chaotic attractor. Int J Bifurcation Chaos 9(7):1465–1466
3. Lu J, Chen G (2002) A new chaotic attractor coined. Int J Bifurcation Chaos 12(3):659–661
4. Yu Y, Zhang S (2003) Controlling uncertain Lu system using backstepping design. Chaos Solitons Fractals 15(5):897–902
5. Vanecek A, Celikovsky S (1996) Control systems: from linear analysis to synthesis of chaos. Prentice Hall, London
6. Celikovsky S, Chen G (2002) On a generalized Lorenz canonical form of chaotic systems. Int J Bifurcation Chaos 12(08):1789–1812
7. Addabbo T, Fort A, Kocarev L, Rocchi S, Vignoli V (2011) Pseudo-chaotic lossy compressors for true random number generation. IEEE Trans Circuits Syst Regul Pap 58(8):1897–1909
8. Zhu F, Xu J, Chen M (2012) The combination of high-gain sliding mode observers used as receivers in secure communication. IEEE Trans Circuits Syst Regul Pap 59(11):2702–2712
9. Wang F-Q, Liu C-X (2006) Hyperchaos evolved from the Liu chaotic system. Chin Phys 15(5):963–968
10. Iu HHC, Yu DS, Fitch AL, Sreeram V, Chen H (2011) Controlling chaos in a memristor based circuit using a twin-T notch filter. IEEE Trans Circuits Syst Regul Pap 58(6):1337–1344
11. Li W (2012) Tracking control of chaotic coronary artery system. Int J Syst Sci 43(1):21–30
12. Zhang Y, Xiao Z, Guo D, Mao M, Yin, Y (2015) Singularity-conquering tracking control of a class of chaotic systems using Zhang-gradient dynamics. IET Control Theory Appl 9(6):871–881
13. Li Y, Tang W, Chen G (2005) Generating hyperchaos via state feedback control. Int J Bifurcation Chaos 15(10):3367–3375
14. Xu Y, Zhou W, Deng L, Lu H (2008) Modified projective synchronization among three modified Chen chaotic systems with unicoupled response system. In: Proceedings of international conference on young computer scientists, pp 2903–2907
15. Ott E, Grebogi C, Yorke JA (1990) Controlling chaos. Phys Rev Lett 64(11):1196–1199
16. Slotine JE, Li W (1991) Applied nonlinear control. Prentice Hall, New Jersey
17. Hauser J, Sastry S, Kokotovic P (1992) Nonlinear control via approximate input-output linearization: the ball and beam example. IEEE Trans Automat Contr 37(3):392–398
18. Kulkarni A, Purwar S (2009) Wavelet based adaptive backstepping controller for a class of nonregular systems with input constraints. Expert Syst Appl 36(3):6686–6696
19. Tomlin CJ, Sastry SS (1998) Switching through singularities. Syst Control Lett 35(3):145–154
20. Hirschorn RM (2008) Output tracking through singularities. SIAM J Control Optim 40(4):993–1010
21. Zhang Y, Yi C (2011) Zhang neural networks and neural-dynamic method. Nova Science Publishers, New York

22. Xiao L, Zhang Y (2013) Solving time-varying nonlinear inequalities using continuous and discrete-time Zhang dynamics. Int J Comput Math 90(5):1114–1127
23. Zhang Y, Chen K, Tan H (2009) Performance analysis of gradient neural network exploited for online time-varying matrix inversion. IEEE Trans Autom Control 54(8):1940–1945
24. Zhang Y, Yang Y, Ruan G (2011) Performance analysis of gradient neural network exploited for online time-varying quadratic minimization and equality-constrained quadratic programming. Neurocomputing 74(10):1710–1719
25. Zhang Y, Wang J (2002) Global exponential stability of recurrent neural networks for synthesizing linear feedback control systems via pole assignment. IEEE Trans Neural Netw 13(3):633–644
26. Chu S, Metcalf F (1967) On Gronwall's inequality. Proc Amer Math Soc 18(3):439–440
27. Abramowitz M, Stegun IA (1972) Handbook of mathematical functions with formulas, graphs, and mathematical tables. Dover, New York

Chapter 3
ZG Synchronization of Lu and Chen Chaotic Systems

Abstract Recently, the synchronization of two chaotic systems is attracting more and more attention because of its potential applications in communication security, aerospace industry and many other fields. In this chapter, the ZG method is investigated for chaos synchronization with multiple inputs (i.e., three or two inputs). Based on the ZG method, the traditional three-input chaos synchronization problem can be successfully solved with desirable convergence rate and satisfactory accuracy. In addition, by taking advantage of the coupling property of chaotic system, an important extension of the ZG method is investigated to solve the thorny two-input chaos synchronization problem. Simulation results illustrate that the controller groups designed by the ZG method not only achieve satisfactory synchronization accuracy and exponential convergence rate on the three-input chaos synchronization problem but also successfully solve the chaos synchronization problem with only two inputs.

3.1 Introduction

Chaotic behaviors can be observed in many real-world physical systems, such as chemical reactors, feedback control devices, and laser systems. As introduced in the previous chapter, the first chaotic system was found by Lorenz in 1963, and much effort has been devoted to the study of chaotic system. As one of the major branches in the study of chaotic system, chaos synchronization plays an important role in the secure communication, encryption and many other fields [1–3], which is investigated in this chapter [4].

In recent years, ZD method [5] has been developed to solve time-varying problems. On the other hand, GD method [6] is intrinsically feasible and efficient to solve time-invariant problems and has been generalized to solve time-varying problems. By combining the ZD and GD methods together, an effective controller-design method termed ZG method was developed for solving the tracking control problem, which possesses both advantages of ZD and GD in accuracy and convergence performance [7]. Compared with other controller-design methods such as the IOL

Y. Zhang et al., *Zhang-Gradient Control*,
https://doi.org/10.1007/978-981-15-8257-8_3

method and the backstepping design technique, the ZG method does not need to introduce any Lyapunov function during the design process and it does not need to define virtual control, which make it a very easy way to design the controllers. Some special controllers based on the ZG method have already been designed for practical systems [8–10] such as the ship course-tracking system, and the corresponding results have illustrated the feasibility and superiority of the ZG method.

In this chapter [4], the ZG method is developed for solving the chaos synchronization problem with multiple inputs. Specifically, by combining the ZG method and the coupling property of the chaotic system, an important extension of the ZG method to solve the chaos synchronization problem with only two inputs is investigated detailedly. The corresponding simulation results show that this ZG method works well for both three-input and two-input chaos synchronization problems.

3.2 ZG Control via Three Inputs

To illustrate the feasibility and efficacy of the ZG method for solving the chaos synchronization problem, the synchronization between the Lu chaotic system and the Chen chaotic system with three inputs is considered in this section. Without loss of generality, the Chen chaotic system is supposed to be the response system, while the Lu chaotic system is supposed to be the drive system.

3.2.1 Problem Description

The Chen chaotic system with three inputs as the response system is presented as follows:

$$\begin{cases} \dot{x}_{1\mathrm{r}} = f_1(x_{1\mathrm{r}}, x_{2\mathrm{r}}, u_1) = a_\mathrm{r}(x_{2\mathrm{r}} - x_{1\mathrm{r}}) + u_1, \\ \dot{x}_{2\mathrm{r}} = f_2(x_{1\mathrm{r}}, x_{2\mathrm{r}}, x_{3\mathrm{r}}, u_2) = (c_\mathrm{r} - a_\mathrm{r})x_{1\mathrm{r}} - x_{1\mathrm{r}}x_{3\mathrm{r}} + c_\mathrm{r}x_{2\mathrm{r}} + u_2, \\ \dot{x}_{3\mathrm{r}} = f_3(x_{1\mathrm{r}}, x_{2\mathrm{r}}, x_{3\mathrm{r}}, u_3) = x_{1\mathrm{r}}x_{2\mathrm{r}} - b_\mathrm{r}x_{3\mathrm{r}} + u_3, \end{cases} \tag{3.1}$$

where u_1, u_2 and u_3 are the control inputs used to make the response system synchronize with the drive system. For notational convenience, the subscripts "d" and "r" are used to distinguish the drive system and the response system. Then, the corresponding synchronization errors can be defined as $e_1 = x_{1\mathrm{d}} - x_{1\mathrm{r}}$, $e_2 = x_{2\mathrm{d}} - x_{2\mathrm{r}}$ and $e_3 = x_{3\mathrm{d}} - x_{3\mathrm{r}}$. The objective is to design controllers such that e_1, e_2 and e_3 asymptotically approach zero (or near zero in practice).

3.2.2 Design of ZD and ZG Controller Groups

To monitor and control the synchronization process, the first ZF is defined as $z_1 = x_{1d} - x_{1r}$. By following the ZD method, the ZD design formula $\dot{z}_1 = -\lambda z_1$ is employed, where the design parameter $\lambda \in \mathbb{R}^+$ is used to scale the convergence rate of the ZD solution [10]. With the first equation of system (3.1) used, we expand the above ZD design formula and then obtain

$$-\lambda(x_{1d} - x_{1r}) = a_d(x_{2d} - x_{1d}) - (a_r(x_{2r} - x_{1r}) + u_1). \tag{3.2}$$

Thus, a controller in the form of u_1 for system (3.1) is obtained from (3.2), i.e.,

$$u_1 = a_d(x_{2d} - x_{1d}) - a_r(x_{2r} - x_{1r}) + \lambda(x_{1d} - x_{1r}). \tag{3.3}$$

Since ZD is exploited once and GD is not exploited during the design of controller (3.3), such a controller is termed z1g0 controller. Similarly, by defining $z_2 = x_{2d} - x_{2r}$ and $z_3 = x_{3d} - x_{3r}$ as well as applying the ZD design formula, two controllers in the forms of u_2 and u_3 are derived as

$$\begin{cases} u_2 = \lambda(x_{2d} - x_{2r}) + (a_r - c_r)x_{1r} + x_{1r}x_{3r} - c_r x_{2r} - x_{1d}x_{3d} + c_d x_{2d}, \\ u_3 = \lambda(x_{3d} - x_{3r}) - x_{1r}x_{2r} + b_r x_{3r} + x_{1d}x_{2d} - b_d x_{3d}. \end{cases} \tag{3.4}$$

Furthermore, from (3.3), we can define

$$\phi_1 = a_d(x_{2d} - x_{1d}) - a_r(x_{2r} - x_{1r}) - \lambda(x_{1r} - x_{1d}) - u_1.$$

Then, by following the GD method, an EF is defined as

$$\epsilon_1 = \frac{\phi_1^2}{2},$$

and the GD design formula is adopted as follows:

$$\dot{u}_1 = -\gamma \frac{\partial \epsilon_1}{\partial u_1}, \tag{3.5}$$

where the design parameter $\gamma \in \mathbb{R}^+$ is used to scale the convergence rate of the GD solution [10]. Thus, a controller in the form of $\dot{u}_1$ is derived from (3.5):

$$\dot{u}_1 = \gamma\phi_1 = \gamma(a_d(x_{2d} - x_{1d}) - a_r(x_{2r} - x_{1r}) - \lambda(x_{1r} - x_{1d}) - u_1). \tag{3.6}$$

Since ZD is exploited once and GD is also exploited once during the design of controller (3.6), such a controller is termed z1g1 controller. Similarly, two controllers in the forms of $\dot{u}_2$ and $\dot{u}_3$ are designed as

$$\begin{cases} \dot{u}_2 = \gamma(\lambda(x_{2d} - x_{2r}) + (a_r - c_r)x_{1r} + x_{1r}x_{3r} - c_r x_{2r} - x_{1d}x_{3d} + c_d x_{2d} - u_2), \\ \dot{u}_3 = \gamma(\lambda(x_{3d} - x_{3r}) - x_{1r}x_{2r} + b_r x_{3r} + x_{1d}x_{2d} - b_d x_{3d} - u_3). \end{cases} \tag{3.7}$$

In summary, by combining (3.3) and (3.4), a group of z1g0 controllers (or say, a z3g0 controller group) in the forms of u_1, u_2 and u_3 can be designed as

$$\begin{cases} u_1 = a_d(x_{2d} - x_{1d}) - a_r(x_{2r} - x_{1r}) + \lambda(x_{1d} - x_{1r}), \\ u_2 = \lambda(x_{2d} - x_{2r}) + (a_r - c_r)x_{1r} + x_{1r}x_{3r} - c_r x_{2r} - x_{1d}x_{3d} + c_d x_{2d}, \\ u_3 = \lambda(x_{3d} - x_{3r}) - x_{1r}x_{2r} + b_r x_{3r} + x_{1d}x_{2d} - b_d x_{3d}. \end{cases} \tag{3.8}$$

In addition, by combining (3.6) and (3.7), a z3g3 controller group in the forms of $\dot{u}_1$, $\dot{u}_2$ and $\dot{u}_3$ can be designed as

$$\begin{cases} \dot{u}_1 = \gamma(a_d(x_{2d} - x_{1d}) - a_r(x_{2r} - x_{1r}) - \lambda(x_{1r} - x_{1d}) - u_1), \\ \dot{u}_2 = \gamma(\lambda(x_{2d} - x_{2r}) + (a_r - c_r)x_{1r} + x_{1r}x_{3r} - c_r x_{2r} - x_{1d}x_{3d} + c_d x_{2d} - u_2), \\ \dot{u}_3 = \gamma(\lambda(x_{3d} - x_{3r}) - x_{1r}x_{2r} + b_r x_{3r} + x_{1d}x_{2d} - b_d x_{3d} - u_3). \end{cases} \tag{3.9}$$

So far, we design two different types of controller groups (i.e. z3g0 and z3g3 controller groups) via the combined ZG method for solving the chaos synchronization problem with three inputs. Corresponding simulation results are presented in the ensuing subsection to illustrate the efficacy of such two controller groups.

3.2.3 Simulation and Verification

Without loss of generality, both systems (2.1) and (3.1) in the simulations are started with randomly generated initial states. In addition, $u_1(0)$, $u_2(0)$ and $u_3(0)$ for controller group (3.9) are set as zero. Besides, parameters $\lambda = 1$ and $\gamma = 10^5$.

The synchronization results between Chen chaotic system (3.1) equipped with controller group (3.8) and Lu chaotic system (2.1) are illustrated in Fig. 3.1. Specifically, Fig. 3.1a–c show that the trajectory of each state variable of Chen chaotic system (3.1) rapidly converges to the corresponding state variable of Lu chaotic system (2.1). In addition, Fig. 3.1d presents a three-dimensional view of the entire synchronization process. Besides, the synchronization error of each state variable shown in Fig. 3.1e converges to zero in an exponential manner. Moreover, Fig. 3.1f shows that the absolute synchronization errors (i.e., the absolute values

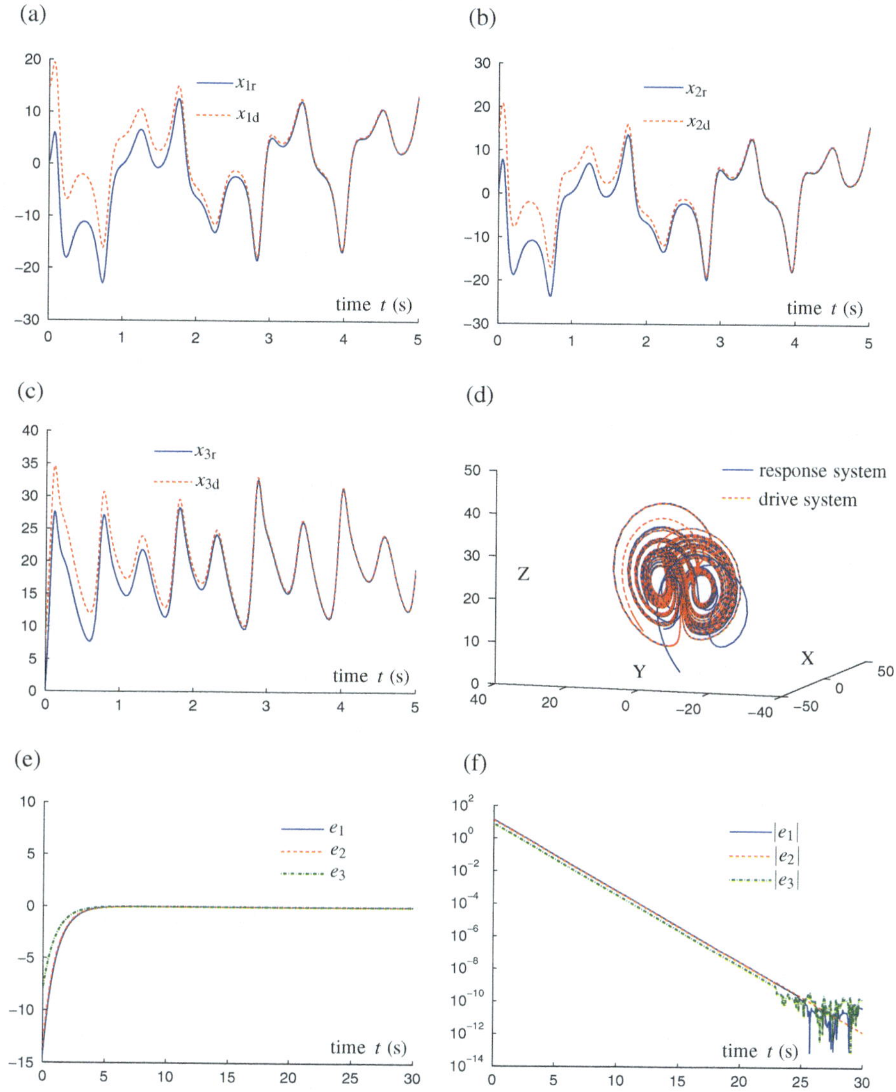

Fig. 3.1 Synchronization performance between Lu chaotic system (2.1) and Chen chaotic system (3.1) equipped with three inputs and using z3g0 controller group (3.8). (**a**) Trajectories of x_{1r} and x_{1d}. (**b**) Trajectories of x_{2r} and x_{2d}. (**c**) Trajectories of x_{3r} and x_{3d}. (**d**) Three-dimensional trajectories. (**e**) Synchronization errors. (**f**) Orders of $|e_1|$, $|e_2|$ and $|e_3|$

of synchronization errors) keep below the order of 10^{-9} after entering the steady state, which substantiates the high accuracy of the ZG method. Similar results can be easily found from Fig. 3.2; i.e., z3g3 controller group (3.9) has the same efficacy as z3g0 controller group (3.8) with tight error bound and exponential convergence

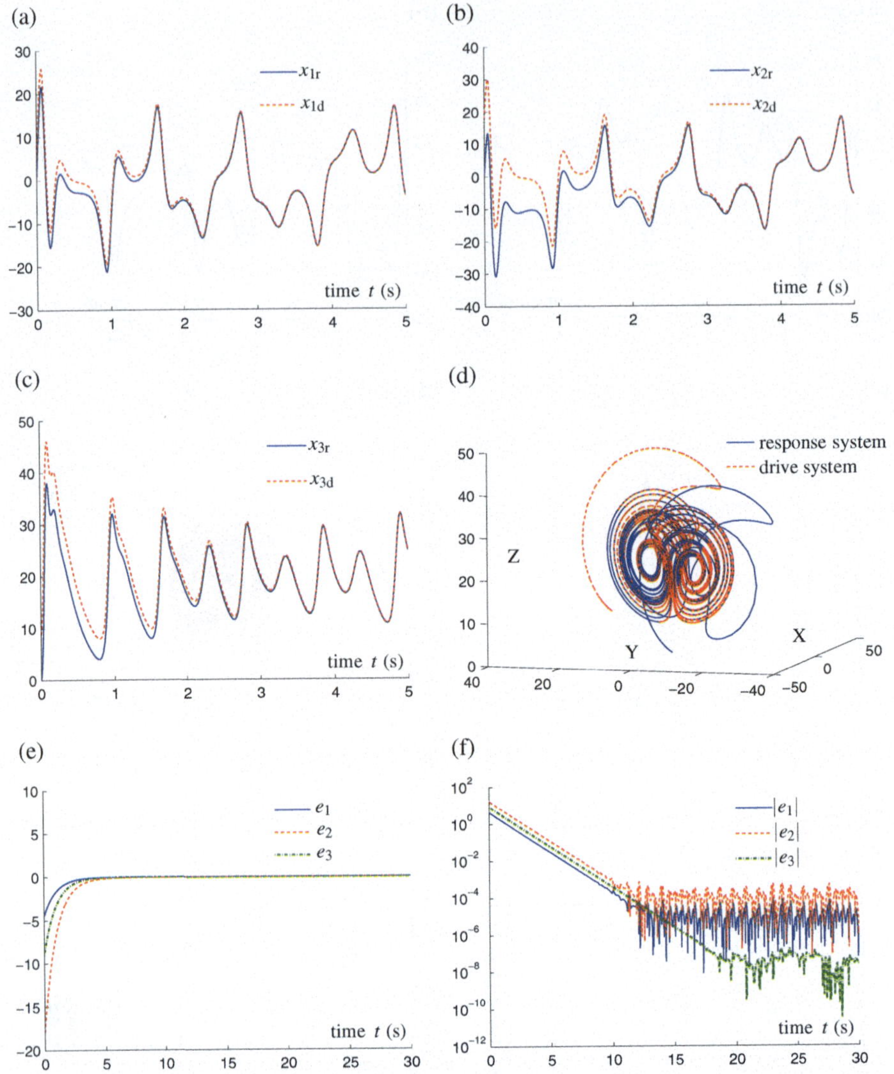

Fig. 3.2 Synchronization performance between Lu chaotic system (2.1) and Chen chaotic system (3.1) equipped with three inputs and using z3g3 controller group (3.9). (**a**) Trajectories of x_{1r} and x_{1d}. (**b**) Trajectories of x_{2r} and x_{2d}. (**c**) Trajectories of x_{3r} and x_{3d}. (**d**) Three-dimensional trajectories. (**e**) Synchronization errors. (**f**) Orders of $|e_1|$, $|e_2|$ and $|e_3|$

rate. These simulation results illustrate the efficacy of the ZG method for solving chaos synchronization problem between Lu chaotic system (2.1) and Chen chaotic system (3.1) with three inputs.

3.3 ZG Control via Two Inputs

In this section, a group of synchronization controllers with only two inputs is designed via the extension of the ZG method. Without loss of generality, the response system and the drive system are defined the same as those in Sect. 3.2 (i.e., the Chen chaotic system is the response system and the Lu chaotic system is the drive system). Simulation results are given to show the satisfactory convergence performance and the high accuracy of this effective extension method.

3.3.1 *Problem Description*

As we know, it is difficult to achieve synchronization between two different chaotic systems with only two inputs. In this section, by combining the ZG method and the coupling property of the chaotic system, a type of controller is designed and presented for solving this difficult problem. The Chen chaotic system with two inputs is shown as below (note that the control inputs can be added to any two states of the response system):

$$\begin{cases} \dot{x}_{1\mathrm{r}} = f_1(x_{1\mathrm{r}}, x_{2\mathrm{r}}, u_1) = a_\mathrm{r}(x_{2\mathrm{r}} - x_{1\mathrm{r}}) + u_1, \\ \dot{x}_{2\mathrm{r}} = f_2(x_{1\mathrm{r}}, x_{2\mathrm{r}}, x_{3\mathrm{r}}, u_2) = (c_\mathrm{r} - a_\mathrm{r})x_{1\mathrm{r}} - x_{1\mathrm{r}}x_{3\mathrm{r}} + c_\mathrm{r}x_{2\mathrm{r}} + u_2, \\ \dot{x}_{3\mathrm{r}} = f_3(x_{1\mathrm{r}}, x_{2\mathrm{r}}, x_{3\mathrm{r}}) = x_{1\mathrm{r}}x_{2\mathrm{r}} - b_\mathrm{r}x_{3\mathrm{r}}, \end{cases} \tag{3.10}$$

where u_1 and u_2 are the control inputs to be designed for making Chen chaotic system (3.10) synchronize with Lu chaotic system (2.1). Similar to Sect. 3.2, the synchronization errors are defined as $e_1 = x_{1\mathrm{d}} - x_{1\mathrm{r}}$, $e_2 = x_{2\mathrm{d}} - x_{2\mathrm{r}}$ and $e_3 = x_{3\mathrm{d}} - x_{3\mathrm{r}}$. The objective is to design controllers u_1 and u_2 such that e_1, e_2 and e_3 converge to zero as soon as possible.

3.3.2 *Design of ZG Controller Group*

Similar to the design procedure shown in Sect. 3.2, we apply the ZD design formulas and obtain

$$\begin{cases} \lambda(x_{1\mathrm{d}} - x_{1\mathrm{r}}) = -a_\mathrm{d}(x_{2\mathrm{d}} + x_{1\mathrm{d}}) + a_\mathrm{r}(x_{2\mathrm{r}} - x_{1\mathrm{r}}) + u_1, \\ \lambda(x_{2\mathrm{d}} - x_{2\mathrm{r}}) = x_{1\mathrm{d}}x_{3\mathrm{d}} - c_\mathrm{d}x_{2\mathrm{d}} + ((c_\mathrm{r} - a_\mathrm{r})x_{1\mathrm{r}} - x_{1\mathrm{r}}x_{3\mathrm{r}} + c_\mathrm{r}x_{2\mathrm{r}} + u_2), \\ \lambda(x_{3\mathrm{d}} - x_{3\mathrm{r}}) = -x_{1\mathrm{d}}x_{2\mathrm{d}} + b_\mathrm{d}x_{3\mathrm{d}} + (x_{1\mathrm{r}}x_{2\mathrm{r}} - b_\mathrm{r}x_{3\mathrm{r}}). \end{cases} \tag{3.11}$$

Since there does not exist u_1 or u_2 in the third equation of (3.11) explicitly, the following ZF is further constructed:

$$z_1 = x_{1d}x_{2d} - b_d x_{3d} - (x_{1r}x_{2r} - b_r x_{3r}) + \lambda(x_{3d} - x_{3r}).$$

By applying the following ZD design formula:

$$\dot{z}_1 = -\lambda z_1,$$

the following result can be obtained:

$$\begin{aligned}\dot{x}_d x_{2d} + x_{1d}\dot{x}_{2d} - b_d\dot{x}_{3d} - (\dot{x}_{1r}x_{2r} + \dot{x}_{2r}x_{1r} - b_r\dot{x}_{3r}) + \lambda(\dot{x}_{3d} - \dot{x}_{3r})\\ = -\lambda(x_{1d}x_{2d} - b_d x_{3d} - (x_{1r}x_{2r} - b_r x_{3r}) + \lambda(x_{3d} - x_{3r})).\end{aligned} \tag{3.12}$$

Combining (3.11) and (3.12), we can define

$$\begin{cases}\phi_1 = u_1 + a_r(x_{2r} - x_{1r}) + \lambda(x_{1r} - x_{1d}) - a_d(x_{2d} - x_{1d}),\\ \phi_2 = \lambda(x_{2d} - x_{2r}) - (c_r - a_r)x_{1r} + x_{1r}x_{3r} - c_r x_{2r} - x_{1d}x_{3d} + c_d x_{2d} - u_2,\\ \phi_3 = -\,(a_r(x_{2r} - x_{1r}) + u_1)x_{2r} - x_{1r}((c_r - a_r)x_{1r} - x_{1r}x_{3r} + c_r x_{2r} + u_2)\\ \quad + b_r(x_{1r}x_{2r} - b_r x_{3r}) + a_d(x_{2d} - x_{1d})x_{2d} + x_{1d}(-x_{1d}x_{3d} + c_d x_{2d})\\ \quad - b_d(x_{1d}x_{2d} - b_d x_{3d}) + \lambda((x_{1d}x_{2d} - b_d x_{3d}) - (x_{1r}x_{2r} - b_r x_{3r}))\\ \quad + \lambda(x_{1d}x_{2d} - b_d x_{3d} - (x_{1r}x_{2r} - b_r x_{3r}) + \lambda(x_{3d} - x_{3r})).\end{cases} \tag{3.13}$$

According to the GD method, the corresponding EFs are defined as $\epsilon_1 = \phi_1^2/2$, $\epsilon_2 = \phi_2^2/2$ and $\epsilon_3 = \phi_3^2/2$. The GD design formulas are constructed as

$$\begin{cases}\dot{u}_1 = -\gamma\frac{\partial(\epsilon_1+\epsilon_3)}{\partial u_1},\\ \dot{u}_2 = -\gamma\frac{\partial(\epsilon_2+\epsilon_3)}{\partial u_2}.\end{cases} \tag{3.14}$$

The above GD design formulas aim at synchronizing two state variables by using only one control input. Since the state variables x_1, x_2 and x_3 are coupled with each other, these controllers can be expected to have good effects. According to (3.14), we can derive another synchronization controller group in the forms of $\dot{u}_1$ and $\dot{u}_2$:

$$\begin{cases}\dot{u}_1 = -\gamma(\phi_1 - \phi_3 x_{2r}),\\ \dot{u}_2 = -\gamma(-\phi_2 - \phi_3 x_{1r}),\end{cases} \tag{3.15}$$

where ϕ_1, ϕ_2 and ϕ_3 are defined in (3.13). Since ZD is exploited four times and GD is exploited two times during the design procedure, controller group (3.15) is termed z4g2 controller group.

In summary, by extending the GD design formula into two directions, a type of controller group termed z4g2 controller group is derived, which can synchronize two different chaotic systems by using only two inputs.

3.3.3 *Simulation and Verification*

In this subsection, the initial values are set the same as those in Sect. 3.2.3. Besides, the parameters are set as $\lambda = 10$ and $\gamma = 10^5$.

From Fig. 3.3a–c, we can observe that the Chen chaotic system is completely synchronized with the Lu chaotic system after very short-time oscillation. In addition, Fig. 3.3d provides a three-dimensional view of two chaotic systems, from which we can find that, in spite of the oscillation of the Chen chaotic system in the beginning, the Chen chaotic system achieves the synchronization with the Lu chaotic system. Moreover, Fig. 3.3e shows that the corresponding synchronization errors converge to zero rapidly. Besides, Fig. 3.3f shows the high accuracy of this ZG method with the absolute synchronization errors keeping below the order of 10^{-3} after a short period of time. Therefore, the feasibility and efficacy of z4g2 controller group (3.15) for solving the chaos synchronization problem between Lu chaotic system (2.1) and Chen chaotic system (3.10) with two control inputs are well substantiated.

3.4 Chapter Summary

In this chapter, the simple and effective ZG method has been developed for solving the chaos synchronization problem with three or two inputs. Specifically, through the ZG method, z3g0 controller group (3.8) and z3g3 controller group (3.9) have been designed, both of which have fulfilled the chaos synchronization tasks between the Lu chaotic system and Chen chaotic system with three inputs effectively. Besides, z4g2 controller group (3.15) based on the extension of the ZG method has been designed and presented for the chaos synchronization with only two inputs. Finally, corresponding simulation results have substantiated the feasibility and efficacy of the presented ZG controller groups for chaos synchronization with multiple inputs (i.e., three or two inputs).

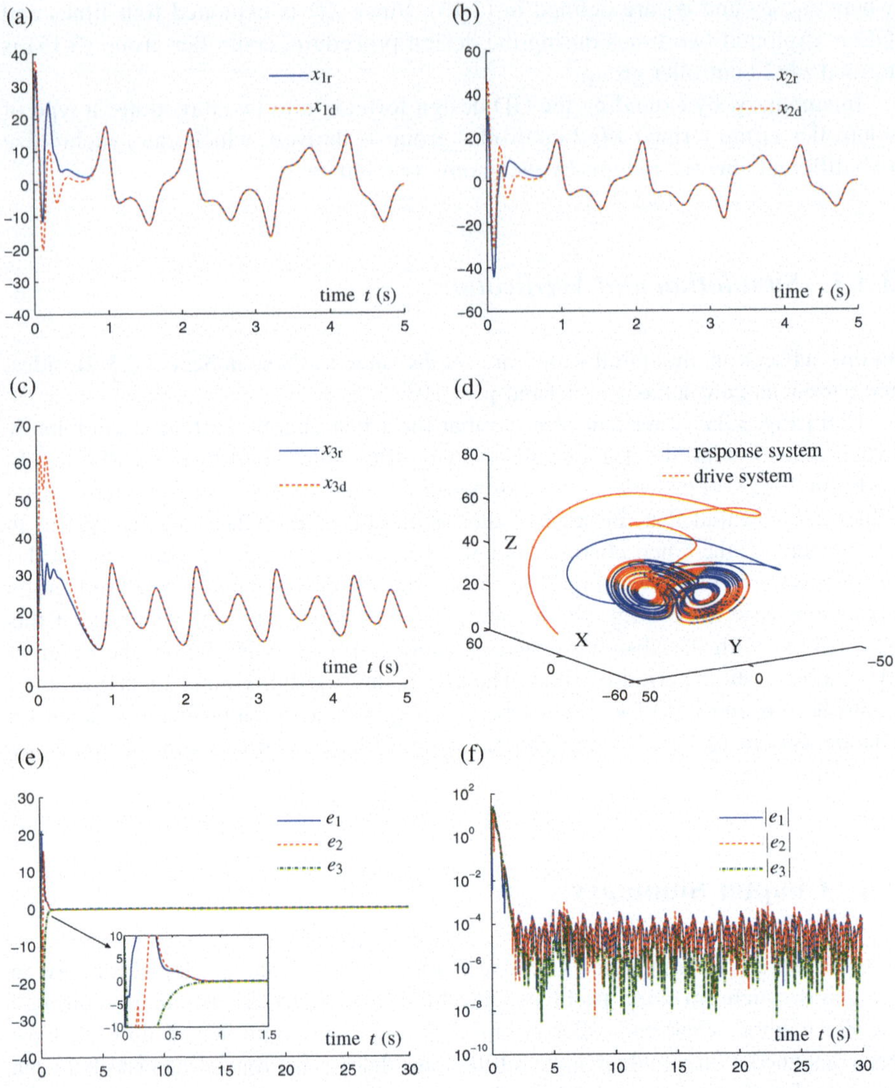

Fig. 3.3 Synchronization performance between Lu chaotic system (2.1) and Chen chaotic system (3.10) equipped with two inputs and using z4g2 controller group (3.15). (**a**) Trajectories of $x_{1\mathrm{r}}$ and $x_{1\mathrm{d}}$. (**b**) Trajectories of $x_{2\mathrm{r}}$ and $x_{2\mathrm{d}}$. (**c**) Trajectories of $x_{3\mathrm{r}}$ and $x_{3\mathrm{d}}$. (**d**) Three-dimensional trajectories. (**e**) Synchronization errors. (**f**) Orders of $|e_1|$, $|e_2|$ and $|e_3|$

References

1. Ott E, Grebogi C, Yorke J (1990) Controlling chaos. Phys Rev Lett 64(11):1196–1199
2. Feki M (2003) An adaptive chaos synchronization scheme applied to secure communication. Chaos Solitons Fractals 18(1):141–148
3. Agiza HN, Yassen MT (2001) Synchronization of Rossler and Chen chaotic dynamical systems using active control. Phys Lett A 278(4):191–197
4. Zhang Y, Liu M, Jin L, Zhang Y, Tan H (2015) Synchronization of two chaotic systems with three or two inputs via ZG method. In: Proceedings of the 34th Chinese control conference, pp 563–568
5. Zhang Y, Yi C (2011) Zhang neural networks and neural-dynamic method. Nova Science Publishers, New York
6. Zhang Y, Yi C, Guo D, Zheng J (2011) Comparison on Zhang neural dynamics and gradient-based neural dynamics for online solution of nonlinear time-varying equation. Neural Comput Appl 20(1):1–7
7. Zhang Y, Yin Y, Wu H, Guo D (2012) Zhang dynamics and gradient dynamics with tracking-control application. In: Proceedings of the 5th international symposium on computational intelligence and design, pp 235–238
8. Yin Y, Xie Q, Wang Y, Chen D, Zhang Y (2013) ZG control for ship course tracking with singularity considered and solved. In: Proceedings of the 11th IEEE international conference on dependable, autonomic and secure computing, pp 352–357
9. Zhang Y, Peng C, Yu X, Yin Y, Ling Y (2013) ZD and ZG controllers for explicit and implicit tracking of pendulum with singularity finally conquered. In: Proceedings of international conference on machine learning and cybernetics, pp 777–782
10. Zhang Y, Yu X, Yin Y, Peng C, Fan Z (2014) Singularity-conquering ZG controllers of z2g1 type for tracking control of the IPC system. Int J Control 87(9):1729–1746

Chapter 4
ZG Tracking Control of Modified Lorenz Nonlinear System

Abstract The tracking-control problem of a special nonlinear system (i.e., the extension of a modified Lorenz chaotic system) with additive control input or the mixture of additive and multiplicative control inputs is considered in this chapter. It is worth pointing out that, with the parameters fixed at some particular values, the modified Lorenz nonlinear system degrades to the modified Lorenz chaotic system. Due to the existence of the DBZ problem at which the nonlinear system fails to have a well-defined relative degree, the IOL method and the backstepping design technique cannot effectively solve the tracking-control problem. By combining the ZD and GD methods together, a simple and effective controller-design method, termed ZG method, is presented for the tracking control of the modified Lorenz nonlinear system. With the DBZ problem conquered, this ZG method can solve the tracking-control problem of the modified Lorenz nonlinear system via additive control input or mixed control inputs (i.e., the mixture of additive and multiplicative control inputs). Both theoretical analyses and simulative verifications substantiate that the tracking controllers based on the ZG method with additive control input or mixed control inputs not only achieve satisfactory tracking accuracy but also successfully conquer the DBZ problem encountered during the tracking-control process.

4.1 Introduction

The chaos theory is considered to be one of the emerging research fields widely encountered in a variety of scientific and engineering fields. As a system of three ordinary differential equations, the first chaotic system was developed in 1963 for atmospheric convection by Lorenz [1], which is shown as follows:

$$\begin{cases} \dot{x}_1 = f_1(x_1, x_2) = a(x_2 - x_1), \\ \dot{x}_2 = f_2(x_1, x_2, x_3) = cx_1 - x_1x_3 - x_2, \\ \dot{x}_3 = f_3(x_1, x_2, x_3) = x_1x_2 - bx_3, \end{cases} \tag{4.1}$$

Y. Zhang et al., *Zhang-Gradient Control*,
https://doi.org/10.1007/978-981-15-8257-8_4

where x_1, x_2 and x_3 are the system states with constants $a = 10$, $b = 8/3$ and $c = 28$. A new chaotic system modified from Lorenz chaotic system (4.1) with fewer terms was presented in [2], which is simpler than those of existing seven-term or six-term ones and shown as follows:

$$\begin{cases} \dot{x}_1 = f_1(x_1, x_2) = a(x_2 - x_1), \\ \dot{x}_2 = f_2(x_1, x_3) = -x_1 x_3, \\ \dot{x}_3 = f_3(x_1, x_2) = x_1 x_2 - b, \end{cases} \tag{4.2}$$

where $a = 1$ and $b = 1$ for additive control input or $b = 1$ with a being a variable of mixed control inputs in this chapter [3].

Two main directions are frequently considered for the investigations of chaotic systems. The first direction includes: (1) utilizing the chaotic behaviors for practical applications, such as true random number generators and encryption and secure communication [4, 5]; (2) generating chaotic system with more chaotic behaviors, such as hyperchaotic systems [6]. The other direction is to get rid of chaotic behaviors existing in physical systems such as electrical circuits, fluid dynamics and mechanical devices, since the systems may perform unexpected behaviors under the influence of chaos [7]. Both the investigations of the above directions involve control, especially the tracking control of chaos. In this chapter [3], the tracking control of modified Lorenz chaotic system (4.2) using additive control input is investigated. In view of the fact that a chaotic system only remains chaotic with particularly fixed parameters and may become a general nonlinear system with parameters changed, the ZG controller group using the mixture of additive and multiplicative control inputs is also applied to the tracking control of a more general nonlinear system (i.e., the modified Lorenz nonlinear system).

IOL is an effective method for solving nonlinear tracking-control problems [8, 9]. However, with the existence of DBZ point(s) at which the nonlinear system fails to have a well-defined relative degree, the IOL method fails to solve the corresponding tracking-control problem. With some nonlinear terms that may generate DBZ points neglected, a method for constructing the approximate nonlinear systems that are input-output linearizable was presented in [10], which attempted to overcome the DBZ drawback. In [11], a switching scheme provided an idea of switching between approximate and exact IOL, so as to avoid the DBZ points and improve the tracking performance. However, this approach costs much in terms of the implementation since it requires two or more controllers for solving a single problem. The backstepping design technique, being a kind of systematic synthetic technique to controller, is a recursive procedure that combines the choice of a Lyapunov function with the design of feedback control. The aforementioned methods generally bring in the operation of division, which leads to the potential possibility of generating DBZ points.

Recently, a model-free optimal control strategy [12] and fuzzy-neural-network based control methods [13, 14] have been presented, in addition to control methods aiming at particular fractional order chaotic systems [15, 16], which all show that there exist more novel and effective control methods to be explored for nonlinear systems. ZD method has been developed to solve time-varying nonlinear equation [17], time-varying matrix pseudoinversion [18], time-varying complex Sylvester equation [19], and so on. On the other hand, GD method is intrinsically feasible and efficient to solve time-invariant problems and has been generalized to solve time-varying problems, such as time-varying matrix inversion [20], time-varying quadratic minimization and equality-constrained quadratic programming [21], and time-varying nonlinear equation [22]. The previous studies generally exploited such ZD and GD methods individually and comparatively, and other researchers rarely consider combining them for the problem solving or discover the superiority of their combination.

In this chapter [3], the ZG method is presented to solve the tracking-control problem of modified Lorenz chaotic system (4.2) in a division-free manner, which can effectively conquer the DBZ problem. Corresponding theoretical analyses and results are provided. Computer simulations are conducted to further substantiate the feasibility and efficacy of the controllers.

4.2 ZG Control via Additive Input

In this section, the tracking-control problem of modified Lorenz chaotic system (4.2) equipped with a single additive control input u is presented as an illustrative example. A tracking controller based on the IOL method is presented to point out the DBZ problem encountered during the tracking-control process. Afterwards, the detailed procedures for designing the ZG controller to solve the DBZ problem in tracking control together with the corresponding theoretical analyses are provided.

Consider the following modified Lorenz chaotic system equipped with a single additive control input u:

$$\begin{cases} \dot{x}_1 = f_1(x_1, x_2) = a(x_2 - x_1), \\ \dot{x}_2 = f_2(x_1, x_3, u) = -x_1x_3 + u, \\ \dot{x}_3 = f_3(x_1, x_2) = x_1x_2 - b. \end{cases} \tag{4.3}$$

Besides, $y = x_3$ denotes the output of system (4.3). The objective of tracking control is to design a controller such that y tracks the smooth and bounded desired trajectory y_d, with the tracking error $e = y - y_d$ asymptotically approaching zero (or say, near zero in practice).

According to the IOL method [8, 9], when $x_1 \neq 0$, the following conventional IOL controller can be obtained to solve the tracking-control problem of modified Lorenz nonlinear system (4.3):

$$u = \frac{1}{x_1}\left(x_1^2x_3 + (a - k_1 - k_0)x_1x_2 - ax_2^2 + \alpha_1\right), \tag{4.4}$$

where $\alpha_1 = -k_1k_0x_3 + k_1b + k_0b + \ddot{y}_\mathrm{d} + (k_0 + k_1)\dot{y}_\mathrm{d} + k_1k_0y_\mathrm{d}$, with k_0 and k_1 satisfying the condition that all roots of the characteristic polynomial $P(s) = s^2 + (k_0 + k_1)s + k_1k_0$ are in the open left-half complex plane. Evidently, this IOL controller (4.4) may encounter the DBZ problem. Specifically, when the value of x_1 is close or even equal to zero, the magnitude of controller (4.4) becomes extremely large or even infinite. This DBZ situation, for which controller (4.4) cannot tackle well, may lead to system crash during the tracking-control process.

4.2.1 Design of ZG Controller

To monitor and control the tracking process, the following first ZF is constructed:

$$z_1 = y - y_\mathrm{d} = x_3 - y_\mathrm{d}.$$

According to the ZD method [18], the following design formula is adopted:

$$\dot{z}_1 = \frac{\mathrm{d}z_1}{\mathrm{d}t} = -\lambda z_1, \tag{4.5}$$

where the design parameter $\lambda \in \mathbb{R}^+$ is used to scale the convergence rate of the ZD solution. It is worth pointing out that different values of λ can affect the convergence performance of the tracking controller. Expanding (4.5), we thus obtain $\dot{x}_3 - \dot{y}_\mathrm{d} = -\lambda(x_3 - y_\mathrm{d})$. It follows from the third equation of system (4.3) that

$$x_1x_2 - b + \lambda x_3 - \dot{y}_\mathrm{d} - \lambda y_\mathrm{d} = 0. \tag{4.6}$$

Since there does not explicitly contain u in (4.6), the second ZF is further constructed as $z_2 = x_1x_2 - b + \lambda x_3 - \dot{y}_\mathrm{d} - \lambda y_\mathrm{d}$. Applying the ZD method again, we have the design formula $\dot{z}_2 = -\lambda z_2$, and then $\dot{x}_1x_2 + x_1\dot{x}_2 + \lambda\dot{x}_3 - \ddot{y}_\mathrm{d} - \lambda\dot{y}_\mathrm{d} = -\lambda(x_1x_2 - b + \lambda x_3 - \dot{y}_\mathrm{d} - \lambda y_\mathrm{d})$. It follows from system (4.3) that

$$x_1^2x_3 + (a - 2\lambda)x_1x_2 - ax_2^2 - x_1u + \alpha_2 = 0, \tag{4.7}$$

where $\alpha_2 = -\lambda^2x_3 + 2\lambda b + \ddot{y}_\mathrm{d} + 2\lambda\dot{y}_\mathrm{d} + \lambda^2y_\mathrm{d}$. Thus, a tracking controller in the form of u for modified Lorenz nonlinear system (4.3) is directly obtained from (4.7), i.e.,

$$u = \frac{1}{x_1}\left(x_1^2x_3 + (a - 2\lambda)x_1x_2 - ax_2^2 + \alpha_2\right). \tag{4.8}$$

Note that, for controller (4.8), there also exist DBZ points when $x_1 = 0$.

Evidently, the possibility of DBZ points mostly lies in the division operation, such as controllers (4.4) and (4.8). To conquer these DBZ points when $x_1 = 0$, the basic idea in this chapter [3] is to transform direct control (4.8) into a time-varying minimization problem. Therefore, the GD method is further exploited for the design of a controller in the form of $\dot{u}$ based on (4.8).

From (4.8), we define $\phi = x_1^2x_3 + (a - 2\lambda)x_1x_2 - ax_2^2 - x_1u + \alpha_2$. Based on the GD method [20, 20, 22], the square-based EF is defined as

$$\epsilon = \frac{1}{2}\phi^2. \tag{4.9}$$

Then, we adopt the following GD design formula:

$$\dot{u} = -\gamma\frac{\partial\epsilon}{\partial u}, \tag{4.10}$$

where the design parameter $\gamma \in \mathbb{R}^+$ is used to scale the convergence rate of the GD solution. From (4.9) and (4.10), the tracking controller in the form of $\dot{u}$ for the modified Lorenz nonlinear system (4.3) is finally obtained as

$$\dot{u} = \gamma x_1\phi = \gamma x_1(x_1^2x_3 + (a - 2\lambda)x_1x_2 - ax_2^2 - x_1u + \alpha_2). \tag{4.11}$$

Note that, as compared with controller (4.8), ZG controller (4.11) is division-free and can conquer the DBZ points when $x_1 = 0$. To sum up, the controller design of ZG method contains two procedures: (1) an elementary controller can be designed by constructing a series of ZFs; (2) based on the elementary controller, the GD method is incorporated for a division-free controller.

4.2.2 Convergence Performance Analyses on ZG Controller

In this subsection, we analyze the tracking-error bound and convergence rate of ZG controller (4.11) for the modified Lorenz nonlinear system (4.3) with the DBZ problem conquered.

4.2.2.1 Preliminary

For conquering the DBZ problem, the basic idea of ZG controller (4.11) is to transform direct control (4.8) into a time-varying minimization problem. That is, the direct computational manner of controller (4.8) is transformed to the manner of minimizing time-varying EF (4.9). For further discussion, controller (4.8) is formulated as $u = \xi/x_1$, where $\xi = x_1^2x_3 + (a - 2\lambda)x_1x_2 - ax_2^2 + \alpha_2$. Besides, we have $\phi = \xi - x_1u$. Thus, the time-varying solution of minimizing (4.9) by using the GD method can be rewritten as

$$\dot{u} = \gamma x_1\phi = \gamma x_1(\xi - x_1u). \tag{4.12}$$

4.2.2.2 Tight Error Bound

When ZG controller (4.11) is applied to the tracking control of modified Lorenz nonlinear system (4.3) with additive control input, the following theorem about its tracking-error bound is presented.

Theorem 4.1 *Consider the modified Lorenz nonlinear system (4.3) equipped with ZG controller (4.11) for smooth and bounded desired trajectory* y_{d}. *Starting with bounded initial state* $\mathbf{x}(0) = [x_1(0), x_2(0), x_3(0)]^{\mathrm{T}} \in \mathbb{R}^3$ *and additive control input* $u(0) \in \mathbb{R}$, *the following results are achieved on a large scale for the tracking control of the system.*

- *For the case of* $x_1 \neq 0$ *(i.e., the non-DBZ case), the tracking error of the system converges toward or stays within the error bound* $\chi\varrho/(\lambda^2\gamma\varsigma)$, *provided that (i)* $\sqrt{\varsigma} \leq |x_1| \leq \chi$, $\exists 0 < \sqrt{\varsigma} \leq \chi < \infty$, *and (ii)* $|\dot{u}^*| = |\mathrm{d}u^*/\mathrm{d}t| \leq \varrho$, $\exists 0 \leq \varrho < +\infty$.
- *For the case of* $x_1 = 0$ *(i.e., the DBZ case), the tracking error of the system is bounded.*

Proof We analyze the following two cases.

(i) For the case of $x_1 \neq 0$ (i.e., the non-DBZ case).

Let us define a solution error of ZG controller (4.11) or (4.12) for solving the time-varying minimization problem of ϵ as $\psi = u - u^*$ with $u^* = \xi/x_1$ being the desired time-varying solution. Then, we have $u = \psi + u^*$ and its time derivative as

$$\dot{u} = \dot{\psi} + \dot{u}^*. \tag{4.13}$$

Substituting (4.12) into (4.13) yields $\gamma x_1(\xi - x_1 u) = \dot{\psi} + \dot{u}^*$ and further $\dot{\psi} = -\gamma x_1^2\psi - \dot{u}^*$. For further analysis, let us define a Lyapunov function candidate as $L = \psi^2/2$. Evidently, L is positive-definite because of $L = \psi^2/2 > 0$ for $\psi \neq 0$ and $L = 0$ for $\psi = 0$ only. Then, taking its time derivative yields

$$\dot{L} = \psi\dot{\psi} = \psi\left(-\gamma x_1^2\psi - \dot{u}^*\right) = -\gamma x_1^2\psi^2 - \psi\dot{u}^*. \tag{4.14}$$

There are two terms in the time derivative $\dot{L}$, i.e., $-\gamma x_1^2\psi^2$ and $-\psi\dot{u}^*$. Let us handle these two terms individually. For the first term $-\gamma x_1^2\psi^2$, we have

$$-\gamma x_1^2\psi^2 \leq -\gamma\varsigma\psi^2, \tag{4.15}$$

where $\varsigma > 0$ is defined previously by $x_1^2 \geq \varsigma$. On the other hand, for the second term $-\psi\dot{u}^*$, we have the following result based on Cauchy inequality [23]:

$$-\psi\dot{u}^* \leq |\psi||\dot{u}^*| \leq \varrho|\psi|, \tag{4.16}$$

where ϱ is defined previously by $|\dot{u}^*| \leq \varrho$.

Then, substituting (4.15) and (4.16) into (4.14) yields

$$\dot{L} = -\gamma x_1^2 \psi^2 - \psi \dot{u}^* \leq -\gamma \varsigma \psi^2 + \varrho|\psi| = -|\psi|(\gamma \varsigma |\psi| - \varrho). \tag{4.17}$$

During the time evolution of solution error ψ, (4.17) falls into one of the following three situations: (i) $\gamma\varsigma|\psi| - \varrho > 0$; (ii) $\gamma\varsigma|\psi| - \varrho = 0$; (iii) $\gamma\varsigma|\psi| - \varrho < 0$. These three situations are analyzed in detail as follows.

- In the first situation (i.e., $|\psi| > \varrho/(\gamma\varsigma)$), $\dot{L} < 0$, which implies that ψ approaches zero (i.e., u approaches u^*) as time evolves.
- In the second situation (i.e., $|\psi| = \varrho/(\gamma\varsigma)$, a so-called ball surface), $\dot{L} \leq 0$, which implies that ψ approaches zero (i.e., u approaches u^*) or ψ stays on the ball surface with $|\psi| = \varrho/(\gamma\varsigma)$ (i.e., $|u - u^*| = \varrho/(\gamma\varsigma)$), in view of $\dot{L} \leq 0$ containing sub-situations $\dot{L} < 0$ and $\dot{L} = 0$, respectively. That is, ψ will not go outside the ball of $\varrho/(\gamma\varsigma)$ in this situation.
- In the third situation (i.e., $|\psi| < \varrho/(\gamma\varsigma)$, inside the ball), it follows from (4.17) that $\dot{L}$ is less than a positive scalar (containing sub-situations $\dot{L} \leq 0$ and $\dot{L} > 0$), and thus ψ may not decrease again. Let us analyze the worst case, i.e., $\dot{L} > 0$. In this case, with $\gamma\varsigma|\psi| - \varrho < 0$, L and $|\psi|$ would increase, which implies that $\gamma\varsigma|\psi| - \varrho$ increases as well. Evidently, there exists a certain time instant such that $\gamma\varsigma|\psi| - \varrho = 0$, which returns to the second situation, i.e., $\dot{L} \leq 0$.

Summarizing the above three situations, one can conclude that the solution error of controller (4.11) or (4.12) for solving the time-varying minimization problem of ϵ is upper bounded by $\varrho/(\gamma\varsigma)$ when the solving process enters the steady state, i.e.,

$$\limsup_{t\to+\infty} |\psi| = \limsup_{t\to+\infty} |u - u^*| \leq \frac{\varrho}{\gamma\varsigma}, \tag{4.18}$$

where $t \geq t_{\mathrm{e}}$ with t_{e} being large enough.

From the derivation process of ZG controller (4.11), we have $\phi = -\dot{z}_2 - \lambda z_2 = -\ddot{z}_1 - 2\lambda\dot{z}_1 - \lambda^2 z_1 = -\ddot{e} - 2\lambda\dot{e} - \lambda^2 e = \xi - x_1 u = x_1(u^* - u) = -x_1\psi$. Together with (4.18), when $t \geq t_{\mathrm{e}}$, we have

$$\begin{cases} -\chi\varrho/(\gamma\varsigma) \leq -|x_1||\psi| \leq \ddot{e} + 2\lambda\dot{e} + \lambda^2 e, \\ \ddot{e} + 2\lambda\dot{e} + \lambda^2 e \leq |x_1||\psi| \leq \chi\varrho/(\gamma\varsigma). \end{cases} \tag{4.19}$$

Then, based on Gronwall inequality [24], the bound constraint

$$-\tilde{c}_1 \exp(-\lambda t) - \tilde{c}_2 t \exp(-\lambda t) - \frac{\chi\varrho}{\lambda^2\gamma\varsigma} \leq e \leq \bar{c}_1 \exp(-\lambda t) + \bar{c}_2 t \exp(-\lambda t) + \frac{\chi\varrho}{\lambda^2\gamma\varsigma} \tag{4.20}$$

holds true when $t \geq t_e$, where $\tilde{c}_1$, $\tilde{c}_2$, $\bar{c}_1$, and $\bar{c}_2$ are constants. Note that the proof of (4.20) is presented in the Appendix. Furthermore, we have

$$|e| \leq |c_1| \exp(-\lambda t) + |c_2| t \exp(-\lambda t) + \frac{\chi \varrho}{\lambda^2 \gamma \varsigma}, \tag{4.21}$$

with $t \geq t_e$. Besides, $|c_1| = \max\{|\tilde{c}_1|, |\bar{c}_1|\}$ and $|c_2| = \max\{|\tilde{c}_2|, |\bar{c}_2|\}$. In view of [25] and $\lambda > 0$, there exist $\bar{c} > 0$ and $\bar{\lambda} > 0$ such that $z_1 = |c_1| \exp(-\lambda t) + |c_2| t \exp(-\lambda t) \leq \bar{c} \exp(-\bar{\lambda} t)$. That is, z_1 exponentially converges toward zero. Therefore, we come to the conclusion that, starting with bounded initial state $\mathbf{x}(0)$, the right side of (4.21) exponentially converges toward $\chi \varrho / (\lambda^2 \gamma \varsigma)$ on a large scale. That is, we have

$$\limsup_{t \to +\infty} |e| \leq \frac{\chi \varrho}{\lambda^2 \gamma \varsigma}. \tag{4.22}$$

Thus, the tracking error of system (4.3) equipped with ZG controller (4.11) converges toward or stays within the error bound $\chi \varrho / (\lambda^2 \gamma \varsigma)$ in this case.

(ii) For the case of $x_1 = 0$ (i.e., the DBZ case).

For system (4.3) equipped with ZG controller (4.11) tracking a time-varying desired trajectory y_d, it can be readily derived that $\lim_{t \to t_s} \dot{u} = \lim_{x_1 \to 0} \dot{u} = 0$ in view of $\dot{u} = \gamma x_1 (\xi - x_1 u)$. Thus, the control input at t_s is the same as that at the previous time instant t_{s-}, which implies that $u(t_s) = u(t_{s-})$. Similarly, at t_{s+} (which is the time instant after the DBZ point), $u(t_s) = u(t_{s+})$. Then, we have the result that $u(t_{s-}) = u(t_s) = u(t_{s+})$ and they are bounded. For a bounded control input, the output of system (4.3) is bounded. Since the desired trajectory y_d is bounded, the tracking error is thus bounded at the time instants t_{s-}, t_s, and t_{s+}. After getting through the DBZ point (i.e., for the time instants after t_{s+}), the tracking error converges toward an error bound again, which implies that the modified Lorenz nonlinear system (4.3) equipped with ZG controller (4.11) finally conquers the DBZ problem.

By the above analysis, the proof is thus completed. □

4.2.2.3 Exponential Convergence Rate

In Theorem 4.1, $|\psi|$ is just asymptotic convergence, and thus it requires an infinitely long time period to guarantee the tight bound of $|e|$, which may be less acceptable in practice. Via further investigation, we obtain the following results on the exponential convergence of ZG controller (4.11) to a relatively loose tracking-error bound.

Theorem 4.2 *Consider the modified Lorenz nonlinear system (4.3) equipped with ZG controller (4.11) for smooth and bounded desired trajectory* y_d. *Starting with bounded initial state* $\mathbf{x}(0) \in \mathbb{R}^3$ *and additive control input* $u(0) \in \mathbb{R}$, *the following results are achieved on a large scale for the tracking control of the system.*

- *For the case of $x_1 \neq 0$ (i.e., the non-DBZ case), the tracking error of the system exponentially converges toward or stays within the error bound $\chi\varrho/(\omega\lambda^2\gamma\varsigma)$ with loosening parameter $\omega \in (0, 1)$, provided that (i) $\sqrt{\varsigma} \leq |x_1| \leq \chi$, $\exists 0 < \sqrt{\varsigma} \leq \chi < +\infty$, and (ii) $|\dot{u}^*| \leq \varrho$, $\exists 0 \leq \varrho < +\infty$.*
- *For the case of $x_1 = 0$ (i.e., the DBZ case), the tracking error of the system is bounded.*

Proof For the case of $x_1 \neq 0$ (i.e., the non-DBZ case), from inequality (4.17), we have $\dot{L} \leq -\gamma\varsigma\psi^2 + \varrho|\psi|$ and then

$$\dot{L} \leq -(1-\omega)\gamma\varsigma\psi^2 + (-\omega\gamma\varsigma\psi^2 + \varrho|\psi|), \tag{4.23}$$

where the loosening parameter $\omega \in (0, 1)$. Evidently, on the right side of (4.23), the first term $-(1-\omega)\gamma\varsigma\psi^2 \leq 0$. Then, the analysis of solution error ψ contains the following two situations.

- For solution error ψ satisfying $-\omega\gamma\varsigma\psi^2 + \varrho|\psi| \leq 0$, (i.e., $|\psi| \geq \varrho/(\omega\gamma\varsigma)$, outside or on the surface of new ball $\varrho/(\omega\gamma\varsigma)$), it follows from (4.23) that

  $$\dot{L} \leq -(1-\omega)\gamma\varsigma\psi^2 + (-\omega\gamma\varsigma\psi^2 + \varrho|\psi|) \leq -(1-\omega)\gamma\varsigma\psi^2 = -2(1-\omega)\gamma\varsigma L.$$

  Then, we have $L \leq L(0)\exp(-2(1-\omega)\gamma\varsigma t)$ and

  $$|\psi| \leq |\psi(0)|\exp(-(1-\omega)\gamma\varsigma t), \forall t \in [0, t_c], \tag{4.24}$$

  where the exponential convergence rate is $(1-\omega)\gamma\varsigma$, and the convergence time $t_c = \ln(\omega\gamma\varsigma|\psi(0)|/\varrho)/((1-\omega)\gamma\varsigma)$ in view of $|\psi(0)|\exp(-(1-\omega)\gamma\varsigma t_c) = \varrho/(\omega\gamma\varsigma)$.
- For solution error ψ satisfying $-\omega\gamma\varsigma\psi^2 + \varrho|\psi| > 0$, (i.e., $|\psi| < \varrho/(\omega\gamma\varsigma)$, inside the new ball $\varrho/(\omega\gamma\varsigma)$), ψ stays within the new ball of $\varrho/(\omega\gamma\varsigma)$. The analysis is similar to that of the third situation of (4.17) in Theorem 4.1.

Thus, with the loosening parameter $\omega \in (0, 1)$ selected, we have the following facts.

- If $|\psi(0)| \geq \varrho/(\omega\gamma\varsigma)$, then

  $$|\psi| \begin{cases} \leq |\psi(0)|\exp(-(1-\omega)\gamma\varsigma t), \forall t \in [0, t_c], \\ \leq \varrho/(\omega\gamma\varsigma), \forall t \in [t_c, +\infty). \end{cases} \tag{4.25}$$

- If $|\psi(0)| \leq \varrho/(\omega\gamma\varsigma)$, then $|\psi| \leq \varrho/(\omega\gamma\varsigma)$, $\forall t \in [0, +\infty)$.

Evidently, even in the worst case, the exponential convergence rate is $(1-\omega)\gamma\varsigma$.

Similar to the derivation of (4.19), when $t \geq t_c$, we have

$$-\frac{\chi\varrho}{\omega\gamma\varsigma} \leq \ddot{e} + 2\lambda\dot{e} + \lambda^2 e \leq \frac{\chi\varrho}{\omega\gamma\varsigma}.$$

Then, similar to the derivation of (4.21), we have

$$|e| \leq |\hat{c}_1| \exp(-\lambda t) + |\hat{c}_2| t \exp(-\lambda t) + \frac{\chi \varrho}{\omega \lambda^2 \gamma \varsigma} \tag{4.26}$$

when $t \geq t_c$, where $\hat{c}_1$ and $\hat{c}_2$ are constants. Then, similar to the analysis on (4.21), the right side of (4.26) exponentially converges toward a new error bound $\chi\varrho/(\omega\lambda^2\gamma\varsigma)$.

Thus, we have the conclusion that the tracking error of modified Lorenz nonlinear system (4.3) equipped with ZG controller (4.11) exponentially converges toward or stays within the error bound $\chi\varrho/(\omega\lambda^2\gamma\varsigma)$ on a large scale.

For the case of $x_1 = 0$ (i.e., the DBZ case), according to the proof of Theorem 4.1, we have the same result that the tracking error of modified Lorenz nonlinear system (4.3) equipped with ZG controller (4.11) is bounded.

By the above analyses, the proof is thus completed. □

4.2.3 *Simulation, Verification and Comparison on ZG Controller*

In this subsection, the simulations are conducted to verify the efficacy and superiority of ZG controller (4.11) with additive control input for the tracking control. Without loss of generality, design parameters are set as $\lambda = 10$ and $\gamma = 10^5$. Besides, the initial state $\mathbf{x}(0)$ and initial control input $u(0)$ are randomly generated.

The tracking control of modified Lorenz nonlinear system (4.3) equipped with the conventional IOL controller (4.4) is conducted for comparison, of which the simulation results are visualized in Fig. 4.1. As seen from the figure, the conventional IOL controller (4.4) cannot complete this tracking-control task because of the existence of DBZ points, and system (4.3) crashes when x_1 approaches zero. These results indicate that a controller that is able to conquer the DBZ problem is needed.

The simulation results of ZG controller (4.11) for the tracking control of modified Lorenz nonlinear system (4.3) with additive control input are shown in Figs. 4.2 and 4.3. Specifically, Fig. 4.2 illustrates the results of system (4.3) equipped with ZG controller (4.11) for desired trajectory $y_d = \sin(t)$. As seen from Fig. 4.2a, all states of system (4.3) are bounded during the whole tracking-control process. As shown in Fig. 4.2c, the system output y rapidly tracks the desired trajectory y_d. In addition, as illustrated in Fig. 4.2d, the tracking error exponentially converges toward a tight error bound that is of order 10^{-5}. These results coincide with the analyses of ZG controller (4.11) in Theorems 4.1 and 4.2, i.e., tight error bound and exponential convergence rate.

More importantly, the DBZ-conquering superiority of ZG controller (4.11) for the tracking control of modified Lorenz nonlinear system (4.3) is shown in Fig. 4.3. As displayed in Fig. 4.3a, b, the system equipped with ZG controller (4.11)

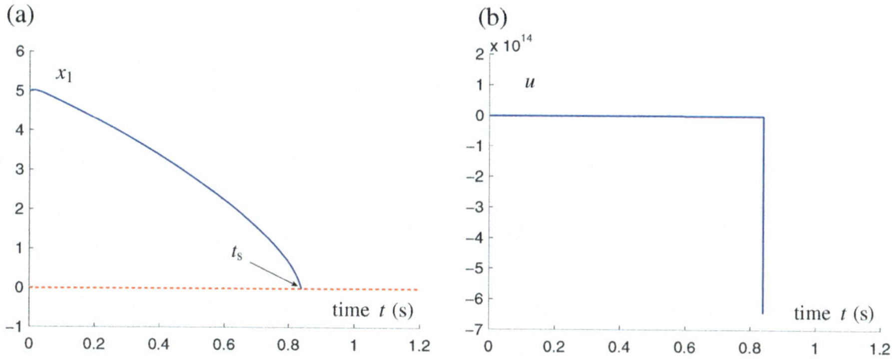

Fig. 4.1 Crash of modified Lorenz nonlinear system (4.3) equipped with conventional IOL controller (4.4) for desired trajectory $y_d = \cos(t)\sin(3t) + 3$. (**a**) Trajectory of x_1. (**b**) Control input

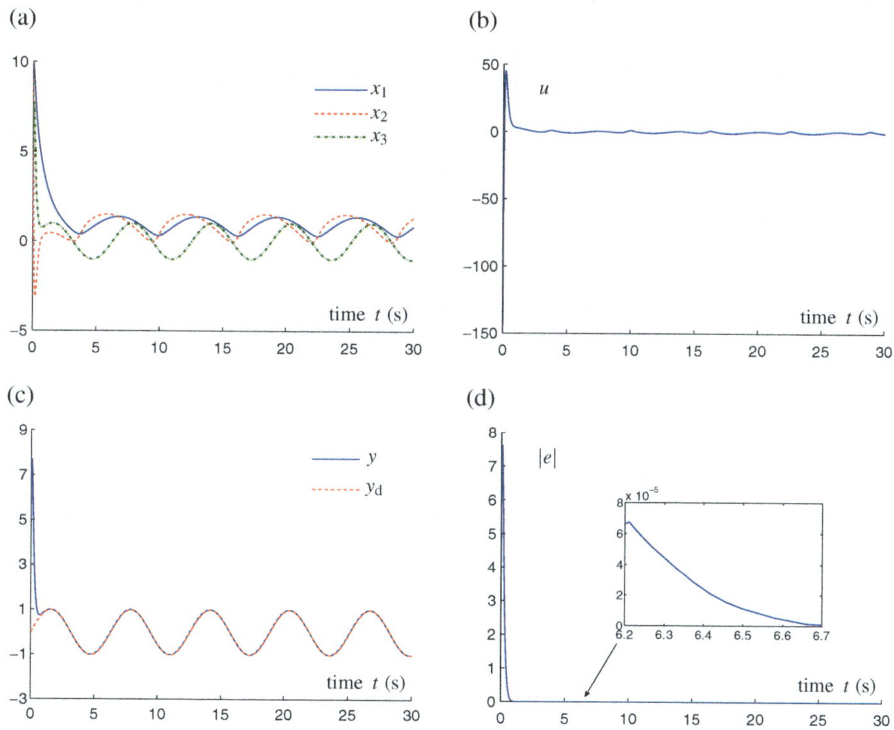

Fig. 4.2 Tracking performance of modified Lorenz nonlinear system (4.3) equipped with ZG controller (4.11) via additive control input for desired trajectory $y_d = \sin(t)$. (**a**) System states. (**b**) Control input. (**c**) Output trajectory and desired trajectory. (**d**) Absolute tracking error

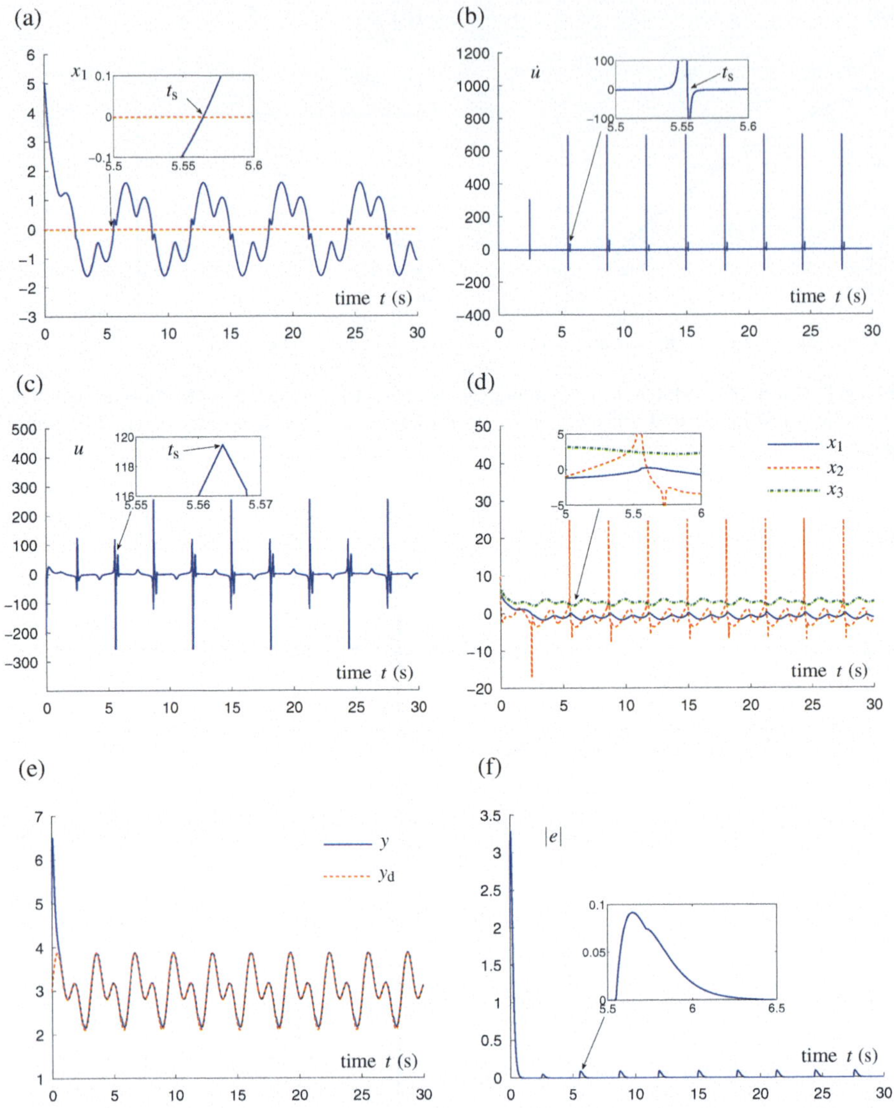

Fig. 4.3 Tracking performance of modified Lorenz nonlinear system (4.3) equipped with ZG controller (4.11) via additive control input for desired trajectory $y_d = \cos(t)\sin(3t) + 3$. (**a**) Trajectory of x_1. (**b**) Control input's time derivative. (**c**) Control input. (**d**) System states. (**e**) Output trajectory and desired trajectory. (**f**) Absolute tracking error

encounters many DBZ points during the tracking-control process but still runs well. In addition, as seen from Fig. 4.3b, c, the $\dot{u}$ and u are bounded at the DBZ points. Moreover, the trajectory of each state shown in Fig. 4.3d is kept bounded. From Fig. 4.3e, we can observe that the actual output y rapidly tracks the desired trajectory y_d with the absolute tracking error $|e|$ shown in Fig. 4.3f slightly increasing in the neighborhood of the DBZ points and then rapidly decreasing to a level of tiny values.

In order to show that the ZG controller is effective in the situation of uncertainties, ZG controller (4.11) is disturbed and described as

$$\dot{u} = \gamma x_1 \phi + \tilde{d} = \gamma x_1 (x_1^2 x_3 + (a - 2\lambda) x_1 x_2 - a x_2^2 - x_1 u + \alpha_2) + \tilde{d}, \tag{4.27}$$

where the design parameters are originally set as $\lambda = 10$ and $\gamma = 1000$, with the disturbance $\tilde{d}$ being 0, 500, 5×10^3 or 5×10^4. The simulation results of absolute tracking error $|e|$ with different disturbances added are shown in Fig. 4.4. It can be seen from Fig. 4.4a–c that, with slightly larger disturbance $\tilde{d}$ added, the absolute tracking error $|e|$ remains within an acceptably small level, i.e., the control method is still effective in controlling the nonlinear system. As seen from Fig. 4.4d, when the disturbance $\tilde{d}$ is too large, the performance of the controller does not satisfy the requirement of tracking precision. However, the controller designed by the ZG method is still able to suppress such considerably large disturbance $\tilde{d}$ by choosing proper (usually larger) design parameters. This is reflected by Fig. 4.4d, f, where the disturbance $\tilde{d}$ is successfully suppressed with larger design parameter γ used; specifically, with λ fixed in this simulation, γ is set as 10^4 and 10^5 in Fig. 4.4e, f, respectively. The efficacy of the presented ZG controller in suppressing disturbance is thus illustrated. The above results coincide with the DBZ-conquering analyses in Theorems 4.1 and 4.2, and substantiate the advantages of ZG controller (4.11) based on the ZG method.

4.3 ZG Control via Mixed Inputs

This section investigates the tracking-control problem of the modified Lorenz chaotic system (4.2) equipped with the mixed control inputs (i.e., the mixture of additive and multiplicative control inputs). The ZG method is used again to design controllers for the tracking control of the system.

Let us consider the following modified Lorenz nonlinear system equipped with a multiplicative control input u_1 and an additive control input u_2:

$$\begin{cases} \dot{x}_1 = f_1(x_1, x_2, u_1) = u_1(x_2 - x_1), \\ \dot{x}_2 = f_2(x_1, x_3, u_2) = -x_1 x_3 + u_2, \\ \dot{x}_3 = f_3(x_1, x_2) = x_1 x_2 - b, \end{cases} \tag{4.28}$$

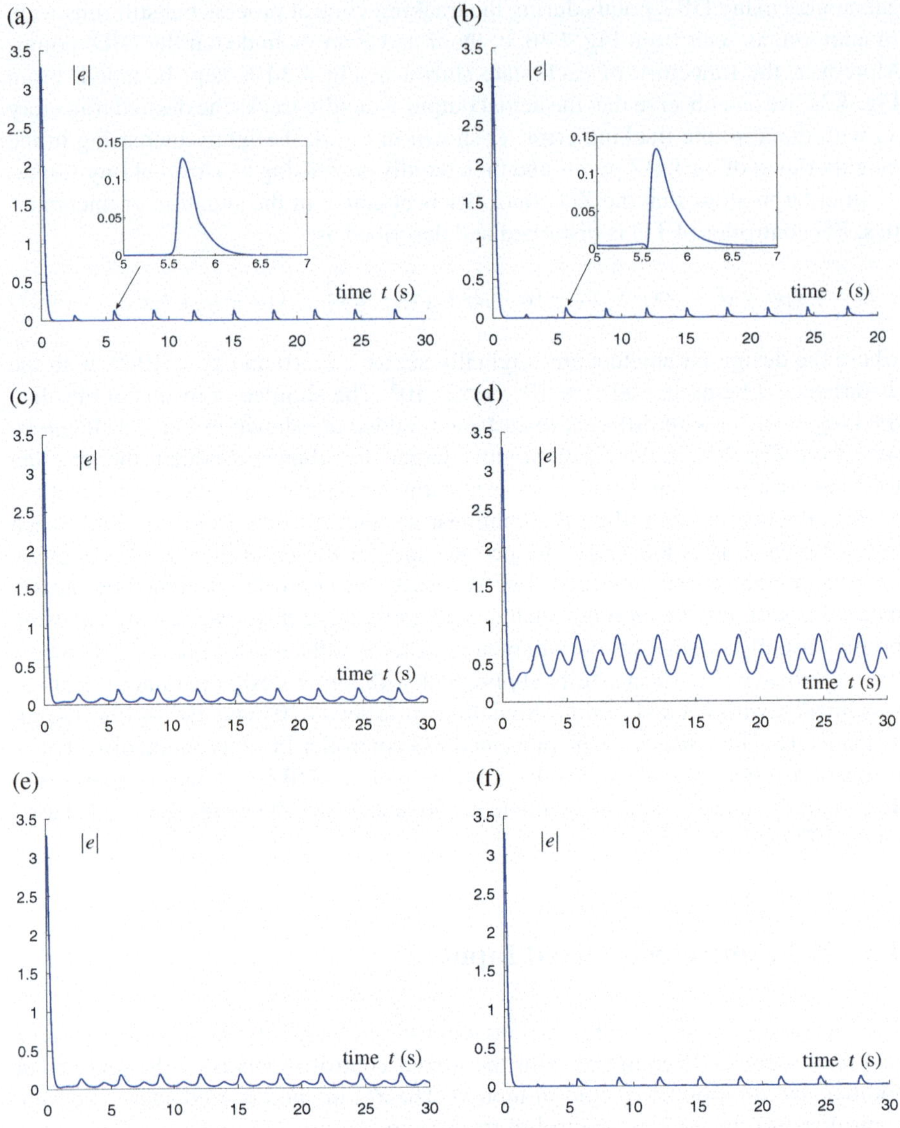

Fig. 4.4 Absolute tracking errors of modified Lorenz nonlinear system (4.3) equipped with ZG controller (4.27) disturbed by $\tilde{d}$ via additive control input for desired trajectory $y_d = \cos(t)\sin(3t) + 3$, with $|e|$ shown in subfigures (**e**) and (**f**) suppressed by increasing γ value. (**a**) With $\tilde{d} = 0$. (**b**) With $\tilde{d} = 500$. (**c**) With $\tilde{d} = 5 \times 10^3$. (**d**) With $\tilde{d} = 5 \times 10^4$. (**e**) With $\tilde{d} = 5 \times 10^4$ but suppressed. (**f**) With $\tilde{d} = 5 \times 10^4$ but suppressed more

with $y_1 = x_1$ and $y_2 = x_3$ denoting the outputs of system (4.28). The objective is to design controllers such that y_1 and y_2 track the smooth and bounded desired trajectories y_{1d} and y_{2d}, respectively, with the tracking errors $e_1 = y_1 - y_{1d}$ and $e_2 = y_2 - y_{2d}$ asymptotically approaching zero (or say, near zero in practice).

4.3.1 Design of ZG Controller Group

To monitor and control the tracking process, the first ZF is defined as $z_{u_1 1} = y_1 - y_{d1} = x_1 - y_{d1}$. Then, applying ZD design formula (4.5) and system (4.28), we have

$$u_1(x_2 - x_1) - \dot{y}_{1d} + \lambda(x_1 - y_{1d}) = 0. \tag{4.29}$$

Thus, the u_1 controller in the form of u_1 for the modified Lorenz nonlinear system (4.28) is directly obtained from (4.29), i.e.,

$$u_1 = \frac{\dot{y}_{1d} - \lambda(x_1 - y_{1d})}{x_2 - x_1}. \tag{4.30}$$

In terms of controller (4.30), there also exist DBZ points when $x_1 = x_2$. To conquer these DBZ points, the GD method is further exploited to design the u_1 controller in the form of $\dot{u}_1$ based on (4.30). By defining $\phi_1 = u_1(x_2 - x_1) - \dot{y}_{1d} + \lambda(x_1 - y_{1d})$ and adopting the GD method [20–22], the u_1 controller in the form of $\dot{u}_1$ for the modified Lorenz nonlinear system (4.28) is finally designed as

$$\dot{u}_1 = \gamma(x_1 - x_2)\phi_1. \tag{4.31}$$

Notably, as compared with controller (4.30), the division-free controller (4.31) can conquer the DBZ points when $x_1 = x_2$.

In addition, via the similar steps presented in Sect. 4.2.1, the u_2 controller in the form of $\dot{u}_2$ for the modified Lorenz nonlinear system (4.28) can be designed as

$$\dot{u}_2 = \gamma x_1 \phi_2, \tag{4.32}$$

where $\phi_2 = x_1^2 x_3 + (u_1 - 2\lambda)x_1 x_2 - u_1 x_2^2 - x_1 u_2 + \alpha_3$ with $\alpha_3 = -\lambda^2 x_3 + 2\lambda b + \ddot{y}_{2d} + 2\lambda \dot{y}_{2d} + \lambda^2 y_{2d}$.

Furthermore, by combining the above two controllers, i.e., (4.31) and (4.32), the controller group with mixed control inputs in the form of $\dot{u}_1$ and $\dot{u}_2$ is presented as

$$\begin{cases} \dot{u}_1 = \gamma(x_1 - x_2)\phi_1 = \gamma(x_1 - x_2)(u_1(x_2 - x_1) - \dot{y}_{1d} + \lambda(x_1 - y_{1d})), \\ \dot{u}_2 = \gamma x_1 \phi_2 = \gamma x_1 (x_1^2 x_3 + (u_1 - 2\lambda)x_1 x_2 - u_1 x_2^2 - x_1 u_2 + \alpha_3). \end{cases} \tag{4.33}$$

As seen from (4.33), the presented ZG controller group can handle and conquer double DBZ points (i.e., the DBZ pair $x_1 - x_2 = 0$ and $x_1 = 0$) simultaneously.

4.3.2 Convergence Performance Analyses on ZG Controller Group

In this subsection, two theorems are presented to guarantee the tracking performance of modified Lorenz nonlinear system (4.28) equipped with ZG controller group (4.33). Note that the corresponding proofs are omitted due to the similarity to those of Theorems 4.1 and 4.2.

Theorem 4.3 *Consider the modified Lorenz nonlinear system (4.28) equipped with ZG controller group (4.33) for smooth and bounded desired trajectories $y_{1\mathrm{d}}$ and $y_{2\mathrm{d}}$. Starting with bounded initial state $\mathbf{x}(0) \in \mathbb{R}^3$ as well as mixed control inputs $u_1(0) \in \mathbb{R}$ and $u_2(0) \in \mathbb{R}$, the following results are achieved on a large scale for the tracking control of the system with $j = 1$ for (4.31) and $j = 2$ for (4.32).*

- *For the case of $x_1 - x_2 \neq 0$ and $x_1 \neq 0$ (i.e., the non-DBZ case), the tracking error $|e_j|$ of the system converges toward or stays within the error bound $\chi_j\varrho_j/(\lambda^2\gamma\varsigma_j)$, provided that (i) $\sqrt{\varsigma_1} \leq |x_2 - x_1| \leq \chi_1$, $\exists 0 < \sqrt{\varsigma_1} \leq \chi_1 < +\infty$, (ii) $\sqrt{\varsigma_2} \leq |x_1| \leq \chi_2$, $\exists 0 < \sqrt{\varsigma_2} \leq \chi_2 < +\infty$, and (iii) $\left|\dot{u}_j^*\right| \leq \varrho_j$, $\exists 0 \leq \varrho_j < +\infty$.*
- *For the case of $x_1 - x_2 = 0$ or $x_1 = 0$ (i.e., the DBZ case), the tracking error $|e_j|$ of the system is bounded.*

Proof It can be generalized from the proof of Theorem 4.1. □

Theorem 4.4 *Consider the modified Lorenz nonlinear system (4.28) equipped with ZG controller group (4.33) for smooth and bounded desired trajectories $y_{1\mathrm{d}}$ and $y_{2\mathrm{d}}$. Starting with bounded initial state $\mathbf{x}(0) \in \mathbb{R}^3$ as well as mixed control inputs $u_1(0) \in \mathbb{R}$ and $u_2(0) \in \mathbb{R}$, the following results are achieved on a large scale for the tracking control of the system with $j = 1$ for (4.31) and $j = 2$ for (4.32).*

- *For the case of $x_1 - x_2 \neq 0$ and $x_1 \neq 0$ (i.e., the non-DBZ case), the tracking error of the system exponentially converges toward or stays within the error bound $\chi_j\varrho_j/(\omega_j\lambda^2\gamma\varsigma_j)$ with $\omega_j \in (0, 1)$, provided that (i) $\sqrt{\varsigma_1} \leq |x_2 - x_1| \leq \chi_1$, $\exists 0 < \sqrt{\varsigma_1} \leq \chi_1 < +\infty$, (ii) $\sqrt{\varsigma_2} \leq |x_1| \leq \chi_2$, $\exists 0 < \sqrt{\varsigma_2} \leq \chi_2 < +\infty$, and (iii) $\left|\dot{u}_j^*\right| \leq \varrho_j$, $\exists 0 \leq \varrho_j < +\infty$.*
- *For the case of $x_1 - x_2 = 0$ or $x_1 = 0$ (i.e., the DBZ case), the tracking error $|e_j|$ of the system is bounded.*

Proof It can be generalized from the proof of Theorem 4.2. □

4.3.3 Simulation and Verification on ZG Controller Group

The DBZ-conquering superiority of ZG controller group (4.33) for the tracking control of modified Lorenz nonlinear system (4.28) is illustrated in Figs. 4.5 and 4.6. As shown in Fig. 4.5a, b, the system equipped with ZG controller group (4.33) encounters many DBZ points during the tracking-control process but still runs well. In addition, it can be observed from Fig. 4.5c, d that the system outputs y_1 and y_2 rapidly track the desired trajectories y_{1d} and y_{2d}, respectively. Moreover, Fig. 4.6a shows that the mixed control inputs are acceptable. From Fig. 4.6b, the error $|e_1|$ slightly increases in the neighborhood of the DBZ points and rapidly decreases to a level of tiny values when the system leaves the DBZ points.

In summary, the efficacy of ZG controller group (4.33) for the tracking control of modified Lorenz nonlinear system (4.28) is verified. Moreover, the capability of ZG controller group (4.33) for conquering the DBZ problem is substantiated. These simulation results show well the feasibility and efficacy of the ZG method for the tracking control of the modified Lorenz nonlinear system with mixed control inputs.

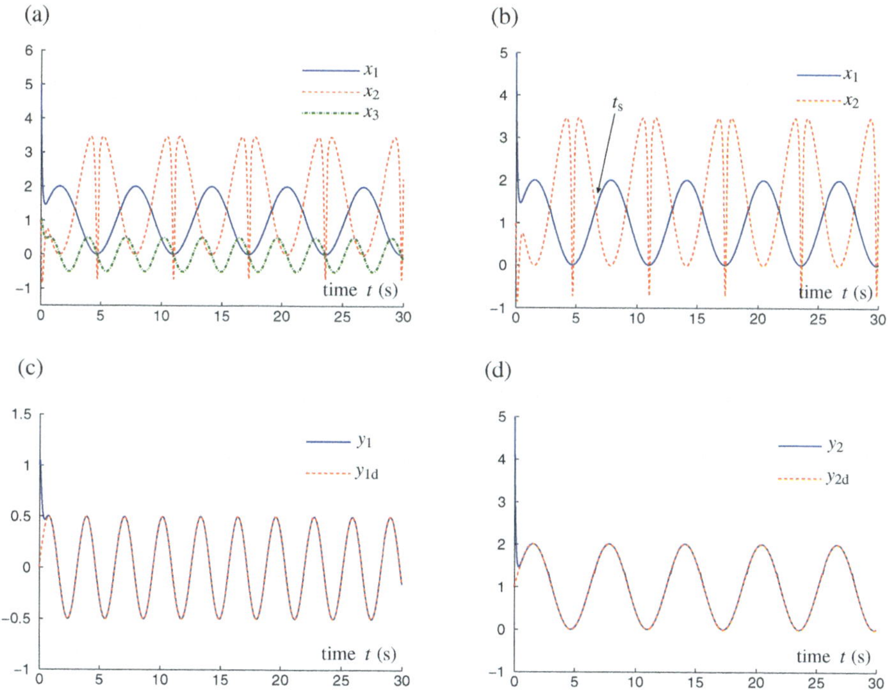

Fig. 4.5 Tracking performance of modified Lorenz nonlinear system (4.28) equipped with ZG controller group (4.33) for desired trajectories $y_{1d} = \sin(t)\cos(t)$ and $y_{2d} = \sin(t) + 1.01$. (**a**) System states. (**b**) Trajectories of x_1 and x_2. (**c**) Output trajectory y_1 and desired trajectory y_{1d}. (**d**) Output trajectory y_2 and desired trajectory y_{2d}

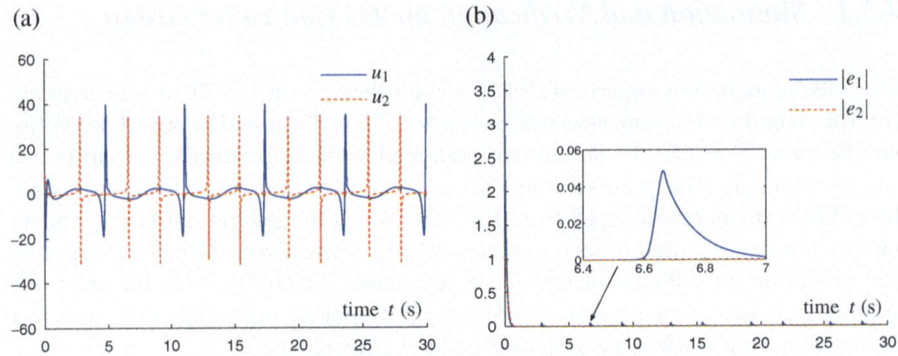

Fig. 4.6 Mixed control inputs and absolute tracking errors of modified Lorenz nonlinear system (4.28) equipped with ZG controller group (4.33) for desired trajectories $y_{1d} = \sin(t)\cos(t)$ and $y_{2d} = \sin(t) + 1.01$. **(a)** Control inputs. **(b)** Absolute tracking errors

4.4 Chapter Summary

In this chapter, by applying the ZG method, the tracking-control problem of the modified Lorenz nonlinear system with the additive control input or the mixture of additive and multiplicative control inputs has been presented and investigated. Such a combined ZG method, together with convergence performance analyses, has been used to solve the tracking-control problem of the modified Lorenz nonlinear system. The simulation results have been shown to further substantiate that the controllers designed via the ZG method can achieve satisfactory tracking accuracy and effectively conquer the DBZ problem encountered during the tracking-control process.

Appendix: Proof of Bound Constraint Depicted in (4.20)

In the Appendix, for completeness, we present the proof of the bound constraint depicted in (4.20).

From (4.19) and the derivation process of (4.11), we have

$$\ddot{e} + 2\lambda\dot{e} + \lambda^2 e = \dot{z}_2 + \lambda z_2 \leq \frac{\chi\varrho}{\gamma\varsigma},$$

where $t \geq t_e$. Making use of Gronwall inequality [24], we have

$$z_2 \leq z_2(t_e)\exp(-\lambda t) + \int_{t_e}^{t} \exp(-\lambda(t-\iota))\frac{\chi\varrho}{\gamma\varsigma}\mathrm{d}\iota \leq \hat{c}\exp(-\lambda t) + \frac{\chi\varrho}{\lambda\gamma\varsigma},$$

where $\hat{c}$ is a constant, and $t \geq t_e$. As $z_2 = \dot{e} + \lambda e$, we have

$$\dot{e} + \lambda e \leq \Omega,$$

where $\Omega = \hat{c}\exp(-\lambda t) + \chi\varrho/(\lambda\gamma\varsigma)$. Making use of Gronwall inequality [24] again, we obtain

$$\begin{aligned} e &\leq e(t_e)\exp(-\lambda t) + \int_{t_e}^{t} \exp(-\lambda(t-\iota))\Omega(\iota)\mathrm{d}\iota \\ &\leq \bar{c}_1\exp(-\lambda t) + \bar{c}_2 t\exp(-\lambda t) + \frac{\chi\varrho}{\lambda^2\gamma\varsigma}, \end{aligned} \tag{4.34}$$

where $\bar{c}_1$ and $\bar{c}_2$ are constants, and $t \geq t_e$.

From (4.19), we also have

$$-\ddot{e} - 2\lambda\dot{e} - \lambda^2 e = -\dot{z}_2 - \lambda z_2 \leq \frac{\chi\varrho}{\gamma\varsigma}.$$

Similar to the derivation process of (4.34), we have

$$e \geq -\tilde{c}_1\exp(-\lambda t) - \tilde{c}_2 t\exp(-\lambda t) - \frac{\chi\varrho}{\lambda^2\gamma\varsigma}, \tag{4.35}$$

where $\tilde{c}_1$ and $\tilde{c}_2$ are constants, and $t \geq t_e$. Combining (4.34) and (4.35) yields (4.20). The proof of (4.20) is thus completed. □

References

1. Lorenz E (1963) Deterministic nonperiodic flows. J Atmos Sci 20:130–141
2. Munmuangsaen B, Srisuchinwong B (2009) A new five-term simple chaotic attractor. Phys Lett A 373(44):4038–4043
3. Jin L, Zhang Y, Qiao T, Tan M, Zhang Y (2016) Tracking control of modified Lorenz nonlinear system using ZG neural dynamics with additive input or mixed inputs. Neurocomputing 196:82–94
4. Stojanovski T, Pihl J, Koncarev L (2001) Chaos-based random number generators-part II: practical realization. IEEE Trans Circuits Syst I Fundam Theor Appl 48(3):382–385
5. Bhatnagar G, Jonathan Wu QM (2015) A novel chaos-based secure transmission of biometric data. Neurocomputing 147(1):444–455
6. Huang Y, Yang X-S (2006) Hyperchaos and bifurcation in a new class of four-dimensional Hopfield neural networks. Neurocomputing 69(13–15):1787–1795
7. Muthuswamy B (2010) Implementing memristor based chaotic circuits. Int J Bifurcat Chaos 20(5):1335–1350
8. Isidori A (1989) Nonlinear control systems: an introduction, 2nd edn. Springer, New York
9. Slotine JE, Li W (1991) Applied nonlinear control. Prentice Hall, New Jersey
10. Hauser J, Sastry S, Kokotovic P (1992) Nonlinear control via approximate input-output linearization: the ball and beam example. IEEE Trans Autom Control 37(3):392–398

11. Tomlin C, Sastry S (1998) Switching through singularities. Syst Control Lett 35(3):145–154
12. Li S, Li Y, Liu B, Murray T (2012) Model-free control of Lorenz chaos using an approximate optimal control strategy. Commu Nonlinear Sci 17(12):4891–4900
13. Lin D, Wang X (2010) Observer-based decentralized fuzzy neural sliding mode control for interconnected unknown chaotic systems via network structure adaptation. Fyzzy Set Syst 161(15):2066–2080
14. Lin D, Wang X, Nian F, Zhang Y (2010) Dynamic fuzzy neural networks modeling and adaptive backstepping tracking control of uncertain chaotic system. Neurocomputing 73(16–18):2873–2881
15. Wang X, Song J (2009) Synchronization of the fractional order hyperchaos Lorenz systems with activation feedback control. Commu Nonlinear Sci 14(8):3351–3357
16. Wang X, He Y, Wang M (2009) Chaos control of a fractional order modified coupled dynamos system. Nonlinear Anal Theory Methods Appl 71(12):6126–6134
17. Xiao L, Lu R (2015) Finite-time solution to nonlinear equation using recurrent neural dynamics with a specially-constructed activation function. Neurocomputing 151:246–251
18. Liao B, Zhang Y (2014) From different ZFs to different ZNN models accelerated via Li activation functions to finite-time convergence for time-varying matrix pseudoinversion. Neurocomputing 133(8):512–522
19. Li S, Li Y (2014) Nonlinearly activated neural network for solving time-varying complex Sylvester equation. IEEE Trans Cybern 44(8):1397–1407
20. Zhang Y, Chen K, Tan H (2009) Performance analysis of gradient neural network exploited for online time-varying matrix inversion. IEEE Trans Autom Control 54(8):1940–1945
21. Zhang Y, Yang Y, Ruan G (2011) Performance analysis of gradient neural network exploited for online time-varying quadratic minimization and equality-constrained quadratic programming. Neurocomputing 74(10):1710–1719
22. Zhang Y, Yi C, Guo D, Zheng J (2011) Comparison on Zhang neural dynamics and gradient-based neural dynamics for online solution of nonlinear time-varying equation. Neural Comput Appl 20(1):1–7
23. Abramowitz M, Stegun IA (1972) Handbook of mathematical functions with formulas, graphs, and mathematical tables. Dover Publications, New York
24. Chu S, Metcalf F (1967) On Gronwall's inequality. Proc Amer Math Soc 18(3):439–440
25. Zhang Z, Zhang Y (2013) Design and experimentation of acceleration-level drift-free scheme aided by two recurrent neural networks. IET Control Theory A 7(1):25–42

Part II
Integrator Systems Using ZG Control

Chapter 5
ZG Tracking Control of Brockett Integrator

Abstract In this chapter, we apply the ZG method to the tracking control of a multiple-input multiple-output (MIMO) nonlinear system (i.e., Brockett integrator). Based on the ZG method, different types of controller groups are designed for tracking control of Brockett integrator. Both theoretical analyses and simulative verifications indicate that the tracking errors are bounded and exponentially convergent. More importantly, comparative simulation results illustrate that the ZG controller group is superior to the ZD controller group in conquering the DBZ problem encountered during the tracking-control process.

5.1 Introduction

Brockett integrator is widely applied in the current-fed inducting motor, Heisenberg flywheel and a variety of robotic and mobile nonholonomic systems [1, 2]. Dynamic tracking-control problems are important issues in nonlinear systems, which is applied in different control fields, such as inverted pendulum systems [3], nonholonomic systems [4], multi-agent systems [5]. So far, a number of methods have been investigated for solving the tracking-control problem, such as feedback linearization [6], sliding mode control [7], and backstepping control [8]. However, most of the aforementioned methods cannot effectively work when encountering the DBZ problem. Recently, by combining the ZD and GD methods, Zhang et al. proposed the simple and effective ZG method to conquer the DBZ problem in the tracking-control process [9–13], which is mainly applied to the single-input single-output nonlinear systems. In this chapter [14], we apply the ZG method to conquer the DBZ problem encountered during the tracking-control process of a multiple-input-multiple-output (MIMO) nonlinear system (i.e., Brockett integrator), and illustrate the efficacy and superiority of ZG controller group.

Y. Zhang et al., *Zhang-Gradient Control*,
https://doi.org/10.1007/978-981-15-8257-8_5

5.2 Preliminaries

Brockett integrator is a kind of nonlinear system with nonholonomic constraint [2, 15]. Mathematically, Brockett integrator can be formulated as

$$\begin{cases} \dot{x}_1 = f_1(u_1) = u_1, \\ \dot{x}_2 = f_2(u_2) = u_2, \\ \dot{x}_3 = f_3(x_1, x_2, u_1, u_2) = x_1 u_2 - x_2 u_1, \end{cases} \tag{5.1}$$

where $\mathbf{x} = [x_1, x_2, x_3]^{\mathrm{T}} \in \mathbb{R}^3$ is the state vector, and $\mathbf{u} = [u_1, u_2]^{\mathrm{T}}$ is the input vector of Brockett integrator. Besides, the third equation of (5.1) is the constraint and a non-integrable one. In this chapter [14], Brockett integrator (5.1) is chosen as a kind of representative MIMO nonlinear system to verify the efficacy and superiority of the presented ZG controller group.

In the tracking-control process, we assume that $y(\mathbf{x})$ is the output of the system and the desired trajectory is y_{d}. Thus, the tracking error of the system should be $z_1 = y(\mathbf{x}) - y_{\mathrm{d}}$. As the ZG method is the combination of ZD and GD methods, we firstly employ the ZD method with the following ZD design formula:

$$\dot{z}_1 = -\lambda z_1,$$

where z_1 also denotes the first ZF, and the design parameter $\lambda \in \mathbb{R}^+$ is used to scale the convergence rate of the ZD solution. With the ZD method used, we can further obtain $\dot{y}(\mathbf{x}) = \dot{y}_{\mathrm{d}} - \lambda(y(\mathbf{x}) - y_{\mathrm{d}})$. If the above equation does not include the explicit expression of u, the second ZF is defined as $z_2 = \dot{z}_1 + \lambda z_1$ and the ZD method is applied again by $\dot{z}_2 = -\lambda z_2$. It does not end in this way until the resultant equation includes the explicit expression of u. In general, we can finally obtain an equation in the form of $h(\mathbf{x})u = g(\mathbf{x})$. Thus, the ZD controller designed by the ZD method is obtained as

$$u = \frac{g(\mathbf{x})}{h(\mathbf{x})}, \tag{5.2}$$

which can also be called zng0 controller, meaning that the ZD method is applied for n times and without using the GD method during the design process.

It can be found that ZD controller (5.2) cannot run well in the situation of $h(\mathbf{x}) = 0$. In order to deal with the DBZ problem, the GD method needs to be used. We can construct an EF as $\epsilon = \phi^2/2$ with $\phi = h(\mathbf{x})u - g(\mathbf{x})$. By applying the GD design formula $\dot{u} = -\gamma \partial\epsilon/\partial u$, the ZG controller is obtained as

$$\dot{u} = -\gamma h(\mathbf{x})\phi = -\gamma h(\mathbf{x})(h(\mathbf{x})u - g(\mathbf{x})), \tag{5.3}$$

where the design parameter $\gamma \in \mathbb{R}^+$ is used to scale the convergence rate of the GD solution. ZG controller (5.3) is also called *zng*1 controller.

5.3 Design of ZD and ZG Controller Groups

In this section, the controller groups are designed for the tracking control of Brockett integrator with different combinations of outputs. When the outputs are set as $y_1 = x_1$ and $y_2 = x_3$ to track the desired trajectories $y_{1\text{d}}$ and $y_{2\text{d}}$, respectively, the ZD controller group is designed by only the ZD method as

$$\begin{cases} u_1 = \dot{y}_{1\text{d}} - \lambda(x_1 - y_{1\text{d}}), \\ u_2 = (x_2 u_1 + \dot{y}_{2\text{d}} - \lambda(x_3 - y_{2\text{d}}))/x_1, \end{cases} \tag{5.4}$$

which is also called z2g0 controller group.

In order to deal with the DBZ problem, the ZG controller group is designed as

$$\begin{cases} u_1 = \dot{y}_{1\text{d}} - \lambda(x_1 - y_{1\text{d}}), \\ \dot{u}_2 = -\gamma x_1 \phi_1, \end{cases} \tag{5.5}$$

where $\phi_1 = x_1 u_2 - x_2 u_1 - \dot{y}_{2\text{d}} + \lambda(x_3 - y_{2\text{d}})$. Controller group (5.5) is also called z2g1 controller group. When other output combinations are set to track desired trajectories $y_{1\text{d}}$ and $y_{2\text{d}}$, the corresponding controller groups are designed and obtained as shown in Table 5.1.

Table 5.1 ZD and ZG Controller groups with different output combinations

Output combination	Controller group	Type
$y_1 = x_1$, $y_2 = x_3$	$u_1 = \dot{y}_{1\text{d}} - \lambda(x_1 - y_{1\text{d}})$, $u_2 = (x_2 u_1 + \dot{y}_{2\text{d}} - \lambda(x_3 - y_{2\text{d}}))/x_1$	z2g0
$y_1 = x_1$, $y_2 = x_3$	$u_1 = \dot{y}_{1\text{d}} - \lambda(x_1 - y_{1\text{d}})$, $\dot{u}_2 = -\gamma x_1(x_1 u_2 - x_2 u_1 - \dot{y}_{2\text{d}} + \lambda(x_3 - y_{2\text{d}}))$	z2g1
$y_1 = x_2$, $y_2 = x_3$	$u_1 = (x_1 u_2 - \dot{y}_{2\text{d}} + \lambda(x_3 - y_{2\text{d}}))/x_2$, $u_2 = \dot{y}_{1\text{d}} - \lambda(x_2 - y_{1\text{d}})$	z2g0
$y_1 = x_2$, $y_2 = x_3$	$\dot{u}_1 = \gamma x_2(x_1 u_2 - x_2 u_1 - \dot{y}_{2\text{d}} + \lambda(x_3 - y_{2\text{d}}))$, $u_2 = \dot{y}_{1\text{d}} - \lambda(x_2 - y_{1\text{d}})$	z2g1
$y_1 = x_1$, $y_2 = x_2$	$u_1 = \dot{y}_{1\text{d}} - \lambda(x_1 - y_{1\text{d}})$, $u_2 = \dot{y}_{2\text{d}} - \lambda(x_2 - y_{2\text{d}})$	z2g0

5.4 Convergence Performance Analyses

In this section, the boundedness and convergence of tracking errors are theoretically analyzed for the tracking control of Brockett integrator (5.1). In view of results' similarity, we only present the theoretical results and analyses on controller groups (5.4) and (5.5). In what follows, $e_1 = z_1 = y_1 - y_{1\mathrm{d}} = x_1 - y_{1\mathrm{d}}$ and $e_2 = z_2 = y_2 - y_{2\mathrm{d}} = x_3 - y_{2\mathrm{d}}$ represents the tracking errors.

Theorem 5.1 *For smooth and bounded desired trajectories $y_{1\mathrm{d}}$ and $y_{2\mathrm{d}}$, starting with bounded initial state $\mathbf{x}(0) = [x_1(0), x_2(0), x_3(0)]^{\mathrm{T}} \in \mathbb{R}^3$, the tracking errors e_1 and e_2 of Brockett integrator (5.1) equipped with z2g0 controller group (5.4) exponentially converge to zero on a large scale, provided that $x_1 \neq 0$, $\forall t \in [0, +\infty)$.*

Proof In view of (5.1) and (5.4), it is not difficult to derive the formulas $\dot{e}_1 = -\lambda e_1$ and $\dot{e}_2 = -\lambda e_2$. Thus, the tracking errors e_1 and e_2 exponentially converge to zero on a large scale. The proof is thus completed. □

Theorem 5.2 *Consider Brockett integrator (5.1) equipped with z2g1 controller group (5.5) for smooth and bounded desired trajectories $y_{1\mathrm{d}}$ and $y_{2\mathrm{d}}$. Starting with bounded initial state $\mathbf{x}(0) \in \mathbb{R}^3$ and control input $\mathbf{u}(0) = [u_1(0), u_2(0)]^{\mathrm{T}} \in \mathbb{R}^2$, the following results are achieved on a large scale for the tracking control of the Brockett integrator.*

- *For the case of $x_1 \neq 0$ (i.e., the non-DBZ case), the tracking error e_2 of the Brockett integrator exponentially converges toward or stays within the error bound $\varrho\sqrt{\eta_2}/(\lambda\gamma\eta_1)$, provided that (i) $\eta_1 \leq x_1^2 \leq \eta_2$, $\exists 0 < \eta_1 \leq \eta_2 < +\infty$, and (ii) $|\dot{u}_2^*| \leq \varrho$, $\exists 0 \leq \varrho < +\infty$.*
- *For the case of $x_1 = 0$ (i.e., the DBZ case), the tracking error e_2 of the Brockett integrator is bounded.*

Proof For the case of $x_1 \neq 0$ (i.e., the non-DBZ case), it can be found that the optimal solution u_2^* is $u_2^* = (x_2 u_1 + \dot{y}_{2\mathrm{d}} - \lambda(x_3 - y_{2\mathrm{d}}))/x_1$. Specifically, a solution error is defined as $\psi = u_2 - u_2^*$. Then, we have its time derivative

$$\dot{\psi} = \dot{u}_2 - \dot{u}_2^*. \tag{5.6}$$

As ϕ_1 can be rewritten as

$$\phi_1 = x_1(u_2 - u_2^*) = x_1\psi, \tag{5.7}$$

the following equation is obtained by substituting (5.5) and (5.7) into (5.6):

$$\dot{\psi} = -\gamma x_1^2 \psi - \dot{u}_2^*.$$

For further analysis, a Lyapunov function candidate is defined as $L = \psi^2/2 \geq 0$. Then, the time derivative of L can be obtained as

$$\dot{L} = \psi\dot{\psi} = -\gamma x_1^2 \psi^2 - \dot{u}_2^* \psi.$$

When $\eta_1 \leq x_1^2 \leq \eta_2$, where η_1 and η_2 are bounded positive real constants, we derive the inequality as

$$\dot{L} \leq -\gamma\eta_1\psi^2 + \varrho|\psi| = -(\gamma\eta_1|\psi| - \varrho)|\psi|. \tag{5.8}$$

During the time evolution of solution error ψ, inequality (5.8) falls into one of the following three situations.

- In the first situation (i.e., $\gamma\eta_1|\psi| > \varrho$), $\dot{L} < 0$, which implies that ψ approaches zero as time evolves.
- In the second situation (i.e., $\gamma\eta_1|\psi| = \varrho$), $\dot{L} \leq 0$, which implies that ψ approaches zero or ψ stays on the ball surface with $|\psi| = |u_2 - u_2^*| = \varrho/(\gamma\eta_1)$.
- In the third situation (i.e., $\gamma\eta_1|\psi| < \varrho$), $\dot{L}$ is less than a positive constant, which consists of sub-situations $\dot{L} \leq 0$ and $0 < \dot{L} \leq -(\gamma\eta_1|\psi| - \varrho)|\psi|$. On the one hand, if $\dot{L} \leq 0$, then it returns to the same case of the second situation. On the other hand, if $0 < \dot{L} \leq -(\gamma\eta_1|\psi| - \varrho)|\psi|$ holds true, then L and ψ would increase, which implies that $\gamma\eta_1|\psi| - \varrho$ increases as well. Thus, since $\gamma\eta_1|\psi| - \varrho < 0$, there exists a certain time instant such that $\gamma\eta_1|\psi| - \varrho = 0$, which would return to the second situation, i.e., $\dot{L} \leq 0$.

Summarizing the above three situations, one can conclude that the solution error synthesized by z2g1 controller group (5.5) is upper bounded by $\varrho/(\gamma\eta_1)$, i.e.,

$$\limsup_{t\to+\infty} |\psi| = \limsup_{t\to+\infty} |u_2 - u_2^*| \leq \frac{\varrho}{\gamma\eta_1}. \tag{5.9}$$

In view of (5.7) and (5.9), the inequality is obtained as

$$\limsup_{t\to+\infty} |\phi_1| = \limsup_{t\to+\infty} (|x_1||\psi|) \leq \frac{\varrho\sqrt{\eta_2}}{\gamma\eta_1}. \tag{5.10}$$

From the definition of ϕ_1 and the design process of u_2, it can be obtained that

$$\phi_1 = \dot{e}_2 + \lambda e_2. \tag{5.11}$$

By substituting (5.11) into (5.10), a new inequality is obtained as

$$\limsup_{t\to+\infty} |\dot{e}_2 + \lambda e_2| \leq \frac{\varrho\sqrt{\eta_2}}{\gamma\eta_1}. \tag{5.12}$$

Furthermore, the solution of inequality (5.12) is the common solution of the following two inequalities:

$$\limsup_{t\to+\infty}(\dot{e}_2+\lambda e_2)\leq\frac{\varrho\sqrt{\eta_2}}{\gamma\eta_1}, \tag{5.13}$$

$$\liminf_{t\to+\infty}(\dot{e}_2+\lambda e_2)\geq-\frac{\varrho\sqrt{\eta_2}}{\gamma\eta_1}. \tag{5.14}$$

Solving (5.13), we obtain the solution

$$\limsup_{t\to+\infty} e_2\leq\frac{\varrho\sqrt{\eta_2}}{\lambda\gamma\eta_1}, \tag{5.15}$$

with its proof presented in Appendix 1. Solving (5.14), we obtain the solution

$$\liminf_{t\to+\infty} e_2\geq-\frac{\varrho\sqrt{\eta_2}}{\lambda\gamma\eta_1}, \tag{5.16}$$

with its proof presented in Appendix 2. In consideration of inequalities (5.15) and (5.16), it can be concluded that the common solution of (5.13) and (5.14) is

$$\limsup_{t\to+\infty}|e_2|\leq\frac{\varrho\sqrt{\eta_2}}{\lambda\gamma\eta_1}.$$

For the case of $x_1=0$ (i.e., the DBZ case), it can be derived that $\lim_{t\to t_{\mathrm{s}}}\dot{u}=\lim_{x_1\to 0}\dot{u}=0$ in view of $\dot{u}=-\gamma x_1\phi_1$, where t_{s} denotes the DBZ time instant. Thus, we have the result that $u(t_{\mathrm{s}^-})=u(t_{\mathrm{s}})=u(t_{\mathrm{s}^+})$ and they are bounded. For a bounded input, the output of Brockett integrator (5.1) is bounded. The tracking error e_2 is thus bounded at the time instants t_{s^-}, t_{s} and t_{s^+}. After getting through the DBZ point, the tracking error e_2 converges toward an error bound again, which implies that Brockett integrator (5.1) equipped with z2g1 controller group (5.5) finally conquers the DBZ problem.

By the above analyses, the proof is thus completed. □

5.5 Simulation, Verification and Comparison

In this section, the simulations are performed to illustrate the feasibility and efficacy of z2g0 controller group (5.4) and z2g1 controller group (5.5). In these simulations, the running time is set as 30 s. Besides, the initial state is set as $\mathbf{x}(0)=[x_1(0),x_2(0),x_3(0)]^{\mathrm{T}}=[-1.5,0.5,0]^{\mathrm{T}}$, the initial control input of z2g1 controller group (5.5) is set as $u_2(0)=0$. In addition, the design parameters are set as $\lambda=50$ and $\gamma=10^3$.

5.5.1 Efficacy of ZD and ZG Controller Groups

In this subsection, the outputs $y_1 = x_1$ and $y_2 = x_3$ are expected to track desired trajectories $y_{1d} = \sin(t) - 2$ and $y_{2d} = \cos(t)$, respectively. The simulation results are shown in Fig. 5.1. Specifically, Fig. 5.1a, c illustrate that the output trajectories of Brockett integrator (5.1) equipped with z2g0 controller group (5.4) tracks well the desired trajectories. As shown in Fig. 5.1a, the output trajectories and the desired trajectories are almost consistent with each other. Besides, Fig. 5.1c shows that the maximal steady-state tracking errors of e_1 and e_2 are both of order 10^{-14}. On the other hand, Fig. 5.1b, d also illustrate that the output trajectories of Brockett integrator (5.1) equipped with z2g1 controller group (5.5) tracks well the desired trajectories. The output trajectories of z2g1 controller group (5.5) track the desired trajectories with the maximal steady-state tracking errors e_1 and e_2 being of orders 10^{-14} and 10^{-5}, respectively. The simulation results show that controller

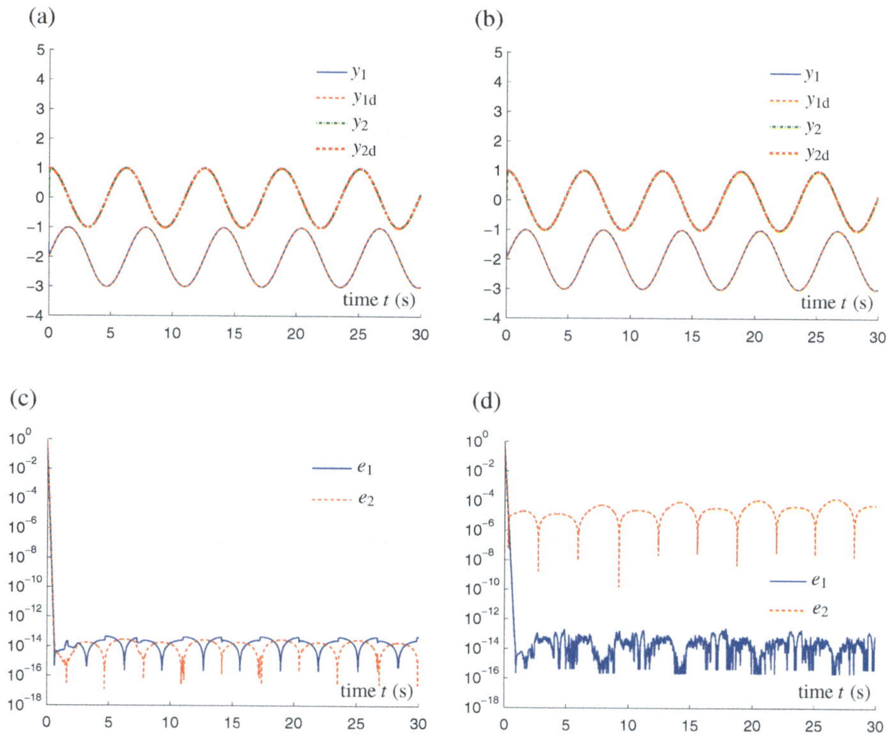

Fig. 5.1 Tracking performance of Brockett integrator (5.1) equipped with z2g0 controller group (5.4) and z2g1 controller group (5.5), respectively, for desired trajectories $y_{1d} = \sin(t) - 2$ and $y_{2d} = \cos(t)$. **(a)** Output trajectories with z2g0 controller group (5.4) and desired trajectories. **(b)** Output trajectories with z2g1 controller group (5.5) and desired trajectories. **(c)** Tracking errors with z2g0 controller group (5.4). **(d)** Tracking errors with z2g1 controller group (5.5)

groups (5.4) and (5.5) are both feasible and effective for the tracking control of Brockett integrator (5.1).

5.5.2 DBZ Conquering of ZG Controller Group

To substantiate the superiority of z2g1 controller group (5.5) over z2g0 controller group (5.4) for conquering the DBZ problem, we select output $y_1 = x_1$ to track desired trajectory $y_{1d} = \sin(t)$ and output $y_2 = x_3$ to track desired trajectory $y_{2d} = \cos(t)\exp(-t/20)$, respectively. The simulation results are illustrated in Fig. 5.2. Specifically, as seen from Fig. 5.2a, c, the simulation stops at time $t = \pi$ s when the trajectory of x_1 passes the DBZ point. The simulation results illustrate that z2g0 controller group (5.4) cannot get through the DBZ point of $x_1 = 0$. More importantly, Fig. 5.2b, d show that the output trajectories of z2g1 controller group (5.5) track well the desired trajectories. Therefore, the simulation results

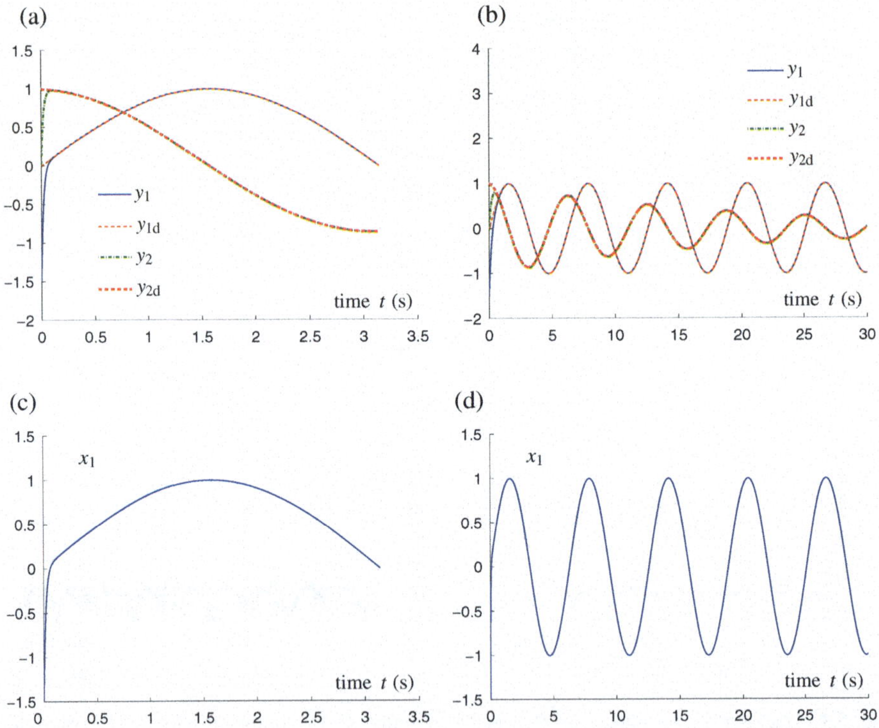

Fig. 5.2 Tracking performance of Brockett integrator (5.1) equipped with z2g0 controller group (5.4) and z2g1 controller group (5.5), respectively, for desired trajectories $y_{1d} = \sin(t)$ and $y_{2d} = \cos(t)\exp(-t/20)$. (**a**) Output trajectories with z2g0 controller group (5.4) and desired trajectories. (**b**) Output trajectories with z2g1 controller group (5.5) and desired trajectories. (**c**) Trajectory of x_1 with z2g0 controller group (5.4). (**d**) Trajectory of x_1 with z2g1 controller group (5.5)

substantiate the superiority of z2g1 controller group (5.5) over z2g0 controller group (5.4) for conquering the DBZ problem.

In order to show the situation of the trajectories in the vicinity of the DBZ point, z2g1 controller group (5.5) is expected to track four desired trajectories $y_{1d} = \sin(t) - \kappa_i$ and $y_{2d} = \cos(t)$, where $\kappa_i = 1.01, 2, 5, 10$ for $i = 1, 2, 3, 4$. Thus, the four trajectories x_1 of the second controller of (5.5) are the nearest points away from the DBZ point of $x_1 = 0$ when time $t = (2m + 1/2)\pi$ s, $m \in \mathbb{N}$, with the nearest distances being 0.01, 1, 4 and 9, respectively. The simulation results are shown in Fig. 5.3. It illustrates the relation between the nearest distance and the maximal steady-state tracking error. Specifically, as shown in Table 5.2, the larger the nearest distance is, the less the maximal steady-state tracking error e_2 is.

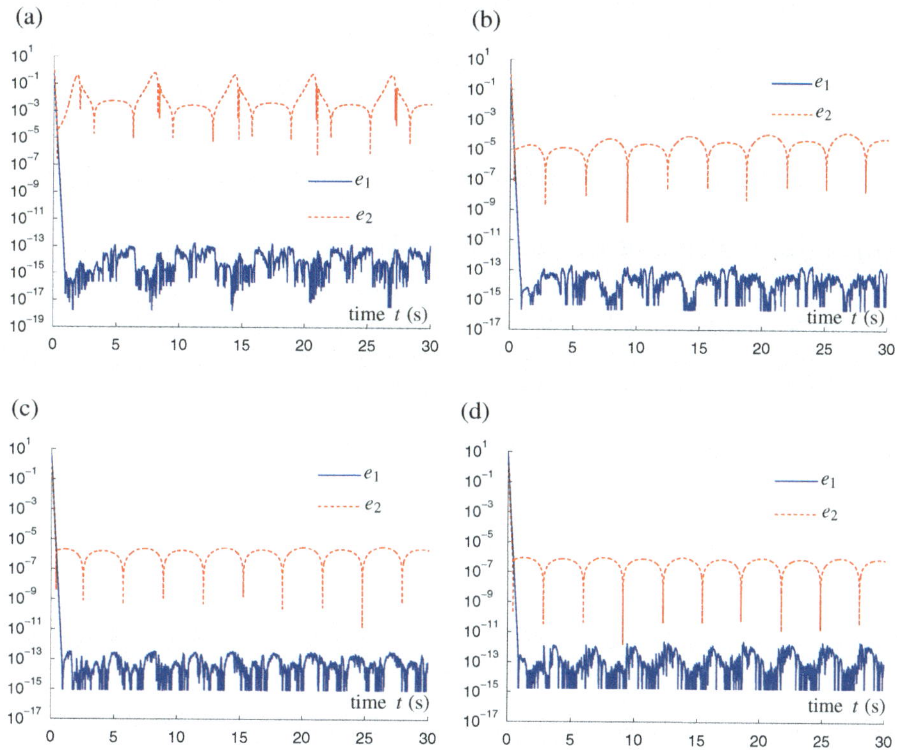

Fig. 5.3 Tracking errors of Brockett integrator (5.1) equipped with z2g1 controller group (5.5) for desired trajectories $y_{1d} = \sin(t) - \kappa_i$, with $i \in \{1, 2, 3, 4\}$, and $y_{2d} = \cos(t)$. (**a**) With $\kappa_1 = 1.01$. (**b**) With $\kappa_2 = 2$. (**c**) With $\kappa_3 = 5$. (**d**) With $\kappa_4 = 10$

Table 5.2 Maximal steady-state tracking errors of Brockett integrator (5.1) equipped with z2g1 controller group (5.5) for desired trajectories $y_{1d} = \sin(t) - \kappa_i$, with $i \in \{1, 2, 3, 4\}$, and $y_{2d} = \cos(t)$

	$\kappa_1 = 1.01$	$\kappa_2 = 2$	$\kappa_3 = 5$	$\kappa_4 = 10$
e_1	6.9×10^{-14}	1.9×10^{-13}	2.6×10^{-13}	1.8×10^{-12}
e_2	0.67	1.1×10^{-4}	2.5×10^{-6}	6.7×10^{-7}

5.6 Chapter Summary

In this chapter, based on the ZG method, two different controller groups, i.e., z2g0 controller group (5.4) and z2g1 controller group (5.5), have been designed and investigated for the tracking control of Brockett integrator (5.1). Theoretical analyses and simulation verifications both have shown the feasibility and efficacy of z2g0 controller group (5.4) and z2g1 controller group (5.5). More importantly, the comparative simulation results have substantiated that z2g1 controller group (5.5) is superior to z2g0 controller group (5.4) in conquering the DBZ problem.

Appendix 1: Proof of Inequality (5.13)

The solution of (5.13) can be obtained by converting the inequality constraint into the equality constraint [16]. The solution of inequality $\dot{e}_2 + \lambda e_2 \leq \varrho\sqrt{\eta_2}/(\gamma\eta_1)$ is equivalent to the solution of the following equation:

$$\dot{e}_2 + \lambda e_2 + \xi_1^2 = \frac{\varrho\sqrt{\eta_2}}{\gamma\eta_1}, \tag{5.17}$$

where time-varying parameter $\xi_1 \in \mathbb{R}$ with $-\infty < \xi_1 < +\infty$. The solution of (5.17) can be expressed as

$$e_2 = e_2(0)\exp(-\lambda t) + \frac{\varrho\sqrt{\eta_2}}{\lambda\gamma\eta_1}(1 - \exp(-\lambda t)) - \int_0^t \xi_1^2 \exp(-\lambda(t - \iota))\mathrm{d}\iota. \tag{5.18}$$

As $0 \leq \xi_1^2 < +\infty$, we have the following inequality:

$$\int_0^t \xi_1^2 \exp(-\lambda(t - \iota)\mathrm{d}\iota \geq 0. \tag{5.19}$$

Then, the inequality of e_2 can be further obtained as

$$e_2 \leq \left(e_2(0) - \frac{\varrho\sqrt{\eta_2}}{\lambda\gamma\eta_1}\right)\exp(-\lambda t) + \frac{\varrho\sqrt{\eta_2}}{\lambda\gamma\eta_1}.$$

In view of $t \to +\infty$, we have

$$e_2(+\infty) \leq \frac{\varrho\sqrt{\eta_2}}{\lambda\gamma\eta_1}.$$

That is,

$$\limsup_{t\to+\infty} e_2 \leq \frac{\varrho\sqrt{\eta_2}}{\lambda\gamma\eta_1}. \tag{5.20}$$

Thus, inequality (5.20) is the solution of (5.13).

Appendix 2: Proof of Inequality (5.14)

The solution of (5.14) can be obtained by converting the inequality constraint into the equality constraint [16]. The solution of inequality $\dot{e}_2 + \lambda e_2 \geq -\varrho\sqrt{\eta_2}/(\gamma\eta_1)$ is equivalent to the solution of the following equation:

$$\dot{e}_2 + \lambda e_2 - \xi_2^2 = -\frac{\varrho\sqrt{\eta_2}}{\gamma\eta_1}, \tag{5.21}$$

where time-varying parameter $\xi_2 \in \mathbb{R}$ with $-\infty < \xi_2 < +\infty$. The solution of (5.21) can be formulated as

$$e_2 = e_2(0)\exp(-\lambda t) - \frac{\varrho\sqrt{\eta_2}}{\lambda\gamma\eta_1}(1 - \exp(-\lambda t)) + \int_0^t \xi_2^2 \exp(-\lambda(t-\iota)\mathrm{d}\iota. \tag{5.22}$$

As $0 \leq \xi_2^2 < +\infty$, we have the following inequality:

$$\int_0^t \xi_2^2 \exp(-\lambda(t-\iota)\mathrm{d}\iota \geq 0. \tag{5.23}$$

Then, the inequality of e_2 can be further obtained as

$$e_2 \geq \left(e_2(0) + \frac{\varrho\sqrt{\eta_2}}{\lambda\gamma\eta_1}\right)\exp(-\lambda t) - \frac{\varrho\sqrt{\eta_2}}{\lambda\gamma\eta_1}.$$

In view of $t \to +\infty$, we have

$$e_2(+\infty) \geq -\frac{\varrho\sqrt{\eta_2}}{\lambda\gamma\eta_1}.$$

That is,

$$\liminf_{t\to+\infty} e_2 \geq -\frac{\varrho\sqrt{\eta_2}}{\lambda\gamma\eta_1}. \tag{5.24}$$

Thus, inequality (5.24) is the solution of (5.14).

References

1. Chwa D (2004) Sliding-mode tracking control of nonholonomic wheeled mobile robots in polar coordinates. IEEE Trans Control Syst Technol 12(4):637–644
2. Bloch A, Drakunov S (1998) Discontinuous stabilization of Brockett's canonical driftless system. In: Essays on mathematical robotics. Springer, New York, pp 169–183
3. Consolini L, Tosques M (2009) On the exact tracking of the spherical inverted pendulum via an homotopy method. Syst Control Lett 58(1):1–6
4. Drakunov SV, Floquet T, Perruquetti W (2005) Stabilization and tracking control for an extended Heisenberg system with a drift. Syst Control Lett 54(5):435–445
5. Meng D, Jia Y, Du J, Yu F (2012) Tracking control over a finite interval for multi-agent systems with a time-varying reference trajectory. Syst Control Lett 61(7):807–818
6. Khalil HK, Grizzle JW (1996) Nonlinear systems. Prentice Hall, New Jersey
7. Drakunov SV, Utkin VI (1992) Sliding mode control in dynamic systems. Int J Control 55(4):1029–1037
8. Fukao T, Nakagawa H, Adachi N (2000) Adaptive tracking control of a nonholonomic mobile robot. IEEE Trans Robot Autom 16(5):609–615
9. Hu C, Wang Y, Kang X, Guo D, Zhang Y (2014) ZG trajectory generation of Van der Pol oscillator in affine-control form with division-by-zero problem handled. In: Proceedings of the 10th international conference on natural computation, pp 1082–1087
10. Hu C, Wang Y, Kang X, Xiao Z, Zhang Y (2014) ZG controller groups for two-output tracking of two-input Brockett integrator. In: Proceedings of the 5th international conference on intelligent control and information, pp 330–335
11. Zhang Y, Zhai K, Chen D, Jin L, Hu C (2016) Challenging simulation practice (failure and success) on implicit tracking control of double-integrator system via Zhang-gradient method. Math Comput Simul 120:104–119
12. Zhang Y, Yu X, Yin Y, Chen P, Fan Z (2014) Singularity-conquering ZG controllers of z2g1 type for tracking control of the IPC system. Int J Control 87(9):1729–1746
13. Zhang Y, Xiao Z, Guo D, Mao M, Yin Y (2015) Singularity-conquering tracking control of a class of chaotic systems using Zhang-gradient dynamics. IET Control Theory Appl 9(6):871–881
14. Hu C, Kang X, Zhang Y (2018) Singularity-conquering Zhang-gradient controller groups for tracking control of Brockett integrator. In: Proceedings of the 30th Chinese control and decision conference, pp 6291–6296
15. Brockett RW (1982) Control theory and singular Riemannian geometry. Springer, New York
16. Guo D, Zhang Y (2014) Zhang neural network for online solution of time-varying linear matrix inequality aided with an equality conversion. IEEE Trans Neural Netw Learn Syst 25(2):370–382

Chapter 6
ZG Tracking Control and Simulation of DI System

Abstract In this chapter, the ZG controllers for explicit and implicit tracking control of a double-integrator (DI) system are designed and presented. In addition, we conduct the corresponding computer simulations with different values of the design parameter λ used to illustrate the efficacy of ZG controllers. However, although the ZG controllers are powerful, there is still a challenge in the simulation practice. Specifically, different settings of simulation options in MATLAB ordinary differential equation (ODE) solvers may lead to different simulation results (e.g., failure and success). For better comparison, the successful and failed simulation results are both presented. The differences in simulation results remind us to pay more attention to MATLAB defaults and options during conducting such simulations.

6.1 Introduction

The tracking control problem commonly arises in many applications [1–5], e.g., the tracking control of optical disc systems [6], real-time navigation of mobile robots [7], and others [8, 9]. In this chapter [10], the ZG method is used to design ZG controllers to solve the tracking control problem of double-integrator (DI) system. The DI system is a typical model in dynamics, which is widely studied in a variety of areas [11, 12], and it can be formulated as

$$\begin{cases} \dot{x}_1 = f_1(x_2) = x_2, \\ \dot{x}_2 = f_2(u) = u, \end{cases} \tag{6.1}$$

where x_1 and x_2 are the states of the DI system, and u denotes the input of the DI system. The tracking control problem is to design a controller of the input u for the system such that the tracking error $y - y_\mathrm{d}$ is kept within an acceptable tolerance, where y is the output trajectory, and y_d is the desired trajectory for y.

The MATLAB is one of the most commonly used computational softwares and has wide applications in various research areas [13–18]. For example, it can

Y. Zhang et al., *Zhang-Gradient Control*,
https://doi.org/10.1007/978-981-15-8257-8_6

be easily found in digital image processing [18], digital signal processing [13], artificial intelligence [16], automatic control [14], and others [15, 17]. The ordinary differential equation (ODE) is often encountered in many fields of science and engineering [19, 20]. Note that, as for the explicit tracking control, e.g., $y = x_1$ or $y = x_2$, the desired output (or say, target output) y_d can uniquely and directly determine the (steady) value of a single state, e.g., x_1 or x_2. Conversely, for the implicit tracking control, e.g., $y = x_1x_2$ or $y = x_1^2+x_2^2$, the desired output y_d cannot explicitly (i.e., uniquely and directly) determine the (steady) value of a single state, such as x_1 or x_2. Besides, both the explicit tracking control and implicit tracking control have their relative advantages and disadvantages. For the explicit tracking control, it can be achieved more easily by the controller, and the corresponding system has the relatively low computational and structural complexities. However, for the reasons that (i) the output of most practical systems may not uniquely and directly relate to a single state, (ii) the output function may generally be nonlinear, and (iii) the states may have to satisfy some coupling constraints, the explicit tracking control is thus not much applicable in practice. On the contrary, the implicit tracking control has a more complicated output and thus has the relatively high computational and structural complexities. Besides, it would be more difficult for a practitioner to design an appropriate controller for such a system with an implicit tracking control purpose. However, the implicit tracking control would have wider applications in practical systems for the above reasons. In this chapter [10], the computer simulations for explicit and implicit tracking control of the DI system using the ZG method are conducted with the MATLAB ODE solvers. However, different settings of simulation options in ODE solvers may lead to different results (e.g., failure and success), which is very interesting and well worthy of further investigation.

6.2 Explicit Tracking Control of DI System via ZG Method (ETC-DI-ZG)

In this section, to show the efficacy of ZG controllers, the design procedure of the ZG controller using the ZG method for explicit tracking control of the DI system is presented with the output being $y = x_1$.

In view of DI system (6.1) and the output $y = x_1$, the whole tracking system can be described as

$$\begin{cases} \dot{x}_1 = f_1(x_2) = x_2, \\ \dot{x}_2 = f_2(u) = u, \\ y = x_1 \rightarrow y_d. \end{cases} \tag{6.2}$$

The ZG controller design process for system (6.2) is presented as below.

Step 1 To monitor and control the tracking process, the first ZF is constructed as

$$z_1 = y - y_\mathrm{d} = x_1 - y_\mathrm{d}. \tag{6.3}$$

Step 2 To make z_1 exponentially converge to zero, based on the ZD method [21], the following ZD design formula is adopted:

$$\dot{z}_1 = -\lambda z_1, \tag{6.4}$$

where the design parameter $\lambda \in \mathbb{R}^+$ is used to scale the convergence rate of the ZD solution. Substituting (6.3) into (6.4), we obtain the following equation:

$$\dot{x}_1 - \dot{y}_\mathrm{d} = -\lambda(x_1 - y_\mathrm{d}). \tag{6.5}$$

Combining (6.2) and (6.5) yields

$$x_2 + \lambda x_1 - \dot{y}_\mathrm{d} - \lambda y_\mathrm{d} = 0. \tag{6.6}$$

Step 3 Constructing the second ZF as

$$z_2 = x_2 + \lambda x_1 - \dot{y}_\mathrm{d} - \lambda y_\mathrm{d},$$

and applying the ZD method again, we further have

$$\dot{x}_2 + \lambda \dot{x}_1 - \ddot{y}_\mathrm{d} - \lambda \dot{y}_\mathrm{d} = -\lambda(x_2 + \lambda x_1 - \dot{y}_\mathrm{d} - \lambda y_\mathrm{d}).$$

Substituting (6.2) into the above equation, we obtain

$$u - \ddot{y}_\mathrm{d} - 2\lambda(\dot{y}_\mathrm{d} - x_2) - \lambda^2(y_\mathrm{d} - x_1) = 0. \tag{6.7}$$

Step 4 From (6.7), we can define

$$\phi_1 = u - \ddot{y}_\mathrm{d} - 2\lambda(\dot{y}_\mathrm{d} - x_2) - \lambda^2(y_\mathrm{d} - x_1),$$

which should be zero theoretically. According to the GD method [22], a square-based nonnegative EF is defined as $\epsilon = \phi_1^2/2$. By adopting the GD design formula

$$\dot{u} = -\gamma \frac{\partial \epsilon}{\partial u},$$

the ZG controller is obtained as

$$\dot{u} = -\gamma \phi_1 = -\gamma(u - \ddot{y}_\mathrm{d} - 2\lambda(\dot{y}_\mathrm{d} - x_2) - \lambda^2(y_\mathrm{d} - x_1)), \tag{6.8}$$

where the design parameter $\gamma \in \mathbb{R}^+$ is used to scale the convergence rate of the GD solution.

6.3 Successful Simulation of ETC-DI-ZG

To illustrate the efficacy of ZG controller (6.8), computer simulations are conducted in this section, of which the results are shown in Fig. 6.1. The running environment is Windows XP 2002 plus MATLAB R2009b. The system running time for simulation is set as 30 s. Without loss of generality, we set the initial values $x_1(0) = 1, x_2(0) = 0.5$ and $u(0) = 0$ for the ZG controller with $\lambda = 5$ and $\gamma = 1000$. The output of the system is $y = x_1$ together with the desired trajectory $y_\mathrm{d} = \sin(t) + \cos(t)$. In addition, the detailed codes are shown in Appendix 1.

The performance of ZG controller (6.8) for the explicit tracking control of the DI system is illustrated in Fig. 6.1. Specifically, Fig. 6.1a shows the output trajectory of the system as well as the desired trajectory. They coincide well with each other. For better illustration, Fig. 6.1b, c present the states and the input of the system. They are both relatively smooth curves. The absolute tracking error $|e| = |y - y_\mathrm{d}|$ shown in the logarithmic scale is illustrated in Fig. 6.1d, which shows that the absolute tracking error decreases within 5 s and then stays relatively stable at the order of

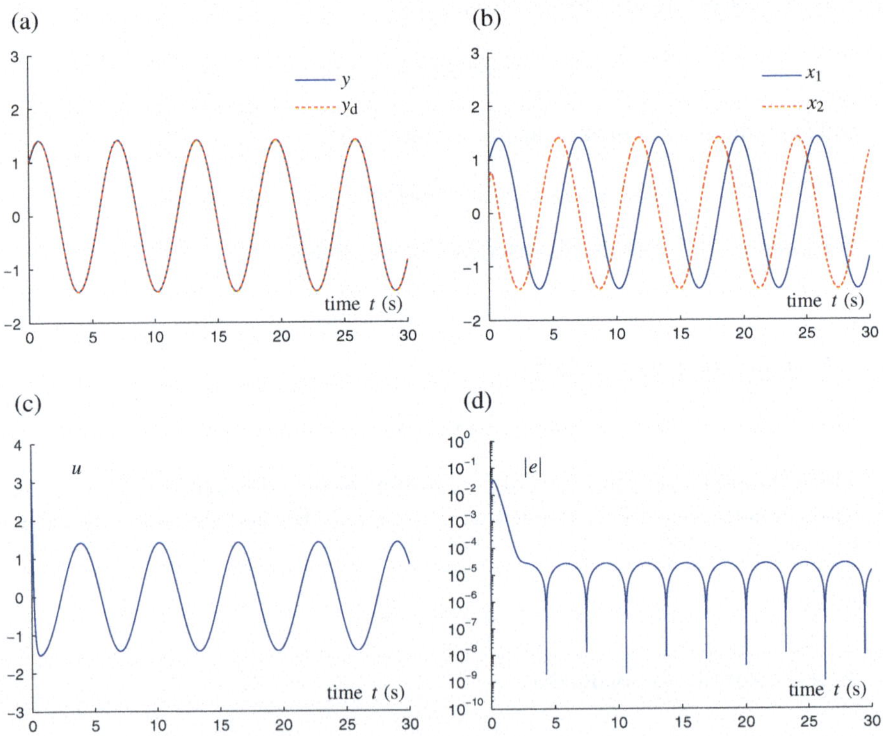

Fig. 6.1 Successful computer simulation with ZG controller (6.8) applied to DI system (6.2) for output $y = x_1$ to track desired trajectory $y_\mathrm{d} = \sin(t) + \cos(t)$. (**a**) Output trajectory and desired trajectory. (**b**) System states. (**c**) Control input. (**d**) Absolute tracking error

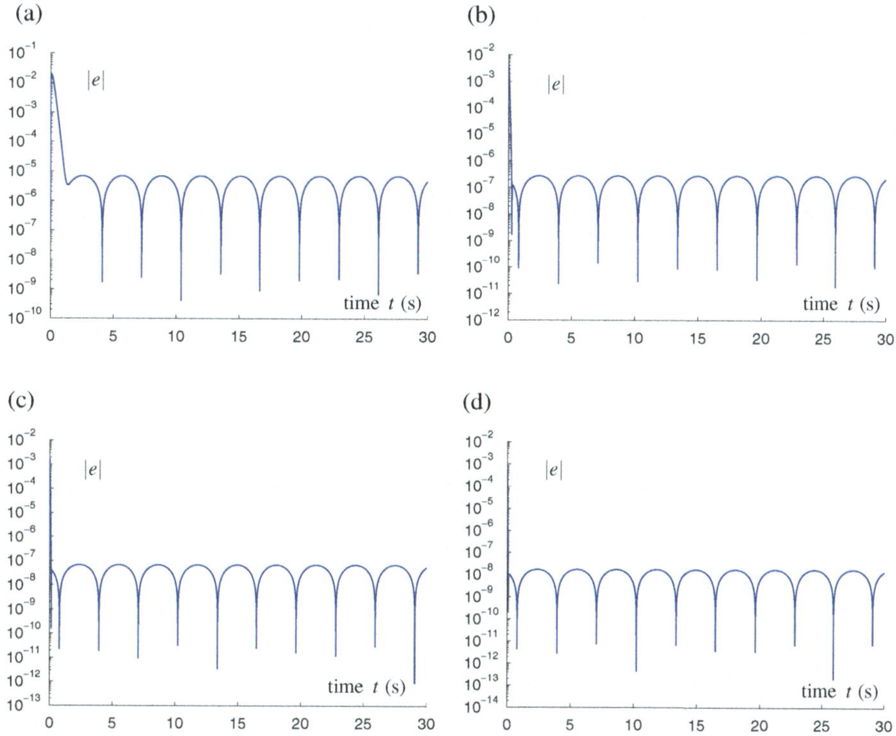

Fig. 6.2 Absolute tracking errors with different values of parameter λ for ZG controller (6.8) applied to DI system (6.2), where output $y = x_1$ tracks desired trajectory $y_d = \sin(t) + \cos(t)$. (**a**) With $\lambda = 10$. (**b**) With $\lambda = 50$. (**c**) With $\lambda = 100$. (**d**) With $\lambda = 200$

10^{-5}. Note that, since a variety of errors such as truncation errors and round-off errors exist in computer simulations, the numerical difference between the output trajectory y and the desired trajectory y_d is inevitable. Hence, the absolute tracking error $|e|$ can hardly be zero.

To better show the performance of ZG controllers, the corresponding computer simulations with different values of λ are conducted. Due to the similarity of the output trajectory y, states x_1 and x_2, and the input u, only the absolute tracking errors are presented. Specifically, Fig. 6.2 illustrates the absolute tracking errors with the value of λ being 10, 50, 100, and 200, respectively. It can be seen from Fig. 6.2 that, as the value of λ increases, the absolute tracking error decreases. Specifically, when the value of λ is increased from 10 to 200, the maximal steady-state absolute tracking error decreases from 10^{-5} to 10^{-8}. Besides, the parameter γ has the same effect on the absolute tracking error, with the related figures omitted here.

Based on the above observations, it can be found that the efficacy of ZG controller (6.8) is illustrated.

6.4 Implicit Tracking Control of DI System via ZG Method (ITC-DI-ZG)

In this section, the implicit tracking controller design is investigated with the output being $y = x_1^2 + x_2^2$, and the design process is shown in detail as follows.

By combining DI system (6.1) and the output $y = x_1^2 + x_2^2$, the whole system can be formulated as

$$\begin{cases} \dot{x}_1 = f_1(x_2) = x_2, \\ \dot{x}_2 = f_2(u) = u, \\ y = x_1^2 + x_2^2 \rightarrow y_d. \end{cases} \tag{6.9}$$

The ZF for (6.9) can be constructed as $z = y - y_d = x_1^2 + x_2^2 - y_d$. Substituting (6.9) into the ZD design formula $\dot{z} = -\lambda z$, we have $2x_1x_2 + 2x_2u - \dot{y}_d + \lambda(x_1^2 + x_2^2 - y_d) = 0$. Then, we define $\phi_2 = 2x_1x_2 + 2x_2u - \dot{y}_d + \lambda(x_1^2 + x_2^2 - y_d)$. According to the GD method, the EF $\epsilon = \phi_2^2/2$ can be defined. By utilizing the GD design formula $\dot{u} = -\gamma\,\partial\epsilon/\partial u$, the ZG controller is finally obtained as

$$\dot{u} = -2\gamma x_2\phi_2 = -2\gamma x_2(2x_1x_2 + 2x_2u - \dot{y}_d + \lambda(x_1^2 + x_2^2 - y_d)). \tag{6.10}$$

6.5 Failed Simulation of ITC-DI-ZG

Sections 6.2 and 6.3 have shown the theoretical derivation and the corresponding computer simulations, which verify the efficacy of ZG controllers. In this section, the ZG controller designed in Sect. 6.4 is simulated with the desired trajectories selected as $y_d = \sin(t)\exp(-0.1t) + 2$ and $y_d = \sin(t) + 2$, respectively. However, the obtained results are not satisfactory.

Specifically, Fig. 6.5 illustrates the tracking performance of ZG controller (6.10) applied to DI system (6.9) with $y = x_1^2 + x_2^2$ and $y_d = \sin(t)\exp(-0.1t) + 2$. The initial values are set as $x_1(0) = 1$, $x_2(0) = 1$, and $u(0) = 0$, together with $\lambda = 5$ and $\gamma = 1000$. In the simulation process, ode15s is utilized as the MATLAB ODE solver with option "AbsTol=1e−8" (i.e., the absolute error tolerance is 10^{-8}). As seen from this figure, in the beginning, the output trajectory fits well with the desired trajectory; however, when the tracking time is over 4 s, there exists a relatively large error (i.e., near 10^0) and the output trajectory becomes a straight line, which cannot track the desired trajectory any more. With the desired trajectory $y_d = \sin(t) + 2$, the initial values $x_1(0) = 0.5$, $x_2(0) = 0.5$ and $u(0) = 0$, and parameters $\lambda = 100$ and $\gamma = 1000$, the tracking performance of ZG controller (6.10) for output $y = x_1^2 + x_2^2$ is presented in Fig. 6.6. The MATLAB ODE solver in the simulation is ode45 with option "RelTol=1e−8" (i.e., the relative error tolerance is 10^{-8}). As

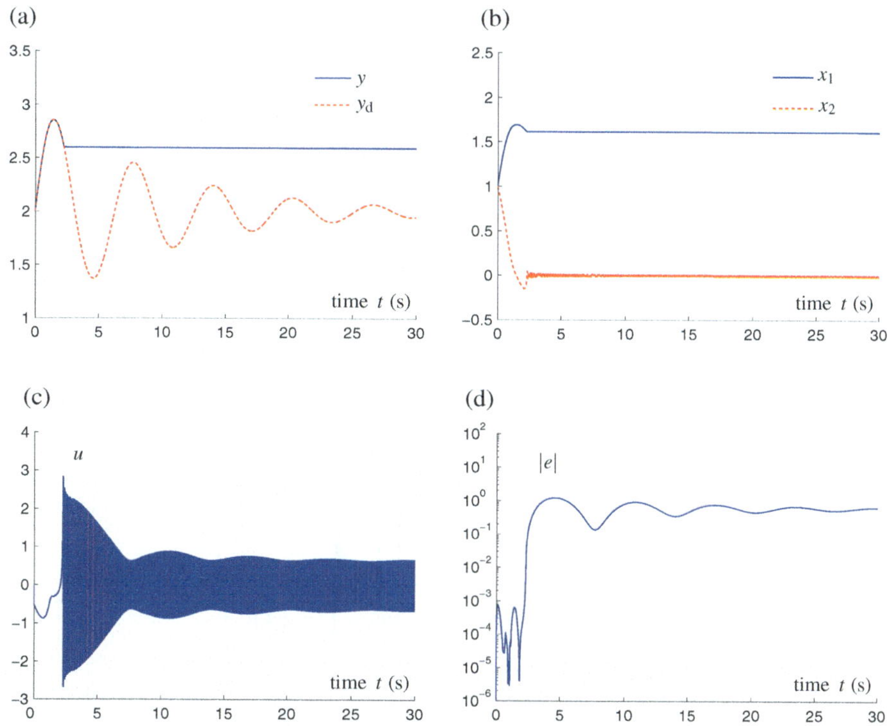

Fig. 6.3 Successful computer simulation with ZG controller (6.10) applied to DI system (6.9) for output $y = x_1^2 + x_2^2$ to track desired trajectory $y_d = \sin(t)\exp(-0.1t) + 2$ using ode15s with option "RelTol=1e−8" in comparison with Fig. 6.5. (**a**) Output trajectory and desired trajectory. (**b**) System states. (**c**) Control input. (**d**) Absolute tracking error

seen from Fig. 6.6, the output trajectory cannot track the desired trajectory during time intervals [15, 20] s and [23, 24] s or so, which means that the simulation of the tracking process is failed.

6.6 Finally Successful Simulation of ITC-DI-ZG

The failed simulations have been presented in Sect. 6.5. However, if the options are slightly changed, the results turn out to be quite different. Part of the corresponding program is attached in Appendix 2.

In order to show the differences, the results of two successful simulations are presented in this section and compared with the corresponding simulations in Sect. 6.5. As shown in Fig. 6.3, the output of the system tracks well the desired trajectory, which illustrates that the designed controller is effective. However, in comparison with Fig. 6.5, it is only the options' change from "AbsTol=1e−8" to

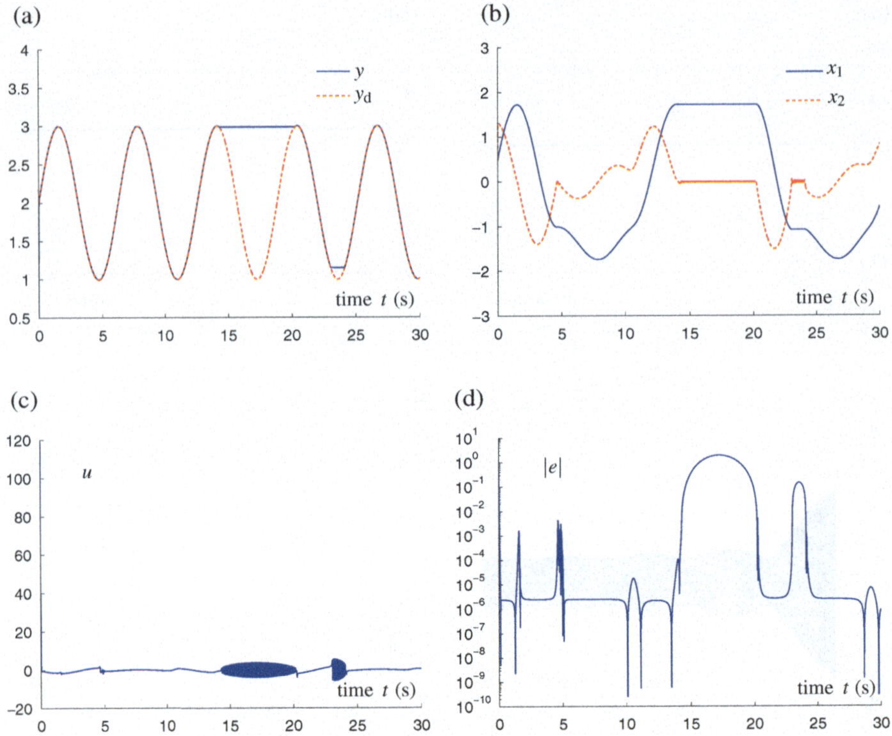

Fig. 6.4 Successful computer simulation with ZG controller (6.10) applied to DI system (6.9) for output $y = x_1^2 + x_2^2$ to track desired trajectory $y_d = \sin(t) + 2$ using ode45 with option "MaxStep=1e−3" (i.e., upper bound on solver stepsize is 10^{-3}) in comparison with Fig. 6.6. (**a**) Output trajectory and desired trajectory. (**b**) System states. (**c**) Control input. (**d**) Absolute tracking error

"RelTol=1e−8". Besides, the successful simulation of the control of system (6.9) for the output $y = x_1^2 + x_2^2$ tracking the desired trajectory $y_d = \sin(t) + 2$ is shown in Fig. 6.4. Specifically, from Fig. 6.4d, we can observe that the maximal steady-state absolute tracking error is below 10^{-2}, with the output trajectory tracking the desired trajectory perfectly as shown in Fig. 6.4a. Such results illustrate the efficacy of the presented controller once again. Similar to the relationship between Figs. 6.5 and 6.3, only the options are different (i.e., changing from "RelTol=1e−8" to "MaxStep=1e−3") between the simulations in Figs. 6.6 and 6.4 but lead to the great differences in the performance and simulation results. To show the differences more clearly, the options, performance and simulation results are comparatively listed in Table 6.1. Based on a large number of simulation tests, it can be readily found from the table that different desired trajectories of output require different settings of ODE options. Besides, a "MaxStep" option setting with a relatively tiny value is more suitable for dealing with the tracking control challenge. In addition, a "RelTol" option setting is more desirable in terms of computational complexity.

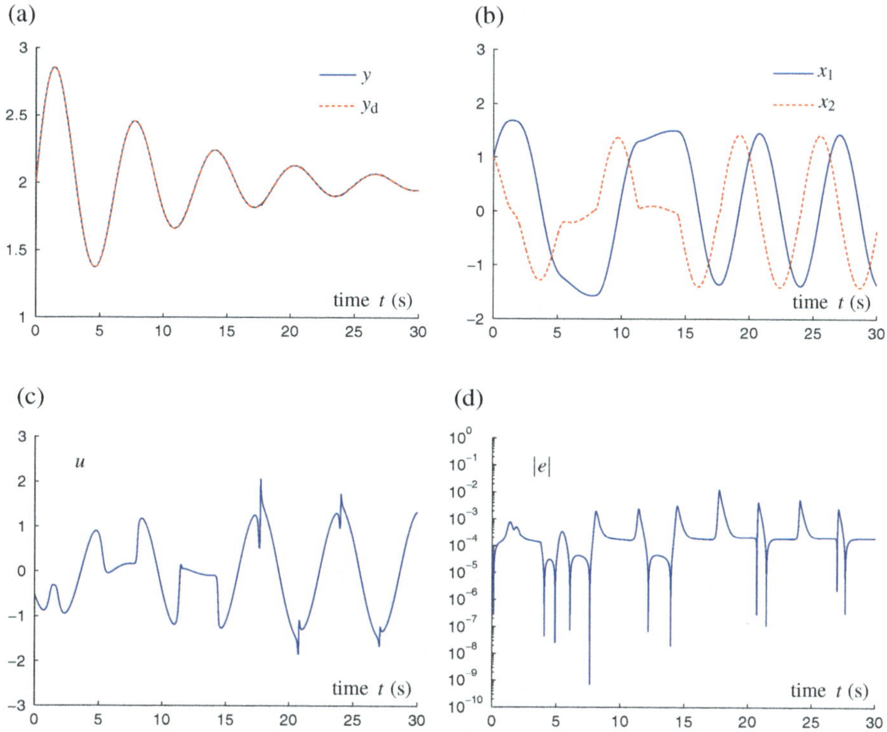

Fig. 6.5 Failed computer simulation with ZG controller (6.10) applied to DI system (6.9) for output $y = x_1^2 + x_2^2$ to track desired trajectory $y_{\rm d} = \sin(t)\exp(-0.1t) + 2$ using ode15s with option "AbsTol=1e−8". (**a**) Output trajectory and desired trajectory. (**b**) System states. (**c**) Control input. (**d**) Absolute tracking error

In short, we have the conclusion that the different settings of simulation options in MATLAB ODE solvers may lead to different simulation results (e.g., failure and success). As seen from Table 6.1, the least computing time of a successful tracking test is 0.660021 s. As compared with the task duration 30 s, the computing time is applicable in most of the real-time control applications. Besides, almost all of the computing times of successful tracking tests (except the "MaxStep=1e−4" ones) are smaller than the task duration (i.e., 30 s), which implies that the general software or hardware implementation of the ZG controller can satisfy the real-time computation requirement. Furthermore, we can choose the suitable values of design parameters (i.e., λ and γ) so as to have a relatively smaller computing time. Moreover, in order to clarify the reason why the simulation challenge happens both in principle and programming levels, the following two remarks are provided.

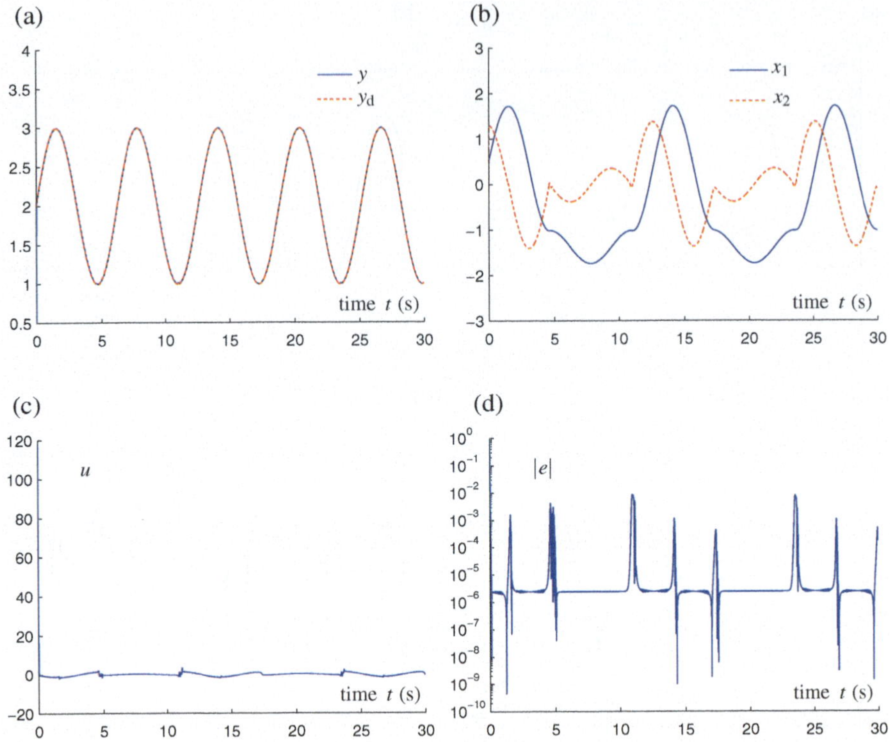

Fig. 6.6 Failed computer simulation with ZG controller (6.10) applied to DI system (6.9) for output $y = x_1^2 + x_2^2$ to track desired trajectory $y_d = \sin(t) + 2$ using ode45 with option "RelTol=1e−8". (**a**) Output trajectory and desired trajectory. (**b**) System states. (**c**) Control input. (**d**) Absolute tracking error

Remark 6.1 Although the ordinary differential equation can be solved by MATLAB ODE solvers effectively, we may still have the challenge in the simulation practice of system-output tracking control (via the ZG method). Many reasons account for the challenge and they can be divided into two levels, i.e., principle and programming levels. From the principle perspective, the types of system to be handled and the design trajectories of output have great impacts on the tracking-performance simulation of ZG controller. For example, the DI system investigated in this work encounters the simulation challenge, while the ZG method based tracking control of other systems (e.g., ship course system, Lu chaotic system, and robot manipulator system) can be successfully simulated by MATLAB ODE solvers. In addition, design parameters (e.g., λ and γ of the ZG method) and initial values (e.g., $x_1(0)$, $x_2(0)$ and $u(0)$) may also have influences on the tracking results. From the programming perspective, the simulation environment and the simulation settings such as stepsize are the important factors of the simulation results. Please see, compare and generalize Table 6.1.

Table 6.1 Statistics of failed and successful simulation tests

RelTol			AbsTol			MaxStep		
Value	Computing time (s)	Category	Value	Computing time (s)	Category	Value	Computing time (s)	Category
$\lambda = 5 \quad \gamma = 1000 \quad x_1(0) = 1 \quad x_2(0) = 1 \quad u(0) = 0 \quad y_d = \sin(t)\exp(-0.1t) + 2$								
Option settings for solver ode15s								
1e−1	0.405966	Failure	1e−1	0.263174	Failure	1e−1	4.172982	Failure
1e−2	0.689756	Failure	1e−2	0.530255	Failure	1e−2	1.039064	Success
1e−3	3.425832	Failure	1e−3	3.502439	Failure	1e−3	6.063296	Success
1e−4	0.740600	Success	1e−4	0.660021	Success	1e−4	59.911112	Success
1e−5	0.854406	Success	1e−5	4.268113	Failure	–	–	–
1e−6	1.047260	Success	1e−6	3.433776	Failure	–	–	–
1e−7	1.003816	Success	1e−7	0.676923	Success	–	–	–
1e−8	1.019171	Success	1e−8	3.423199	Failure	–	–	–
1e−9	0.970222	Success	1e−9	3.734611	Failure	–	–	–
1e−10	0.994165	Success	1e−10	3.067046	Failure	–	–	–
$\lambda = 100 \quad \gamma = 1000 \quad x_1(0) = 0.5 \quad x_2(0) = 0.5 \quad u(0) = 0 \quad y_d = \sin(t) + 2$								
Option settings for solver ode45								
1e−1	17.331623	Failure	1e−1	17.030226	Success	1e−1	19.344821	Failure
1e−2	14.100663	Failure	1e−2	21.308804	Failure	1e−2	24.138390	Success
1e−3	17.203062	Failure	1e−3	17.483012	Failure	1e−3	29.970807	Success
1e−4	20.402083	Failure	1e−4	22.299325	Failure	1e−4	148.908840	Success
1e−5	21.067568	Failure	1e−5	21.926064	Failure	–	–	–
1e−6	31.316403	Failure	1e−6	18.137608	Failure	–	–	–
1e−7	37.883427	Failure	1e−7	10.676167	Failure	–	–	–
1e−8	27.004442	Failure	1e−8	13.809725	Failure	–	–	–
1e−9	36.195654	Failure	1e−9	16.945451	Failure	–	–	–
1e−10	35.605396	Failure	1e−10	16.598676	Failure	–	–	–

Remark 6.2 When we choose the MATLAB ODE solvers and their corresponding option settings for the ZG controller, the following rules can be considered.

- The MATLAB ODE solver ode15s with the "RelTol" setting is more appropriate for simulating the time-varying systems and real-time practical systems.
- A "MaxStep" option setting with a relatively tiny value is more suitable and effective for dealing with the tracking control challenge, no matter whether the MATLAB ODE solver is ode15s or ode45.
- A relatively tiny value of "RelTol" option setting for ode15s would increase the possibility of the successful simulation practice of ZG control.

At the end of the main body of this chapter [10], it is worth pointing out that the ZG method could have other extended applications. For better understanding and for completeness, another type of output function (i.e., $y = x_1x_2$) is exploited and investigated in Appendix 3 for design and substantiation of ZG controller.

6.7 Chapter Summary

This chapter has shown the design procedures of explicit and implicit tracking controllers by using the ZG method. The corresponding computer simulations have been performed to show the efficacy of ZG controllers. Although the ZG controllers are powerful, there is still a challenge in the simulation practice. After conducting abundant simulation practices, we have the conclusion that different settings of simulation options in ODE solvers may lead to different simulation results (e.g., failure and success). For better comparison, the successful and failed simulation results have been presented in this chapter, which reminds us to pay more attention to the option settings in the simulation program.

Appendix 1: MATLAB Codes Related to Sect. 6.2

(I) The following is the main body of the program, which calls the function "outputPath" and function "systemDI" to calculate the output, error, etc, and plot corresponding figures.

```
clear all; close all; clc; format long;
global path_index lambda gamma;
lambda=5; gamma=1000; T=30; %set task duration
path_index=3;
switch path_index
    case 1
        path_name='sin(t)';
    case 2
        path_name='cos(t)';
    case 3
        path_name='sin(t)+cos(t)';
end
xu0=[1,0.5,0]; %set initial values
%now set the options
options=odeset('MaxStep',1e-3);
%options=odeset('RelTol',1e-8);
%options=odeset('AbsTol',1e-8);
%now set the ODE solver
[t,xu]=ode15s('systemDI',[0,T],xu0,options);
%[t,xu]=ode45('systemDI',[0,T],xu0,options);
t_length=length(t); yd=zeros(t_length,1);
for j=1:t_length
    [yd(j),yd1,yd2]=outputPath(path_index,t(j));
end
```

```
error=abs(xu(:,1)-yd);
tracking_plot=figure(1);
plot(t,xu(:,1),'r',t,yd,'--b');
error_plot=figure(2); semilogy(t,error);
x1x2_plot=figure(3);
plot(t,xu(:,1),'g',t,xu(:,2),'--b');
u_plot=figure(4); plot(t,xu(:,3));
```

(II) The following user-defined function "outputPath" is used to calculate y_d, $\dot{y}_\mathrm{d}$ and $\ddot{y}_\mathrm{d}$, and then assign them to yd, yd1 and yd2.

```
function [yd,yd1,yd2]=outputPath(pi,t)
switch pi
    case 1
        yd=sin(t);
        yd1=cos(t);
        yd2=-sin(t);
    case 2
        yd=cos(t);
        yd1=-sin(t);
        yd2=-cos(t);
    case 3
        yd=sin(t)+cos(t);
        yd1=cos(t)-sin(t);
        yd2=-sin(t)-cos(t);
end
end
```

(III) The following "systemDI" function is defined to achieve DI system (6.2) equipped with ZG controller (6.8).

```
function dotxu=systemDI(t,xu)
global path_index lambda gamma;
dotxu=zeros(3,1);
dotxu(1)=xu(2);
dotxu(2)=xu(3);
[yd,yd1,yd2]=outputPath(path_index,t);
dotxu(3)=-2*gamma*(lambda^2*xu(1)+2*lambda*xu(2)+...
xu(3)-lambda^2*yd-2*lambda*yd1-yd2);
t
end
```

Appendix 2: MATLAB Code Related to Sect. 6.4

Since the main program and the "outputPath" function are similar to those in Appendix 1, only the "systemDI" function is shown as below. That is, the following "systemDI" function is defined to achieve DI system (6.9) equipped with ZG controller (6.10).

```
function dotxu=systemDI(t,xu)
global path_index lambda gamma;
dotxu=zeros(3,1);
dotxu(1)=xu(2);
dotxu(2)=xu(3);
[yd,yd1,yd2]=outputPath(path_index,t);
h=2*xu(1)*xu(2)+2*xu(2)*xu(3)+lambda*xu(1)^2+...
lambda*xu(2)^2-yd1-lambda*yd;
dotxu(3)=-4*gamma*xu(2)*h;
t
end
```

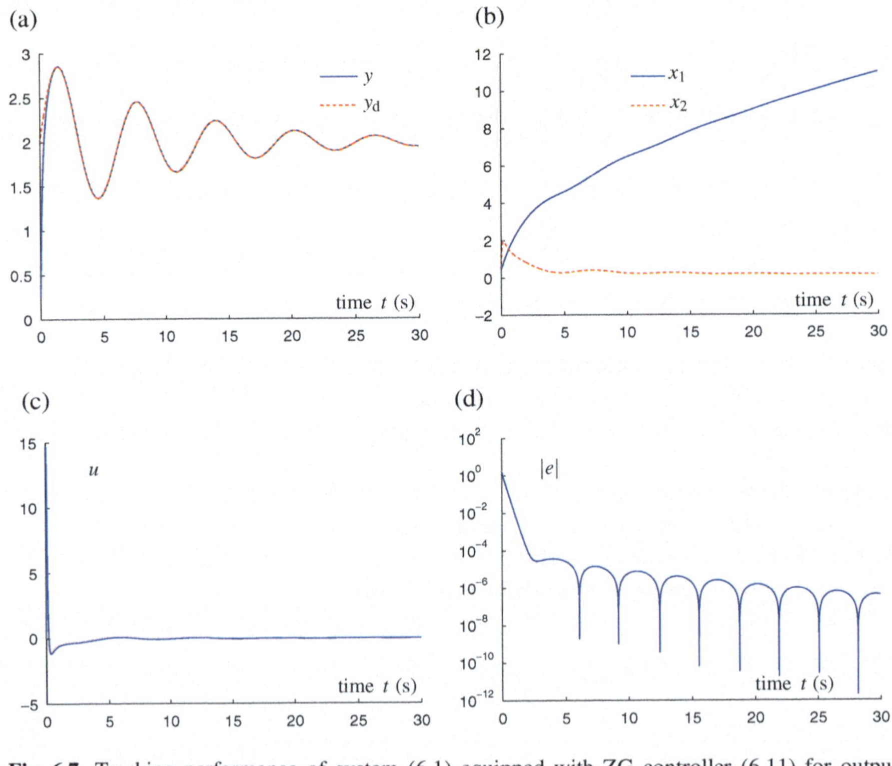

Fig. 6.7 Tracking performance of system (6.1) equipped with ZG controller (6.11) for output $y = x_1 x_2$ to track desired trajectory $y_d = \sin(t)\exp(-0.1t) + 2$. (**a**) Output trajectory and desired trajectory. (**b**) System states. (**c**) Control input. (**d**) Absolute tracking error

Appendix 3: Simulation on Another Type of Output Function via ZG Controller

In order to show the generalization of the ZG method for the implicit tracking control of the DI system, we present another type of output function, i.e., $y = x_1x_2$ as an illustrative example. The design process of the ZG controller for DI system (6.1) with output function $y = x_1x_2$ can be generalized from Sect. 6.4, and thus is omitted here due to the similarity. The ZG controller for DI system (6.1) with output function $y = x_1x_2$ is obtained as

$$\dot{u} = -2\gamma x_1(x_2^2 + x_1u - \dot{y}_\mathrm{d} + \lambda(x_1x_2 - y_\mathrm{d})). \tag{6.11}$$

The corresponding simulation results are shown in Fig. 6.7, where the desired trajectory is set as $y_\mathrm{d} = \sin(t)\exp(-0.1t) + 2$ again. As seen from this figure, the output y quickly tracks the desired trajectory y_d, and the absolute tracking error $|e|$ converges to a level of tiny values within a short time. Such simulation results substantiate that ZG controller (6.11) can still possess effective tracking performance even when the system exploits a different type of output function.

References

1. Frechard J, Knittel D, Dessagne P, Pelle JS, Gaudiot G, Caspar JC, Heitz G (2013) Modelling and fast position control of a new unwinding-winding mechanism design. Math Comput Simul 90:116–131
2. Li C (2012) Tracking control and generalized projective synchronization of a class of hyperchaotic system with unknown parameter and disturbance. Commun Nonlinear Sci Numer Simul 17(1):405–413
3. Qiu J, Lu J, Cao J, He H (2011) Tracking analysis for general linearly coupled dynamical systems. Commun Nonlinear Sci Numer Simul 16(4):2072–2085
4. Xiao X, Zhou L, Zhang Z (2014) Synchronization of chaotic Lur'e systems with quantized sampled-data controller. Commun Nonlinear Sci Numer Simul 19(6):2039–2047
5. Xin X, Liu Y (2014) Trajectory tracking control of variable length pendulum by partial energy shaping. Commun Nonlinear Sci Numer Simul 19(5):1544–1556
6. Miyazaki T, Ohishi K, Shibutani I, Yoshida Y, Koide D, Tokumaru H (2007) Perfect tracking control based on prediction of reference for high speed optical disc system. In: Proceedings of the 33rd annual conference of IEEE industrial electronics society, pp 345–350
7. Yang S, Zhu A, Yuan G, Meng M (2012) A bioinspired neurodynamics-based approach to tracking control of mobile robots. IEEE Trans Ind Electron 59(8):3211–3220
8. Chen M, Ge SS, Choo YS (2009) Neural network tracking control of ocean surface vessels with input saturation. In: Proceedings of IEEE international conference on automation and logistics, pp 85–89
9. Gao H, Chen T (2008) Network-based H_∞ output tracking control. IEEE Trans Autom Control 53(3):655–667
10. Zhang Y, Zhai K, Chen D, Jin L, Hu C (2015) Challenging simulation practice (failure and success) on implicit tracking control of double-integrator system via Zhang-gradient method. Math Comput Simul 120:104–119

11. Hao H, Barooah P (2013) Stability and robustness of large platoons of vehicles with double-integrator models and nearest neighbor interaction. Int J Robust Nonlinear Control 23(18):2097–2122
12. Zhou B, Duan G, Lin Z (2010) Global stabilization of the double integrator system with saturation and delay in the input. IEEE Trans Circuits Syst Regul Pap 57(6):1371–1383
13. Arshak K, Jafer E, McDonagh D (2007) Modelling and simulation of a wireless microsensor data acquisition system using PCM techniques. Simul Model Pract Theory 15(7):764–785
14. Quaglia D, Muradore R, Bragantini R, Fiorini P (2012) A SystemC/MATLAB co-simulation tool for networked control systems. Simul Model Pract Theory 23:71–86
15. Gelen A, Yalcinoz T (2010) An educational software package for Thyristor Switched Reactive Power Compensators using MATLAB/Simulink. Simul Model Pract Theory 18(3):366–377
16. Haidar A, Mohamed A, Milano F (2010) A computational intelligence-based suite for vulnerability assessment of electrical power systems. Simul Model Pract Theory 18(5):533–546
17. Lim CL, Jones NB, Spurgeon SK, Scott JJA (2003) Modelling of knee joint muscles during the swing phase of gait—a forward dynamics approach using MATLAB/Simulink. Simul Model Pract Theory 11(2):91–107
18. Wu L, Yu C (2012) Powder particle size measurement with digital image processing using MATLAB. Adv Mater Res 443–444:589–593
19. Zhang Z, Un An D, Kim H, Chong KT (2009) Comparative study of matrix exponential and Taylor series discretization methods for nonlinear ODEs. Simul Model Pract Theory 17(2):471–484
20. Migoni G, Bortolotto M, Kofman E, Cellier FE (2013) Linearly implicit quantization-based integration methods for stiff ordinary differential equations. Simul Model Pract Theory 35:118–136
21. Zhang Y, Yi C, Guo D, Zheng J (2011) Comparison on Zhang neural dynamics and gradient-based neural dynamics for online solution of nonlinear time-varying equation. Neural Comput Appl 20(1):1–7
22. Zhang Y, Ke Z, Xu P, Yi C (2010) Time-varying square roots finding via Zhang dynamics versus gradient dynamics and the former's link and new explanation to Newton-Raphson iteration. Inf Process Lett 110(24):1103–1109

Chapter 7
ZG Tracking Control of MI Systems

Abstract In this chapter, the tracking-control problems of multiple-integrator (MI) systems are investigated by using the ZG method. Several types of ZG controllers are presented for the tracking control of MI systems, e.g., triple-integrator (TI) systems. As an example, the design procedures of ZG controllers for TI systems with a linear output function (LOF) and a nonlinear output function (NOF) are presented. Corresponding theoretical analyses are given to guarantee the convergence performance of both z3g0 controllers and z3g1 controllers for TI systems. Computer simulations concerning the tracking control of MI systems with different types of output functions are further performed to substantiate the feasibility and efficacy of ZG controllers for tracking-control problem solving. Moreover, comparative simulation results for tracking control of MI systems with NOFs substantiate that the $zmg1$ controllers can effectively conquer the DBZ problem with m being the times of using the ZD method.

7.1 Introduction

Multiple-integrator (MI) systems have many practical applications [1, 2]. For example, one of those systems can be treated as the linearized model of the inverted pendulum. For this reason, such systems have attracted a lot of attention in the past decades [3–5]. The nth-order MI system [4] can be formulated as below:

$$\begin{cases} \dot{x}_i = f_i(x_{i+1}) = x_{i+1}, \text{ with } i = 1, 2, \cdots, n-1, \\ \dot{x}_n = f_n(u) = u. \end{cases} \tag{7.1}$$

As a mathematical abstraction or idealization, linear systems, which form the cornerstone of much of modern day electrical engineering, find important applications in nature, electrical engineering and economics [6, 7]. Besides, the tracking-control problem is very important and has appeared in many applications [8–10]. Hence, the tracking control of linear systems with a linear output function (LOF) and a nonlinear output function (NOF) is an important study topic in the

Y. Zhang et al., *Zhang-Gradient Control*,
https://doi.org/10.1007/978-981-15-8257-8_7

field of control. There have been many researches aiming at the tracking-control problems for their wide applications in numerous fields of science and engineering [10–12]. The tracking control of the DI system, which is a typical dynamic system, has widely been investigated [1, 5]. However, it is difficult to find literatures concerning the tracking-control problem of higher-order MI systems, especially those systems with NOF that are subject to the DBZ problem. The DBZ problem is a big challenge, which has attracted many researchers. For example, a switching-type control scheme has been developed as one of the main techniques to the DBZ problem [10, 13, 14]. Nevertheless, the switching-type control scheme may cost much in terms of the implementation since it requires two or more controllers for solving a single problem. Another example is the DBZ problem encountered in the path tracking of parallel manipulators. The scheme of solving the DBZ problem has been studied for motion feasibility in the neighborhood of DBZ point in [14]. The above solution avoids passing the DBZ point. Besides, there are many researches concerning the global stabilization of MI systems by bounded controls [15–19]. For example, in [17–19], several types of effectively improved saturation functions and corresponding control laws are considered and researched therein for the saturation problem solving, which is desirable and interesting in linear systems for the achievement of global stabilization by bounded controls. Those researches are meaningful and significant for the deep investigation of MI systems. Comparatively, in this chapter [7], by the presented method. Besides, a relatively different, simple and smooth form of controller is investigated, which can effectively achieve the tracking control of MI systems and can elegantly solve the DBZ problem.

In this chapter [7], an effective way is given for the tracking control of MI systems. ZD and GD are two different types of powerful methods for online problem solving [20–22]. The design of GD, which is usually based on a norm-based or square-based EF, was originally designed for time-invariant problem solving, and has recently been extended for time-varying problem solving [23]. Differing from GD, the design of ZD is based on an indefinite EF and has been exploited for solving online time-varying problems [21, 22, 24]. In general, a ZG controller obtained by adopting the ZD method m times and the GD method n times is called a zmgn controller. It is worth pointing out that a zmg1 controller can conquer the DBZ problem, which is a troublesome problem in the controller design [5, 10]. By combining the ZD and GD methods together, the ZG method is presented to design ZG controllers for solving the tracking-control problems of higher-order MI systems. Those controllers can successfully track the desired trajectory, especially for the systems with NOF that would be subject to the DBZ problem. Besides, corresponding theoretical analyses indicate that z3g0 and z3g1 controllers both possess the exponential convergence for triple-integrator (TI) systems with LOF and NOF, which guarantees the efficacy of ZG controllers in theory. Simulation results also illustrate that MI systems equipped with the designed controllers can accurately track the desired trajectories, thereby showing the feasibility and superiority of the ZG method for MI systems. More importantly, the efficacy of zmg1 controllers in conquering the DBZ problem is also substantiated by comparative simulation results for the tracking control of MI systems with NOF.

7.2 Design of Controllers

Without loss of generality, we take the TI system as an illustrative example of MI systems. The TI system can be formulated as below:

$$\begin{cases} \dot{x}_1 = f_1(x_2) = x_2, \\ \dot{x}_2 = f_2(x_3) = x_3, \\ \dot{x}_3 = f_3(u) = u, \end{cases} \tag{7.2}$$

where x_1, x_2 and x_3 are the states of the TI system, and u is the input of this system. The detailed design procedures of ZG controllers are presented in the following subsections, with the output functions $y = x_1$ and $y = \cos x_1$ taken as illustrative examples. Besides, in the following, the tracking error is defined as $e = y - y_\mathrm{d}$, where y_d denotes the desired output.

7.2.1 *Design of ZD and ZG Controllers for LOF*

Considering TI system (7.2) with the LOF (i.e., $y = x_1$), and applying the ZD method [5, 10] twice, we have the third ZF as follows:

$$z_3 = x_3 - \ddot{y}_\mathrm{d} + 2\lambda(x_2 - \dot{y}_\mathrm{d}) + \lambda^2(x_1 - y_\mathrm{d}),$$

where the design parameter $\lambda \in \mathbb{R}^+$ is used to scale the convergence rate of the ZD solution. By applying the ZD method the third time, based on TI system (7.2) and the above equation, the z3g0 controller in the form of u is finally obtained as

$$u = \lambda^3(y_\mathrm{d} - x_1) + 3\lambda^2(\dot{y}_\mathrm{d} - x_2) + 3\lambda(\ddot{y}_\mathrm{d} - x_3) + \dddot{y}_\mathrm{d}. \tag{7.3}$$

From Eq. (7.3), we can define $\phi_1 = u - \dddot{y}_\mathrm{d} - 3\lambda(\ddot{y}_\mathrm{d} - x_3) - 3\lambda^2(\dot{y}_\mathrm{d} - x_2) - \lambda^3(y_\mathrm{d} - x_1)$. Note that ϕ_1 should theoretically be zero. Then, based on the GD method [5, 10], a corresponding EF in the form of $\epsilon_1 = \phi_1^2/2$ can be defined, and the GD design formula is adopted as $\dot{u} = -\gamma(\partial\epsilon_1/\partial u)$, where the design parameter $\gamma \in \mathbb{R}^+$ is used to scale the convergence rate of the GD solution. Thus, with the output function of TI system (7.2) being $y = x_1$, a z3g1 controller in the form of $\dot{u}$ can be obtained as

$$\dot{u} = -\gamma\phi_1 = -\gamma\left(u - \dddot{y}_\mathrm{d} - 3\lambda(\ddot{y}_\mathrm{d} - x_3) - 3\lambda^2(\dot{y}_\mathrm{d} - x_2) - \lambda^3(y_\mathrm{d} - x_1)\right). \tag{7.4}$$

Remark 7.1 The control inputs, which are obtained by using the ZG method and given in the form of an ODE equation, can be handled in practice. For example, on the one hand, the integral of Eq. (7.4) can be obtained as

$$u=-\gamma\left(3\lambda^2(x_1-y_\mathrm{d})+\lambda^3\int_0^t(x_1(\iota)-y_\mathrm{d}(\iota))\mathrm{d}\iota+x_3-\ddot{y}_\mathrm{d}+3\lambda(x_2-\dot{y}_\mathrm{d})\right), \tag{7.5}$$

which is actually a generalized proportional-integral-derivative controller with the second-order time derivative considered as well. On the one hand, we can use the traditional proportional-integral-derivative controller method to handle and implement (7.5) in practice. On the other hand, from the viewpoint of the physical implementation based on digital circuits and computers, $\dot{u}$ can be discretized as $\dot{u}(t_k)=(u(t_{k+1})-u(t_k))/\tau$, where $u(t_k)$ denotes the control input at time instant t_k (with $k=1,2,\ldots$), and $\tau>0$ denotes the sampling time interval [10]. Thus, we have

$$\begin{aligned}u(t_{k+1})=&u(t_k)-\gamma\tau(u(t_k)-\dddot{y}_\mathrm{d}(t_k)-3\lambda(\ddot{y}_\mathrm{d}(t_k)-x_3(t_k))\\&-3\lambda^2(\dot{y}_\mathrm{d}(t_k)-x_2(t_k))-\lambda^3(y_\mathrm{d}(t_k)-x_1(t_k))),\end{aligned}$$

which is also feasible in practice. The other control inputs presented in this chapter [7] can also be handled in a similar manner. Thus, it is feasible and effective to apply the presented ZG controllers to the tracking control of MI systems in practice.

7.2.2 *Design of ZD and ZG Controllers for NOF*

For completeness of investigation on the tracking control of TI system (7.2), it is necessary to consider the output function in a complex form. Therefore, $y=\cos x_1$ is further considered. With such a complex form of the output, the tracking control of TI system (7.2) is implicit.

Similar to the design process of ZG controllers for the tracking control of TI system with LOF, by adopting the ZD method thrice and without using the GD method, the following z3g0 controller can readily be obtained, with the output of TI system (7.2) being $y=\cos x_1$:

$$u=\frac{1}{\sin x_1}(\alpha_1-3\lambda\alpha_2-3\lambda^2\alpha_3+\lambda^3\alpha_4), \tag{7.6}$$

where $\alpha_1=x_2^3\sin x_1-3x_2x_3\cos x_1-\dddot{y}_\mathrm{d}$, $\alpha_2=x_2^2\cos x_1+x_3\sin x_1+\ddot{y}_\mathrm{d}$, $\alpha_3=\dot{y}_\mathrm{d}+x_2\sin x_1$ and $\alpha_4=\cos x_1-y_\mathrm{d}$. Note that if the divisor $\sin x_1$ of z3g0 controller (7.6) is zero, the value of u is infinite. In other words, this DBZ problem can probably lead to crash of physical systems in practical applications. By combining the following GD method, the above-mentioned DBZ problem can be elegantly solved.

Defining $\phi_2 = -\sin x_1 u + \alpha_1 - 3\lambda\alpha_2 - 3\lambda^2\alpha_3 + \lambda^3\alpha_4$ and $\epsilon_2 = \phi_2^2/2$, we can similarly have the following z3g1 controller:

$$\dot{u} = \gamma \sin x_1 \phi_2 = -\gamma \sin x_1(\sin x_1 u - \alpha_1 + 3\lambda\alpha_2 + 3\lambda^2\alpha_3 - \lambda^3\alpha_4). \tag{7.7}$$

Note that z3g1 controller (7.7) can conquer the DBZ problem in tracking control since it has no division operation.

7.3 Convergence Performance Analyses on ZD Controllers

In this section, the convergence performance of z3g0 controllers (7.3) and (7.6) is analyzed for the tracking control with LOF and NOF.

7.3.1 Analysis on Tracking Control with LOF

For the tracking control of TI system (7.2) with LOF (i.e., $y = x_1$), the corresponding convergence performance analysis on z3g0 controller (7.3) is presented as the following theorem.

Lemma 7.1 *$\forall\, \tilde{\alpha} > 0$ and $\tilde{\beta} > 0$, with $t > 0$, there exist $\alpha > 0$ and $\beta > 0$ such that*

$$\tilde{\alpha} t^2 \exp(-\tilde{\beta} t) \leq \alpha \exp(-\beta t). \tag{7.8}$$

Proof It can be generalized from [25]. □

Theorem 7.1 *For smooth and bounded desired trajectory y_d, starting with bounded initial state $\mathbf{x}(0) = [x_1(0), x_2(0), x_3(0)]^T \in \mathbb{R}^3$, the tracking error of TI system (7.2) equipped with z3g0 controller (7.3) exponentially converges to zero on a large scale.*

Proof According to the ZD design formulas about z_1, z_2 and z_3, we can readily obtain the following equation:

$$\dddot{z}_1 + 3\lambda\ddot{z}_1 + 3\lambda^2\dot{z}_1 + \lambda^3 z_1 = 0.$$

Then, solving the above equation, we have

$$z_1 = \acute{c}_1 \exp(-\lambda t) + \acute{c}_2 t \exp(-\lambda t) + \acute{c}_3 t^2 \exp(-\lambda t), \tag{7.9}$$

where $\acute{c}_1$, $\acute{c}_2$ and $\acute{c}_3$ are constants, and $t \in [0, +\infty)$. We can derive the expressions of $\dot{z}_1$ and $\ddot{z}_1$ from (7.9), and then the initial conditions $z_1(0)$, $\dot{z}_1(0)$ and $\ddot{z}_1(0)$ are obtained as

$$\begin{cases} z_1(0) = \acute{c}_1, \\ \dot{z}_1(0) = -\lambda\acute{c}_1 + \acute{c}_2, \\ \ddot{z}_1(0) = \lambda^2\acute{c}_1 - 2\lambda\acute{c}_2 + 2\acute{c}_3. \end{cases}$$

Solving the above equations, we have

$$\begin{cases} \acute{c}_1 = z_1(0), \\ \acute{c}_2 = \lambda z_1(0) + \dot{z}_1(0), \\ \acute{c}_3 = \frac{1}{2}(\lambda^2 z_1(0) + 2\lambda\dot{z}_1(0) + \ddot{z}_1(0)). \end{cases}$$

Then, according to Lemma 7.1 and [25], in view of $\lambda > 0$, there exist $\bar{c} > 0$ and $\bar{\lambda} > 0$ such that $z_1 \leq \bar{c}\exp(-\bar{\lambda}t)$, which means that $z_1 = e$ exponentially converges to zero. Therefore, the tracking error e of system (7.2) equipped with controller (7.3) exponentially converges to zero. The proof is thus completed. □

7.3.2 Analysis on Tracking Control with NOF

For the tracking control of TI system (7.2) with NOF (i.e., $y = \cos x_1$), the corresponding convergence performance analysis on z3g0 controller (7.6) is presented as the following theorem.

Theorem 7.2 *For smooth and bounded desired trajectory* y_d, *starting with bounded initial state* $\mathbf{x}(0) = [x_1(0), x_2(0), x_3(0)]^T \in \mathbb{R}^3$, *the tracking error of TI system (7.2) equipped with z3g0 controller (7.6) exponentially converges to zero on a large scale, provided that* $\sin x_1 \neq 0$, $\forall t \in [0, +\infty)$.

Proof It can be generalized from the proof of Theorem 7.1. □

7.4 Convergence Performance Analyses on ZG Controllers

In this section, the convergence performance of z3g1 controllers (7.4) and (7.7) is analyzed for the tracking control of TI system (7.2) with LOF and NOF.

7.4.1 Analyses on Tracking Control with LOF

For the tracking control of TI system (7.2) with LOF (i.e., $y = x_1$), the corresponding convergence performance analysis on z3g1 controller (7.4) is presented as the following theorem.

Theorem 7.3 *Consider TI system (7.2) equipped with z3g1 controller (7.4) for smooth and bounded desired trajectory* y_d. *Starting with bounded initial state* $\mathbf{x}(0) \in \mathbb{R}^3$ *and control input* $u(0) \in \mathbb{R}$, *provided that* $|\dot{u}^*| = |\mathrm{d}u^*/\mathrm{d}t| \leq \varrho$, $\exists 0 \leq \varrho < +\infty$, *the steady-state tracking error of the system is upper bounded tightly as*

$$\limsup_{t\to+\infty} |e| \leq \frac{\varrho}{\lambda^3\gamma}.$$

Proof Let us firstly analyze the convergence performance of solution error for z3g1 controller (7.4). Specifically, a solution error is defined as $\psi = u - u^*$ with u^* being the desired time-varying solution. We have $u = \psi + u^*$ and its time derivative $\dot{u} = \dot{\psi} + \dot{u}^*$. In view of $\dot{u} = -\gamma\psi$, we have $\dot{\psi} = -\gamma\psi - \dot{u}^*$. A Lyapunov function candidate can be defined as $L_1 = \psi^2/2 \geq 0$. Then, we have

$$\dot{L}_1 = \psi\dot{\psi} = -\gamma\psi^2 - \psi\dot{u}^*. \tag{7.10}$$

There are two terms in the expression of $\dot{L}_1$, i.e., $-\gamma\psi^2$ and $-\psi\dot{u}^*$. For the first term, $-\gamma\psi^2 \leq 0$ always holds true. For the second term, we have the following result based on Cauchy inequality [26]:

$$-\psi\dot{u}^* \leq |\psi||\dot{u}^*| \leq \varrho|\psi|.$$

Then, (7.10) can be derived as

$$\dot{L}_1 = -\gamma\psi^2 - \psi\dot{u}^* \leq -\gamma\psi^2 + \varrho|\psi| = -|\psi|(\gamma|\psi| - \varrho). \tag{7.11}$$

During the time evolution of solution error ψ, (7.11) falls into one of the following three situations.

- $\gamma|\psi| - \varrho > 0$ (i.e., $|\psi| > \varrho/\gamma$, outside a so-called ball of radius ϱ/γ). $\dot{L}_1 < 0$, which implies that ψ approaches zero (i.e., u approaches u^*) as time evolves.
- $\gamma|\psi| - \varrho = 0$ (i.e., $|\psi| = \varrho/\gamma$, on the ball surface). $\dot{L}_1 \leq 0$ implies that ψ approaches zero (i.e., u approaches u^*) or stays on the ball surface with $|\psi| = \varrho/\gamma$ as time evolves.
- $\gamma|\psi| - \varrho < 0$ (i.e., $|\psi| < \varrho/\gamma$, inside the ball). According to (7.11), $\dot{L}_1$ is less than a positive scalar, and thus $|\psi|$ may not decrease again. Moreover, if $\dot{L}_1 > 0$, with $\gamma|\psi| - \varrho < 0$, $\dot{L}_1$ and $|\psi|$ would increase, which implies that $\gamma|\psi| - \varrho$ increases as well. Evidently, there exists a certain time instant such that $2\gamma|\psi| - \varrho = 0$, which returns to the second situation, i.e., $\dot{L}_1 \leq 0$.

By summarizing the above situations, it can be concluded that the solution error of controller (7.4) is upper bounded by ϱ/γ, i.e.,

$$\limsup_{t\to+\infty} |\psi| = \limsup_{t\to+\infty} |u - u^*| \leq \frac{\varrho}{\gamma}.$$

From the derivation process of controller (7.4), we can obtain $\psi = \dot{z}_3 + \lambda z_3$. Then, we have

$$\limsup_{t\to+\infty} |\dot{z}_3 + \lambda z_3| \leq \frac{\varrho}{\gamma},$$

which means

$$-\frac{\varrho}{\gamma} \leq \dot{z}_3 + \lambda z_3 \leq \frac{\varrho}{\gamma}, \tag{7.12}$$

where $t \geq t_\mathrm{e}$ with t_e being large enough.

Then, focusing on the right part of (7.12), we have

$$\dot{z}_3 + \lambda z_3 \leq \frac{\varrho}{\gamma},$$

where $t \geq t_\mathrm{e}$. Making use of Gronwall inequality [27], we obtain

$$z_3 \leq z_3(t_\mathrm{e}) \exp(-\lambda t) + \int_{t_\mathrm{e}}^{t} \exp(-\lambda(t-\iota))\frac{\varrho}{\gamma}\mathrm{d}\iota \leq \hat{c}\exp(-\lambda t) + \frac{\varrho}{\lambda\gamma},$$

where $t \geq t_\mathrm{e}$ and $\hat{c}$ is a constant. Since $z_3 = \dot{z}_2 + \lambda z_2$, we have

$$\dot{z}_2 + \lambda z_2 \leq \vartheta,$$

where $t \geq t_\mathrm{e}$ and $\vartheta = \hat{c}\exp(-\lambda t) + \varrho/(\lambda\gamma)$. Making use of Gronwall inequality again, we have

$$\begin{aligned} z_2 &\leq z_2(t_\mathrm{e}) \exp(-\lambda t) + \int_{t_\mathrm{e}}^{t} \exp(-\lambda(t-\iota))\vartheta(\iota)\mathrm{d}\iota \\ &\leq \bar{c}_1 \exp(-\lambda t) + \bar{c}_2 t \exp(-\lambda t) + \frac{\varrho}{\lambda^2\gamma}, \end{aligned} \tag{7.13}$$

where $\bar{c}_1$ and $\bar{c}_2$ are constants, and $t \geq t_\mathrm{e}$. Letting $\psi = \bar{c}_1\exp(-\lambda t) + \bar{c}_2 t\exp(-\lambda t) + \varrho/(\lambda^2\gamma)$, in view of $z_2 = \dot{z}_1 + \lambda z_1 = \dot{e} + \lambda e$ and similar to the above derivation process of (7.13), we obtain

$$\begin{aligned} e &\leq e(t_\mathrm{e}) \exp(-\lambda t) + \int_{t_\mathrm{e}}^{t} \exp(-\lambda(t-\iota))\psi(\iota)\mathrm{d}\iota \\ &\leq \check{c}_1 \exp(-\lambda t) + \check{c}_2 t \exp(-\lambda t) + \check{c}_3 t^2 \exp(-\lambda t) + \frac{\varrho}{\lambda^3\gamma}, \end{aligned} \tag{7.14}$$

where $\check{c}_1$, $\check{c}_2$ and $\check{c}_3$ are constants, and $t \geq t_e$. According to the left part of (7.12), similar to the derivation process of (7.14), we have

$$e \geq -\tilde{c}_1 \exp(-\lambda t) - \tilde{c}_2 t \exp(-\lambda t) - \tilde{c}_3 t^2 \exp(-\lambda t) - \frac{\varrho}{\lambda^3 \gamma}, \tag{7.15}$$

where $\tilde{c}_1$, $\tilde{c}_2$ and $\tilde{c}_3$ are constants, and $t \geq t_e$. Combining (7.14) and (7.15) yields

$$|e| \leq |c_1| \exp(-\lambda t) + |c_2| t \exp(-\lambda t) + |c_3| t^2 \exp(-\lambda t) + \frac{\varrho}{\lambda^3 \gamma}, \tag{7.16}$$

where $|c_1| = \max\{|\check{c}_1|, |\tilde{c}_1|\}$, $|c_2| = \max\{|\check{c}_2|, |\tilde{c}_2|\}$ and $|c_3| = \max\{|\check{c}_3|, |\tilde{c}_3|\}$, and $t \geq t_e$. Thus, in view of $\lambda > 0$ and $t_e \to +\infty$, we obtain

$$\limsup_{t \to +\infty} |e| \leq \frac{\varrho}{\lambda^3 \gamma}. \tag{7.17}$$

Therefore, the tracking error of TI system (7.2) equipped with z3g1 controller (7.4) converges toward or stays within the error bound $\varrho/(\lambda^3\gamma)$ in this case. The proof is thus completed. □

Theorem 7.3 shows that the tracking error converges toward the ball of $\varrho/(\lambda^3\gamma)$, which implies that, when γ tends to infinite, the tracking-error bound approaches zero. However, the nature of Theorem 7.3 is asymptotic convergence, which may be less acceptable in practice as it may require an infinitely long time period to accomplish the tracking control. Via further investigation, the tracking error is proven to be exponentially convergent toward a relatively loose tracking-error bound, which is presented as the following theorem.

Theorem 7.4 *Consider TI system (7.2) equipped with z3g1 controller (7.4) for smooth and bounded desired trajectory* y_d. *Starting with bounded initial state* $\mathbf{x}(0) \in \mathbb{R}^3$ *and control input* $u(0) \in \mathbb{R}$, *provided that* $|\dot{u}^*| \leq \varrho$, $\exists 0 \leq \varrho < +\infty$, *the tracking error of the system exponentially converges toward or stays within the error bound* $\varrho/(\omega\lambda^3\gamma)$ *on a large scale, where* $\omega \in (0, 1)$ *is a loosening parameter.*

Proof As shown in Theorem 7.3, the solution error ψ of z3g1 controller (7.4) asymptotically converges toward the ball of ϱ/γ. Thus, we have

$$\dot{L}_1 \leq -\gamma\psi^2 + \varrho|\psi| = -(1-\omega)\gamma\psi^2 + (-\omega\gamma\psi^2 + \varrho|\psi|). \tag{7.18}$$

Based on the theoretical result in [10], even in the worst case, the exponential convergence rate is $(1-\omega)\gamma$. Besides, for any $t \geq t_c$ with $t_c = \ln(\omega\gamma|\psi(0)|/\varrho)/((1-\omega)\gamma)$, $|\psi| \leq \varrho/(\omega\gamma)$. Similar to the proof of Theorem 7.3, the following inequality

$$|e| \leq |\check{c}_1| \exp(-\lambda t) + |\check{c}_2| t \exp(-\lambda t) + |\check{c}_3| t^2 \exp(-\lambda t) + \frac{\varrho}{\omega\lambda^3\gamma} \tag{7.19}$$

holds true when $t \geq t_c$, where $\breve{c}_1$, $\breve{c}_2$ and $\breve{c}_3$ are constants. Then, similar to the analysis on the right side of (7.16), the right side of (7.19) exponentially converges toward the new error bound $\varrho/(\omega\lambda^3\gamma)$.

By the above analysis, the tracking error of TI system (7.2) equipped with z3g1 controller (7.4) exponentially converges toward or stays within the error bound $\varrho/(\omega\lambda^3\gamma)$. The proof is thus completed. □

7.4.2 Analyses on Tracking Control with NOF

For the tracking control of TI system (7.2) with NOF (i.e., $y = \cos x_1$), the corresponding convergence performance analysis on z3g1 controller (7.7) is presented as the following theorem.

Theorem 7.5 *Consider TI system (7.2) equipped with z3g1 controller (7.7) for smooth and bounded desired trajectory y_d. Starting with bounded initial state $\mathbf{x}(0) \in \mathbb{R}^3$ and control input $u(0) \in \mathbb{R}$, the following results are achieved on a large scale for the tracking control of the system.*

- *For the case of $\sin x_1 \neq 0$ (i.e., the non-DBZ case), the tracking error of the system converges toward or stays within the error bound $\varrho/(\lambda^3\gamma\varsigma)$, provided that (i) $\varsigma \leq \sin^2 x_1 \leq 1$, $\forall 0 < \varsigma \leq 1$, and (ii) $|\dot{u}^*| \leq \varrho$, $\exists 0 \leq \varrho < +\infty$.*
- *For the case of $\sin x_1 = 0$ (i.e., the DBZ case), the tracking error of the system is bounded.*

Proof For the case of $\sin x_1 \neq 0$ (i.e., the non-DBZ case), controller (7.6) can be further formulated as $u = \xi/\sin x_1$, where $\xi = \alpha_1 - 3\lambda\alpha_2 - 3\lambda^2\alpha_3 + \lambda^3\alpha_4$. Similar to the proof of Theorem 7.3, a solution error of controller (7.7) is defined as $\psi = u - u^*$ with $u^* = \xi/\sin x_1$ denoting the desired time-varying solution. Then, we have $u = \psi + u^*$ and its time derivative $\dot{u} = \dot{\psi} + \dot{u}^*$. Besides, we have $\phi_2 = \xi - \sin x_1 u$. Thus, the time-varying solution of z3g1 controller (7.7) can be rewritten as

$$\dot{u} = \gamma \sin x_1 \phi_2 = \gamma \sin x_1 (\xi - \sin x_1 u).$$

Then, we have $\dot{\psi} = -\gamma \sin^2 x_1 \psi - \dot{u}^*$. In addition, a Lyapunov function candidate is defined as $L_2 = \psi^2/2 \geq 0$. The time derivative of L_2 is obtained as

$$\dot{L}_2 = \psi\dot{\psi} = -\gamma \sin^2 x_1 \psi^2 - \psi\dot{u}^*. \tag{7.20}$$

With $\varsigma \leq \sin^2 x_1$ considered, $-\gamma \sin^2 x_1 \psi^2 \leq -\gamma\varsigma\psi^2$ always holds true for the first term on the right part of (7.20). Besides, making use of Cauchy inequality [26], we have $-\psi\dot{u}^* \leq |\psi||\dot{u}^*| \leq \varrho|\psi|$ for the second term on the right part of (7.20).

Evidently, (7.20) can be reformulated as

$$\dot{L}_2 = -\gamma \sin^2 x_1 \psi^2 - \psi \dot{u}^* \leq -\gamma \varsigma \psi^2 + \varrho |\psi| = -|\psi|(\gamma \varsigma |\psi| - \varrho). \tag{7.21}$$

Similar to the analysis of (7.11), according to (7.21), we have

$$\limsup_{t\to+\infty} |\psi| = \limsup_{t\to+\infty} |u - u^*| \leq \frac{\varrho}{\gamma \varsigma}. \tag{7.22}$$

According to the design procedure of controller (7.7), similar to the proof of Theorem 7.3, we can obtain

$$\limsup_{t\to+\infty} |e| \leq \frac{\varrho}{\lambda^3 \gamma \varsigma}. \tag{7.23}$$

For the case of $\sin x_1 = 0$ (i.e., the DBZ case), it can be readily obtained that $\lim_{\sin x_1 \to 0} \dot{u} = 0$ in view of $\dot{u} = \gamma \sin x_1 (\xi - \sin x_1 u)$. Therefore, let t_{s} denote the DBZ time instant. The control input at t_{s} is the same as that at the previous time instant $t_{\text{s}-}$, which implies that $u(t_{\text{s}}) = u(t_{\text{s}-})$. Similarly, at $t_{\text{s}+}$ (which is after the DBZ time instant), $u(t_{\text{s}}) = u(t_{\text{s}+})$. Note that, at $t_{\text{s}-}$, from the above proof for the non-DBZ case, we know that the control value $u(t_{\text{s}-})$ is bounded and the tracking error converges toward or stays within an error bound. Then, we have the result that $u(t_{\text{s}-}) = u(t_{\text{s}}) = u(t_{\text{s}+})$ is bounded. For a bounded input, the output of TI system (7.2) is bounded. Since the desired trajectory y_{d} is bounded, the tracking error e is thus bounded at $t_{\text{s}-}$, t_{s} and $t_{\text{s}+}$. After getting through the DBZ point (i.e., for time instants after $t_{\text{s}+}$), the tracking error converges toward an error bound again, which implies that TI system (7.2) equipped with z3g1 controller (7.7) finally conquers the DBZ problem.

By the above analyses, the proof is thus completed. □

Theorem 7.6 *Consider TI system (7.2) equipped with z3g1 controller (7.7) for smooth and bounded desired trajectory* y_{d}. *Starting with bounded initial state* $\mathbf{x}(0) \in \mathbb{R}^3$ *and control input* $u(0) \in \mathbb{R}$, *the following results are achieved on a large scale for the tracking control of the system.*

- *For the case of* $\sin x_1 \neq 0$ *(i.e., the non-DBZ case), the tracking error of the system exponentially converges toward or stays within the error bound* $\varrho/(\omega\lambda^3\gamma\varsigma)$ *with* $\omega \in (0, 1)$, *provided that (i)* $\varsigma \leq \sin^2 x_1 \leq 1, \forall 0 < \varsigma \leq 1$, *and (ii)* $|\dot{u}^*| \leq \varrho, \exists 0 \leq \varrho < +\infty$.
- *For the case of* $\sin x_1 = 0$ *(i.e., the DBZ case), the tracking error of the system is bounded.*

Proof It can be generalized from the proofs of Theorems 7.3, 7.4, and 7.5. □

7.5 Simulation, Verification and Comparison

To illustrate the feasibility and efficacy of the ZG method for the tracking control of MI systems, the simulation results of the ZG controllers for the tracking control of TI system (7.2) are shown in this section. The simulation duration is set as 30 s. Without loss of generality, we set the initial state $\mathbf{x}(0) = [1, 0, 0]^{\mathrm{T}}$ for z3g0 and z3g1 controllers, and the initial input $u(0) = 0$ for z3g1 controllers.

Example 7.1 In this example, we select the desired trajectory $y_{\mathrm{d}} = \sin(2t)\cos(2t)$ and the output function $y = x_1$ for TI system (7.2). The design parameters λ and γ are both set as 10. To achieve the purpose of tracking y_{d}, z3g0 controller (7.3) and z3g1 controller (7.4) are used in the tracking control of TI system (7.2).

Simulation results of this example are shown in Figs. 7.1 and 7.2, which illustrate the tracking performance of TI system (7.2) with z3g0 controller (7.3) and z3g1 controller (7.4). Specifically, Fig. 7.1 shows the output trajectories and control inputs of TI system (7.2) equipped with z3g0 controller (7.3) and z3g1 controller (7.4),

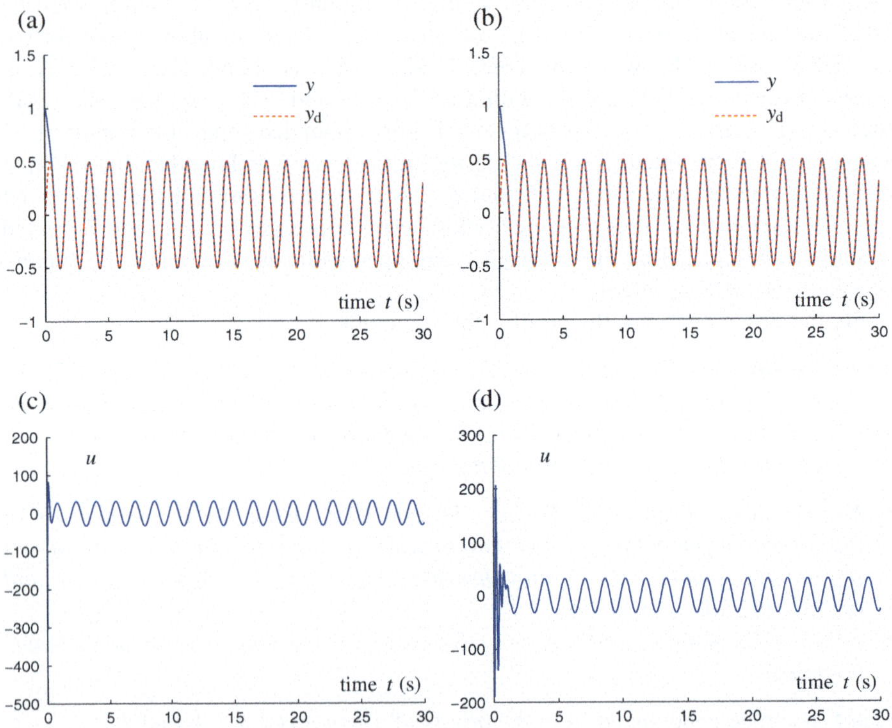

Fig. 7.1 Output trajectories and control inputs of TI system (7.2) equipped with z3g0 controller (7.3) and z3g1 controller (7.4), respectively, for desired trajectory $y_{\mathrm{d}} = \sin(2t)\cos(2t)$. (**a**) Output trajectory with z3g0 controller (7.3) and desired trajectory. (**b**) Output trajectory with z3g1 controller (7.4) and desired trajectory. (**c**) Control input with z3g0 controller (7.3). (**d**) Control input with z3g1 controller (7.4)

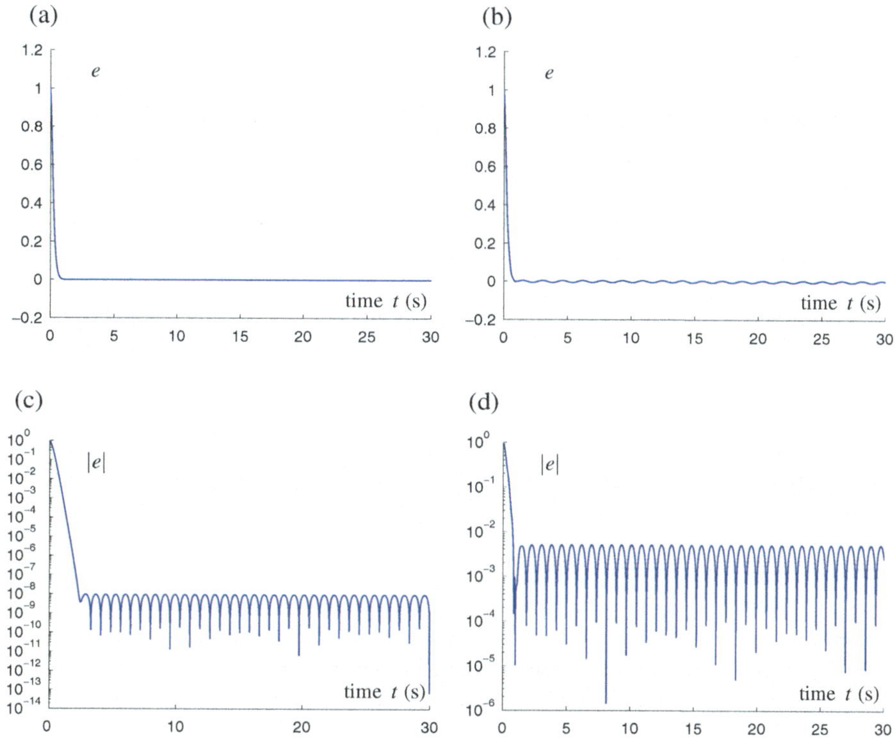

Fig. 7.2 Tracking errors of TI system (7.2) equipped with z3g0 controller (7.3) and z3g1 controller (7.4), respectively, for desired trajectory $y_\mathrm{d} = \sin(2t)\cos(2t)$. (**a**) Tracking error with z3g0 controller (7.3). (**b**) Tracking error with z3g1 controller (7.4). (**c**) Order of $|e|$ with z3g0 controller (7.3). (**d**) Order of $|e|$ with z3g1 controller (7.4)

respectively. Fig. 7.1a, b present the tracking processes by using controllers (7.3) and (7.4), from which we can observe that z3g0 and z3g1 controllers both achieve the tracking-control purpose around 1 s. Besides, Fig. 7.1c, d show the control inputs of TI system (7.2) equipped with z3g0 controller (7.3) and z3g1 controller (7.4), respectively. Furthermore, Fig. 7.2 shows the tracking errors of TI system (7.2) equipped with z3g0 controller (7.3) and z3g1 controller (7.4), respectively. From Fig. 7.2a, it can be found that the tracking error of z3g0 controller (7.3) rapidly approaches zero, which coincides well with the convergence performance analysis in Theorem 7.1. In addition, the tracking error exponentially converges toward an error bound that is of order 10^{-3}, which is consistent with the analyses of z3g1 controller (7.4) in Theorems 7.3 and 7.4. For further illustration, Fig. 7.2c, d show the tracking errors of z3g0 controller (7.3) and z3g1 controller (7.4) in the logarithmic scale, respectively. Note that the tracking error of z3g1 controller (7.4) is small enough for general practical applications, as shown in Fig. 7.2d. From the viewpoint of more accurate control, it seems that z3g0 controller (7.3) is a better choice.

In summary, both z3g0 controller (7.3) and z3g1 controller (7.4) are effective for the tracking control of TI system (7.2) with desired trajectory $y_\mathrm{d} = \sin(2t)\cos(2t)$.

Example 7.2 In this example, we consider and investigate a complex output function of TI system (7.2), i.e., $y = \cos x_1$. Without loss of generality, the desired trajectory is set as $y_\mathrm{d} = \sin(t)$. The design parameters λ and γ are set as 10 and 20, respectively. Then, z3g0 controller (7.6) and z3g1 controller (7.7) are employed for tracking control of TI system (7.2).

The simulation results of this example are illustrated in Fig. 7.3. From Fig. 7.3d, we can observe that z3g1 controller (7.7) successfully tracks the desired trajectory y_d while the tracking control with z3g0 controller (7.6) fails at the DBZ point as shown in Fig. 7.3a. As seen from Fig. 7.3b, the tracking error is small in the beginning, which verifies the efficacy of z3g0 controller (7.6) for the tracking control of system (7.2) to some extent. However, the system crashes around 1.6 s when $\sin x_1 = 0$. In contrast, as shown in Fig. 7.3d, e, the output of the system equipped with z3g1 controller (7.7) tracks the desired trajectory well, with the order of steady-state tracking error being 10^{-14}. This illustrates the superiority of z3g1 controller (7.7) with the ability of conquering the DBZ problem. For further illustration, Fig. 7.3c, f display how z3g0 controller (7.6) fails and how z3g1 controller (7.7) works well in view of control input u.

In summary, the efficacy of z3g0 controller (7.3), z3g1 controller (7.4) and z3g0 controller (7.6) designed via the ZG method is verified. Besides, via comparative results, the capability of z3g1 controller (7.7) in conquering the DBZ problem is illustrated. For better understanding and completeness, more NOFs are exploited for design and substantiation of ZG controllers in the Appendix.

Remark 7.2 It is worth pointing out that, in order to further illustrate the feasibility and efficacy of the ZG method for the tracking control of MI systems, the corresponding simulation results of more MI systems (e.g., quadruple-integrator systems and quintuple-integrator systems) equipped with ZG controllers are presented in the Appendix. The theoretical analyses on the convergence performance of ZG controllers for the above two MI systems with LOF and NOF can be generalized from Sects. 7.3 and 7.4, and thus are omitted here due to the derivation similarity.

Remark 7.3 Based on the design procedures of ZG controllers for MI systems, to some extent, we can observe that the parameters in the expression of u for the nth-order MI system satisfy the Pascal's triangle. Therefore, a general formula of the ZG controller can be given for the tracking control of MI systems. A general continuous output function can be defined as $f_\mathrm{gen}(x_1, \cdots, x_k)$, where x_k denotes the kth element in $\mathbf{x}$ and $k = 1, 2, \cdots, n$. Then, we have $z_1 = f_\mathrm{gen}(x_1, \cdots, x_k) - y_\mathrm{d}$. Note that more ZFs should be constructed successively with the ZD method adopted for more times, until the expansion of a ZD design formula contains the explicit

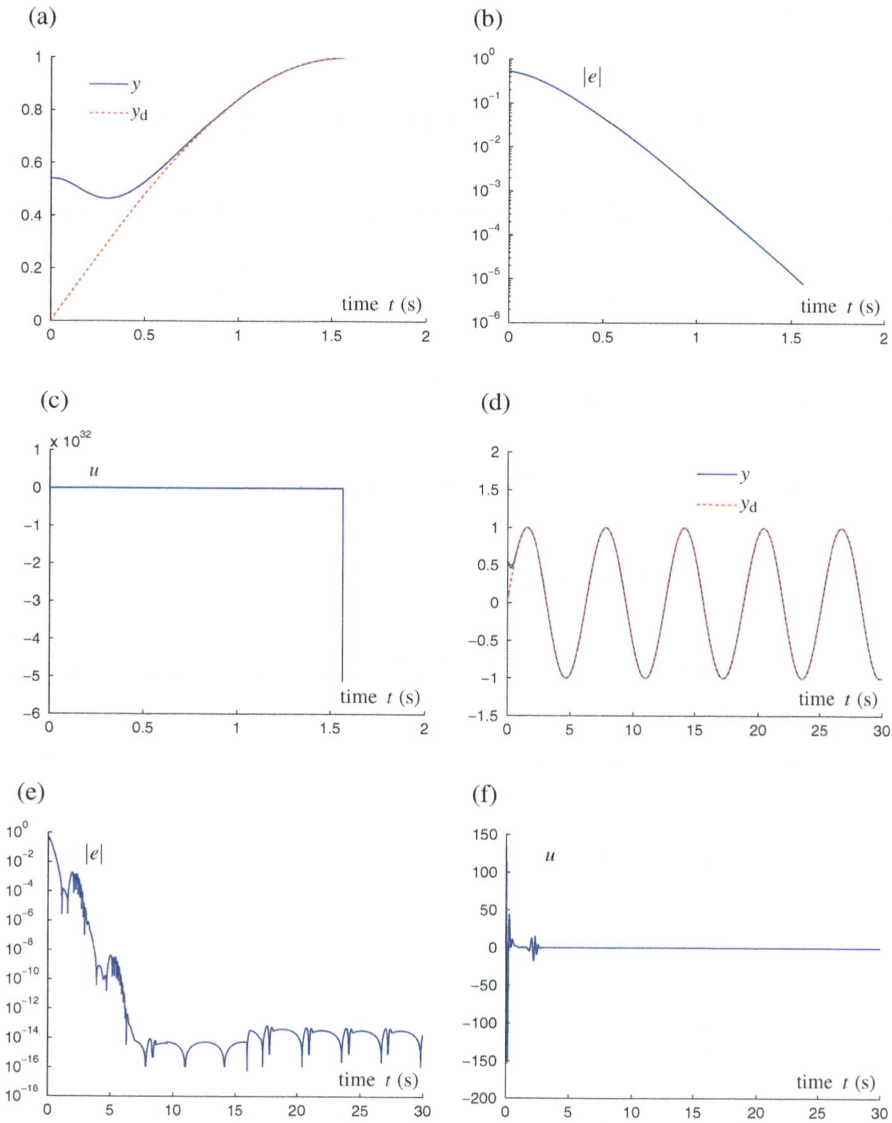

Fig. 7.3 Output trajectories, control inputs and absolute tracking errors of TI system (7.2) equipped with z3g0 controller (7.6) and z3g1 controller (7.7), respectively, for desired trajectory $y_d = \sin(t)$. (**a**) Output trajectory with z3g0 controller (7.6) and desired trajectory. (**b**) Order of $|e|$ with z3g0 controller (7.6). (**c**) Control input with z3g0 controller (7.6). (**d**) Output trajectory with z3g1 controller (7.7) and desired trajectory. (**e**) Order of $|e|$ with z3g1 controller (7.7). (**f**) Control input with z3g1 controller (7.7)

expression of u. Letting $z_1^{(N)}$ be the Nth-order time derivative of z_1 with $N = n - k + 1$, we have

$$z_1^{(N)} = f_{\text{gen}}^{(N)} - y_{\text{d}}^{(N)} = \frac{\partial f_{\text{gen}}^{(N-1)}}{\partial x_n} u + \frac{\partial f_{\text{gen}}^{(N-1)}}{\partial x_{n-1}} x_n + \cdots + \frac{\partial f_{\text{gen}}^{(N-1)}}{\partial x_1} x_2 - y_{\text{d}}^{(N)}.$$

Thus, according to [28], we have the general expression of the zNg0 controller:

$$\begin{aligned} u = & -\frac{1}{g(x_1, \cdots, x_k)} \big(C_N^0 \lambda^N z_1 + C_N^1 \lambda^{N-1} z_1^{(1)} + \cdots + C_N^{k_{\text{ord}}} \lambda^{N-k_{\text{ord}}} z_1^{(k_{\text{ord}})} \\ & + \cdots + C_N^{N-1} \lambda z_1^{(N-1)} + C_N^N h(x_1, \cdots, x_n) \big), \end{aligned}$$

in which $g(x_1, \cdots, x_k) = \partial f_{\text{gen}}^{(N-1)} / \partial x_n$, $h(x_1, \cdots, x_n) = (\partial f_{\text{gen}}^{(N-1)} / \partial x_{n-1}) x_n + \cdots + (\partial f_{\text{gen}}^{(N-1)} / \partial x_1) x_2 - y_{\text{d}}^{(N)}$, and $k_{\text{ord}} = 0, 1, \cdots, N$ with the subscript "ord" corresponding to the word "order". Besides, we can define $\phi_{\text{gen}} = C_N^0 \lambda^N z_1 + C_N^1 \lambda^{N-1} z_1^{(1)} + \cdots + C_N^{k_{\text{ord}}} \lambda^{N-k_{\text{ord}}} z_1^{(k_{\text{ord}})} + \cdots + C_N^{N-1} \lambda z_1^{(N-1)} + C_N^N z_1^{(N)}$. The following z$N$g1 controller can also be obtained by utilizing the GD method once:

$$\dot{u} = -\gamma g(x_1, \cdots, x_k) \phi_{\text{gen}}.$$

Therefore, we can obtain the zNg0 and zNg1 controllers of the Nth-order MI system directly, which can avoid complex derivations and computational difficulties.

7.6 Chapter Summary

The tracking-control problems of MI systems with LOF and NOF have been considered and investigated in this chapter. To achieve the tracking control of TI system (7.2), based on the ZG method, the controllers of z3g0 and z3g1 types for the tracking control of TI system (7.2) have been developed and investigated. Furthermore, the presented controllers not only complete the tracking-control task (for controllers of z3g0 and z3g1 types), but also conquer the DBZ problem (for controllers of z3g1 type). Besides, the theoretical analyses have been provided to guarantee the convergence performance of both z3g0 and z3g1 controllers. The corresponding simulation results have substantiated the feasibility and efficacy of z3g0 and z3g1 controllers for the tracking control of TI system (7.2). More importantly, the comparative simulation results have illustrated the superiority of z3g1 controllers in conquering the DBZ problem. Moreover, we have shown the tracking performance of ZG controllers for more MI systems, and the simulation results have further verified the feasibility and efficacy of the ZG method for solving the tracking-control problem of MI systems.

Appendix: Tracking Control of More ZG Controllers

In the Appendix, for solving the tracking-control problem, the expressions of controllers on the TI system with more NOFs as well as the expressions of controllers on more MI systems are presented, and the corresponding simulation results are given to further show the feasibility and efficacy of the ZG method.

Tracking Control of TI Systems with More NOFs

To better illustrate the advantages and show the generalization of the ZG method for the implicit tracking control of MI systems, we consider and investigate more complex NOFs, e.g., $y = x_1^2 + x_2^2$ and $y = x_1 x_2 x_3$. The design procedures of the ZG controllers for the TI system with output functions $y = x_1^2 + x_2^2$ and $y = x_1 x_2 x_3$ can be generalized from Sect. 7.2, and are omitted here due to the derivation similarity. Specifically, the z2g1 controller in the form of $\dot{u}$ with output function $y = x_1^2 + x_2^2$ is presented as below:

$$\begin{aligned}\dot{u} = &-2\gamma x_2\big(2x_2u + (2x_2^2 + 2x_1x_3 + 2x_3^2 - \ddot{y}_\mathrm{d})\\ &+ 2\lambda(2x_1x_2 + 2x_3x_2 - \dot{y}_\mathrm{d}) + \lambda^2(x_1^2 + x_2^2 - y_\mathrm{d})\big).\end{aligned} \tag{7.24}$$

Then, we present the z1g1 controller in the form of $\dot{u}$ with output function $y = x_1 x_2 x_3$ as below:

$$\dot{u} = -\gamma x_1 x_2\big(x_1x_2u + x_2^2x_3 + x_1x_3^2 - \dot{y}_\mathrm{d} + \lambda(x_1x_2x_3 - y_\mathrm{d})\big). \tag{7.25}$$

The corresponding simulation results are shown in Fig. 7.4. The desired trajectory is selected as $y_\mathrm{d} = \sin(t)\exp(-0.1t) + 2$. The design parameters λ and γ are set as 10 and 20, respectively. As seen from Fig. 7.4, the output trajectory quickly tracks the desired trajectory, and the steady-state tracking error is small enough. These simulation results substantiate well that the ZG controllers with different complex NOFs still possess good tracking performance.

Tracking Control of Quadruple-Integrator Systems

Similar to the design process of ZG controllers for the tracking control of TI systems, we have the ZG controllers for the tracking control of the quadruple-integrator system with LOF $y = x_1$. Specifically, the z4g0 controller in the form

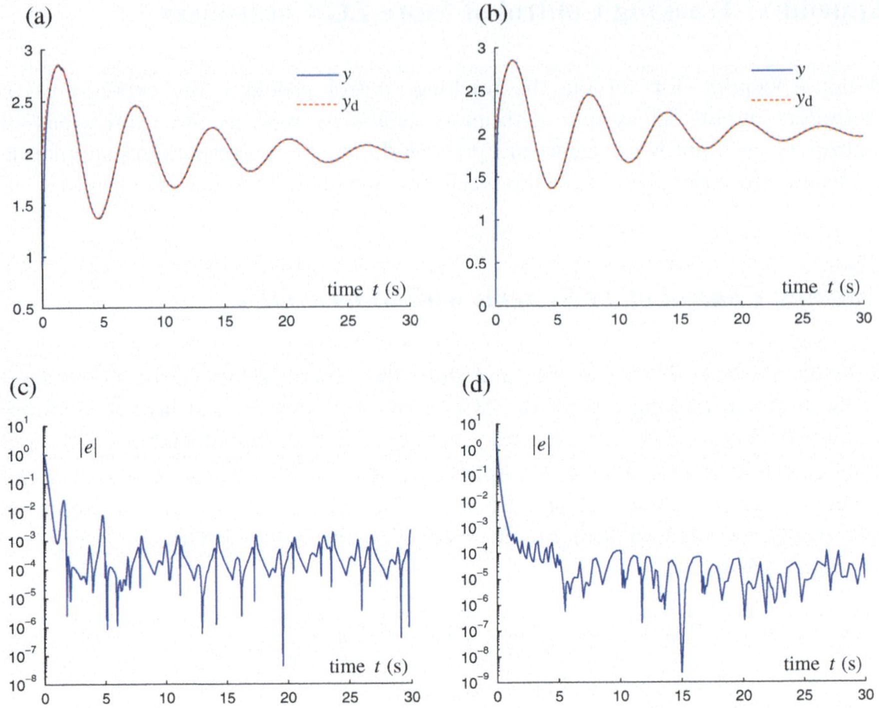

Fig. 7.4 Output trajectories and absolute tracking errors of TI system (7.2) equipped with z2g1 controller (7.24) with $y = x_1^2 + x_2^2$ and z1g1 controller (7.25) with $y = x_1x_2x_3$, respectively, for desired trajectory $y_\text{d} = \sin(t)\exp(-0.1t) + 2$. (**a**) Output trajectory with z2g1 controller (7.24) and desired trajectory. (**b**) Output trajectory with z1g1 controller (7.25) and desired trajectory. (**c**) Order of $|e|$ with z2g1 controller (7.24). (**d**) Order of $|e|$ with z1g1 controller (7.25)

of u can be obtained as below:

$$u = \lambda^4(y_\text{d} - x_1) + 4\lambda^3(\dot{y}_\text{d} - x_2) + 6\lambda^2(\ddot{y}_\text{d} - x_3) + 4\lambda(\dddot{y}_\text{d} - x_4) + \ddddot{y}_\text{d}. \tag{7.26}$$

Then, by defining $\phi_3 = u - \big(\lambda^4(y_\text{d} - x_1) + 4\lambda^3(\dot{y}_\text{d} - x_2) + 6\lambda^2(\ddot{y}_\text{d} - x_3) + 4\lambda(\dddot{y}_\text{d} - x_4) + \ddddot{y}_\text{d}\big)$, the z4g1 controller in the form of $\dot{u}$ can be obtained as below:

$$\dot{u} = -\gamma\phi_3. \tag{7.27}$$

Moreover, we can obtain the ZG controllers for the tracking control of the quadruple-integrator system with NOF $y = \cos x_1$, which are omitted here since they are complicated and can be obtained according to the general expression of the ZG controller for MI systems.

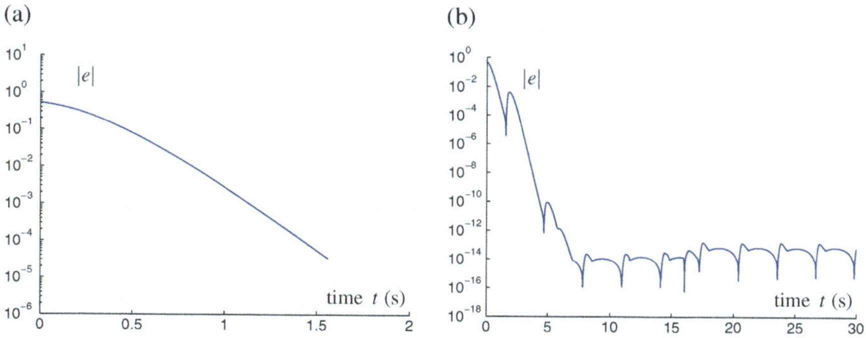

Fig. 7.5 Absolute tracking errors of quadruple-integrator system equipped with z4g0 controller (7.26) and z4g1 controller (7.27), respectively, for desired trajectory $y_\mathrm{d} = \sin(t)$. (**a**) Order of $|e|$ with z4g0 controller (7.26). (**b**) Order of $|e|$ with z4g1 controller (7.27)

In the corresponding simulations, the design parameters are set as $\lambda = 10$ and $\gamma = 1000$, respectively. In addition, the initial state $\mathbf{x}(0)$ and the initial control input $u(0)$ are set as 0 for z4g1 controllers. As shown in Fig. 7.5a, the tracking-control task of the quadruple-integrator system equipped with z4g0 controller (7.26) fails when time t is around 1.6 s. However, as illustrated in Fig. 7.5b, z4g1 controller (7.26) successfully conquers the DBZ problem and tracks well the desired trajectory, with the steady-state tracking error being of order 10^{-14}.

Tracking Control of Quintuple-Integrator Systems

The ZG controllers are also designed for the tracking control of quintuple-integrator system with LOF (i.e., $y = x_1$), all of which can track well the desired trajectory. The design and formulation of z5g0 and z5g1 controllers as well as the corresponding simulation results are omitted here due to the results' similarity. Specially, the controllers for the tracking control of the system and the corresponding simulation results with NOF $y = x_1^2 + x_2^2$ are simply presented here. The z4g0 controller in the form of u can be obtained as below:

$$
\begin{aligned}
u =& \frac{1}{2x_2}\Big(- 6x_3^2 - 8x_2x_4 - 2x_1x_5 - 8x_3x_5 - 6x_4^2 + \ddddot{y}_\mathrm{d} - 4\lambda(6x_2x_3 \\
&+ 2x_1x_4 + 6x_3x_4 + 2x_2x_5 - \dddot{y}_\mathrm{d}) - 6\lambda^2(2x_2^2 + 2x_1x_3 + 2x_2x_4 + 2x_3^2 - \ddot{y}_\mathrm{d}) \\
&- 4\lambda^3(2x_1x_2 + 2x_2x_3 - \dot{y}_\mathrm{d}) - \lambda^4(x_1^2 + x_2^2 - y_\mathrm{d})\Big).
\end{aligned}
\tag{7.28}
$$

Then, similar to the design process of z3g1 controllers, by defining $\phi_4 = 2x_2u - (-6x_3^2 - 8x_2x_4 - 2x_1x_5 - 8x_3x_5 - 6x_4^2 + \ddddot{y}_\mathrm{d} - 4\lambda(6x_2x_3 + 2x_1x_4 + 6x_3x_4 + 2x_2x_5 - \dddot{y}_\mathrm{d}) - 6\lambda^2(2x_2^2 + 2x_1x_3 + 2x_2x_4 + 2x_3^2 - \ddot{y}_\mathrm{d}) - 4\lambda^3(2x_1x_2 + 2x_2x_3 - \dot{y}_\mathrm{d}) - \lambda^4(x_1^2 + x_2^2 - y_\mathrm{d}))$, the z4g1 controller in the form of $\dot{u}$ can be obtained as below:

$$\dot{u} = -2\gamma x_2\phi_4. \tag{7.29}$$

In the corresponding simulations, the design parameters are set as $\lambda = 100$ and $\gamma = 1000$. In addition, the initial states $x_1(0)$, $x_2(0)$, and control input $u(0)$ are set as 1, 1, and 0, respectively. The simulation results are shown in Fig. 7.6, with the desired trajectory $y_\mathrm{d} = \cos(t) + 2$. As seen from this figure, the tracking-control task of the quintuple-integrator system equipped with z4g0 controller (7.28) fails when time t is close to 0.8 s. On the contrary, z4g1 controller (7.29) basically fulfills the tracking-control task and conquers the DBZ problem.

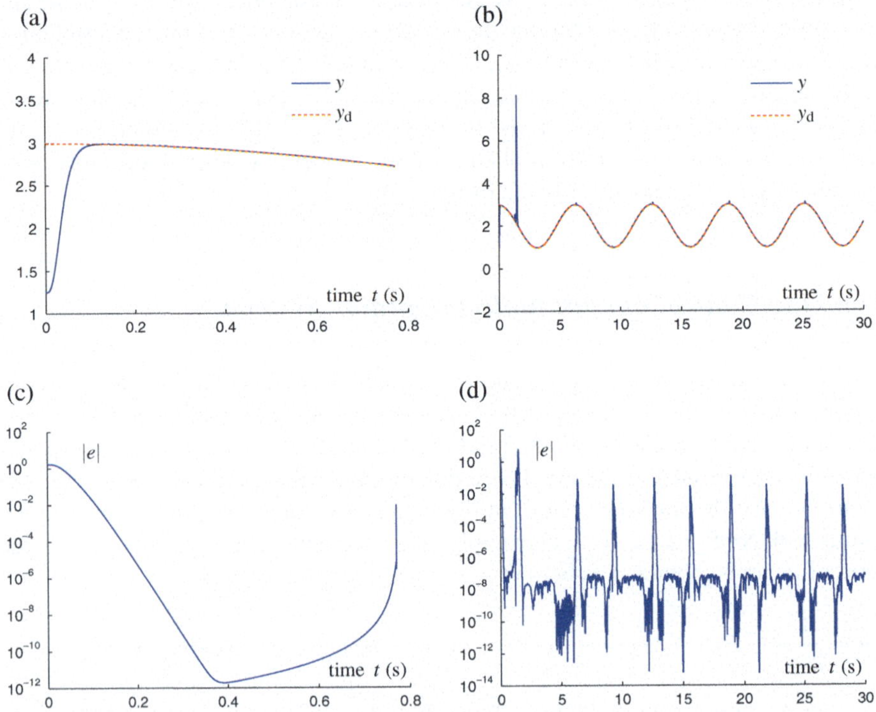

Fig. 7.6 Output trajectories and absolute tracking errors of quintuple-integrator system equipped with z4g0 controller (7.28) and z4g1 controller (7.29), respectively, for desired trajectory $y_\mathrm{d} = \cos(t) + 2$. (**a**) Output trajectory with z4g0 controller (7.28) and desired trajectory. (**b**) Output trajectory with z4g1 controller (7.29) and desired trajectory. (**c**) Order of $|e|$ with z4g0 controller (7.28). (**d**) Order of $|e|$ with z4g1 controller (7.29)

References

1. Rao VG, Bernstein DS (2001) Naive control of the double integrator. IEEE Control Syst Mag 21(5):86–97
2. Marchand N (2003) Further results on global stabilization for multiple integrators with bounded controls. In: Proceedings of the 42nd IEEE conference on decision and control, pp 4440–4444
3. Kaliora G, Astolfi A (2004) Nonlinear control of feedforward systems with bounded signals. IEEE Trans Autom Control 49(11):1975–1990
4. Zhou B, Duan G, Li Z (2008) On improving transient performance in global control of multiple integrators system by bounded feedback. Syst Control Lett 57(10):867–875
5. Zhang Y, Zhai K, Wang Y, Chen D, Peng C (2014) Design and illustration of ZG controllers for linear and nonlinear tracking control of double-integrator system. In: Proceedings of the 33rd Chinese control conference, pp 28–30
6. Xu D, Wang W (2010) Research on applications of linear system theory in economics. In: Proceedings of the 8th IEEE international conference on control and automation. pp 1843–1847
7. Zhang Y, Ding S, Chen D, Mao M, Zhai K (2015) Zhang-gradient controllers for tracking control of multiple-integrator systems. ASME J Dyn Syst Meas Control 137(11):111013
8. Jarzebowska EM (2008) Advanced programmed motion tracking control of nonholonomic mechanical systems. IEEE Trans Robot 24(6):1315–1328
9. Li W (2010) Tracking control of chaotic coronary artery system. Int J Syst Sci 43(1):21–30
10. Zhang Y, Yu X, Yin Y, Peng C, Fan Z (2014) Singularity-conquering ZG controllers of z2g1 type for tracking control of the IPC system. Int J Control 87(9):1729–1746
11. Bialy BJ, Pasiliao CL, Dinh HT, Dixon WE (2014) Tracking Control of limit cycle oscillations in an aero-elastic system. ASME J Dyn Syst Meas Control 136(6):064505
12. Dontchev AL, Krastanov MI, Rockafellar RT, Veliov VM (2014) Neural network-based tracking control of underactuated autonomous underwater vehicles with model uncertainties. ASME J Dyn Syst Meas Control 137(2):021004
13. Hirschorn RM (2002) Output tracking through singularities. SIAM J Control Optim 40(4):993–1010
14. Jui CKK, Qiao S (2005) Path tracking of parallel manipulators in the presence of force singularity. ASME J Dyn Syst Meas Control 127(4):550–563
15. Teel AR (1992) Global stabilization and restricted tracking for multiple integrators with bounded controls. Syst Control Lett 18(3):165–171
16. Marchand N, Hably A (2005) Global stabilization of multiple integrators with bounded controls. Automatica 41(12):2147–2152
17. Zhou B, Duan G (2007) Global stabilization of multiple integrators via saturated controls. IET Control Theory Appl 1(6):1586–1593
18. Zhou B, Duan G (2008) A novel nested nonlinear feedback law for global stabilization of linear systems with bounded controls. Int J Control 81(9):1352–1363
19. Zhou B, Duan G (2009) Global stabilization of linear systems via bounded controls. Syst Control Lett 58(1):54–61
20. Zhang Y, Yi C (2011) Zhang neural networks and neural-dynamic method. Nova Science Publishers, New York
21. Xiao L, Zhang Y (2011) Zhang neural network versus gradient neural network for solving time-varying linear inequalities. IEEE Trans Neural Netw 22(10):1676–1684
22. Zhang Y, Yi C, Guo D, Zheng J (2011) Comparison on Zhang neural dynamics and gradient-based neural dynamics for online solution of nonlinear time-varying equation. Neural Comput Appl 20(1):1–7
23. Zhang Y, Yang Y, Ruan G (2011) Performance analysis of gradient neural network exploited for online time-varying quadratic minimization and equality-constrained quadratic programming. Neurocomputing 74(10):1710–1719

24. Zhang Y, Yin Y, Wu H, Guo D (2012) Zhang dynamics and gradient dynamics with tracking-control application. In: Proceedings of the 5th international symposium on computational intelligence and design, pp 235–238
25. Zhang Z, Zhang Y (2013) Design and experimentation of acceleration-level drift-free scheme aided by two recurrent neural networks. IET Control Theory Appl 7(1):25–42
26. Abramowitz M, Stegun IA (1972) Handbook of mathematical functions with formulas, graphs, and mathematical tables. Dover Publications, New York
27. Chu S, Metcalf F (1967) On Gronwall's inequality. Proc Amer Math Soc 18(3):439–440
28. Zhang Y, Luo F, Yin Y, Liu J, Yu X (2013) Singularity-conquering ZG controller for output tracking of a class of nonlinear systems. In: Proceedings of the 32nd chinese control conference, pp 477–482

Part III
Pendulum Systems Using ZG Control

Chapter 8
ZD and ZG Control of Simple Pendulum System

Abstract In this chapter, we firstly design ZD controllers for the explicit and implicit tracking control of a simple pendulum system. For achieving the DBZ-containing implicit tracking control, ZG controllers are further designed for conquering the DBZ problem. Computer simulations with an explicit tracking example and two implicit tracking examples are conducted. Comparative simulation results have substantiated the superiority of the ZG controllers for the DBZ-containing implicit tracking control of simple pendulum system.

8.1 Introduction

Tracking control is widely encountered in the engineering field [1–5]. Traditionally and generally speaking, the tracking control problem of a system is to design a controller in terms of the input u for the system such that the actual output y tracks the desired output y_d. As an example, simple pendulum system [6] considered in this chapter [7] can be expressed as

$$\begin{cases} \dot{x}_1 = f_1(x_2) = x_2, \\ \dot{x}_2 = f_2(x_1, x_2, u) = -\frac{g}{l}\sin x_1 - \frac{\sigma}{m}x_2 + \frac{1}{ml^2}u, \end{cases} \tag{8.1}$$

where x_1 and x_2 are the state variables with x_1 corresponding to the angle between the rod and the vertical axis through the pivot point. The rod is assumed to be rigid and of length l with zero mass, connecting with a ball of mass m. Besides, g is the acceleration of gravity, and σ is the frictional coefficient of the joint. The control variable u corresponds to a torque applied to the pendulum.

In this chapter [7], the explicit and implicit tracking control problems of simple pendulum system (8.1) are both formulated and investigated. On the one hand, the explicit tracking control of the pendulum system is solved by the ZD method. On the other hand, the ZD and GD methods are combined and exploited together to solve the DBZ-containing implicit tracking control problems of simple pendulum

Y. Zhang et al., *Zhang-Gradient Control*,
https://doi.org/10.1007/978-981-15-8257-8_8

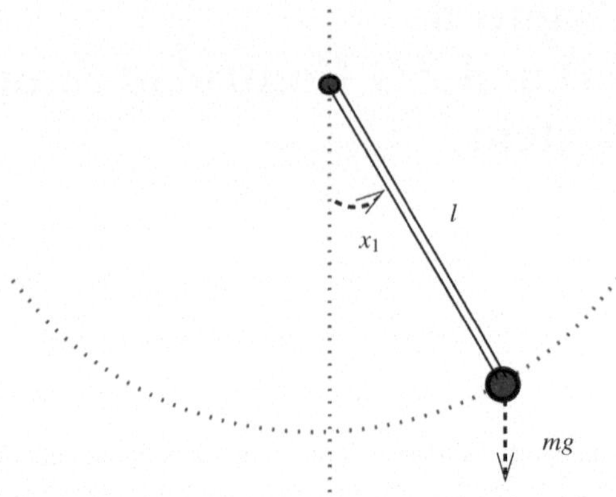

Fig. 8.1 Schematic of simple pendulum system

system (8.1). For better understanding, the schematic of simple pendulum system is shown in Fig. 8.1.

8.2 ZD Controller for Explicit Tracking Control

In the explicit tracking control, we assume the actual output $y = x_1$, which is expected to track a desired output y_d, e.g., $y_d = \sin(t)$ or $y_d = \sin(2t)\cos(t)$. Thus, the first ZF [8–10] can be constructed as

$$z_1 = y - y_d = x_1 - y_d \in \mathbb{R}.$$

Applying the ZD design formula $\dot{z}_1 = -\lambda z_1$ with the design parameter $\lambda \in \mathbb{R}^+$ used to scale the convergence rate of the ZD solution, we have

$$x_2 - \dot{y}_d = -\lambda(x_1 - y_d).$$

Then, by constructing the second ZF $z_2 = x_2 - \dot{y}_d + \lambda(x_1 - y_d)$ and applying the ZD design formula $\dot{z}_2 = -\lambda z_2$, a ZD controller in the form of u for the explicit tracking control of the simple pendulum system (8.1) with $y = x_1$ can be obtained as

$$u = mgl\sin x_1 + (\sigma l^2 - 2\lambda ml^2)x_2 + ml^2\ddot{y}_d + 2\lambda ml^2\dot{y}_d - \lambda^2 ml^2(x_1 - y_d). \tag{8.2}$$

Since controller (8.2) is obtained by applying the ZD method twice and without using the GD method, it is thus termed z2g0 controller. Besides, let us define the tracking error as $e = |y - y_d|$.

8.3 ZG Controller for Implicit Tracking Control

The previous section shows the efficacy of the ZD method individually applied to the explicit tracking control of the simple pendulum system (8.1). In order to solve the DBZ-containing implicit tracking control problem, the ZD method can also be combined with the GD method, which yields the ZG method. The following two examples about implicit tracking control are thus given to show the efficacy and superiority of the ZG method.

For simulation verifications, two desired trajectories are chosen, i.e.,

$$y_d = \sin(t) \tag{8.3}$$

and

$$y_d = \sin(2t)\cos(t). \tag{8.4}$$

The initial states $x_1(0) = x_2(0) = 0$, design parameters $\lambda = 3.9$ and $\gamma = 1000$, and the values of other parameters are given in Table 8.1. The corresponding simulation results are shown in Fig. 8.2. From Fig. 8.2a, b, it can be seen that the actual output y converges to the desired trajectory y_d. In addition, it can be seen from Fig. 8.2c, d that the tracking errors converge to zero, which illustrates the efficacy of z2g0 controller (8.2).

Example 8.1 Let us firstly consider a simple implicit tracking control example, where the actual output $y = x_1x_2 + x_2$, together with the desired trajectory y_d being (8.3) or (8.4) to be tracked. By using the ZG method, a ZG controller is developed below to conquer the DBZ problem.

By constructing the first ZF $z_1 = y - y_d = x_1x_2 + x_2 - y_d$ and applying the ZD design formula $\dot{z}_1 = -\lambda z_1$, function ϕ_1 can be defined as

$$\begin{aligned}\phi_1 &= ml^2(\dot{z}_1 + \lambda z_1)\\ &= (x_1+1)u + ml^2(x_2^2 - \dot{y}_d) - (x_1+1)mgl\sin x_1 - (x_1+1)\sigma l^2 x_2 + \lambda ml^2\big(y - y_d\big),\end{aligned}$$

which should theoretically be zero. Therefore, a z1g0 controller (or say, a conventional controller) can be derived from $\phi_1 = 0$, i.e.,

$$u = -\frac{1}{x_1+1}\big(ml^2(x_2^2 - \dot{y}_d) + \lambda ml^2(y - y_d)\big) + mgl\sin x_1 + \sigma l^2 x_2. \tag{8.5}$$

It is evident that z1g0 controller (8.5) cannot get through the DBZ point $x_1 = -1$. The initial states $x_1(0) = -1.5$ and $x_2(0) = 1$, design parameters $\lambda = 3.9$ and

Table 8.1 Parameter values of simple pendulum system

Parameter	m	l	σ	g
Value	1 kg	1 m	0.5 N s/m	9.81 m/s^2

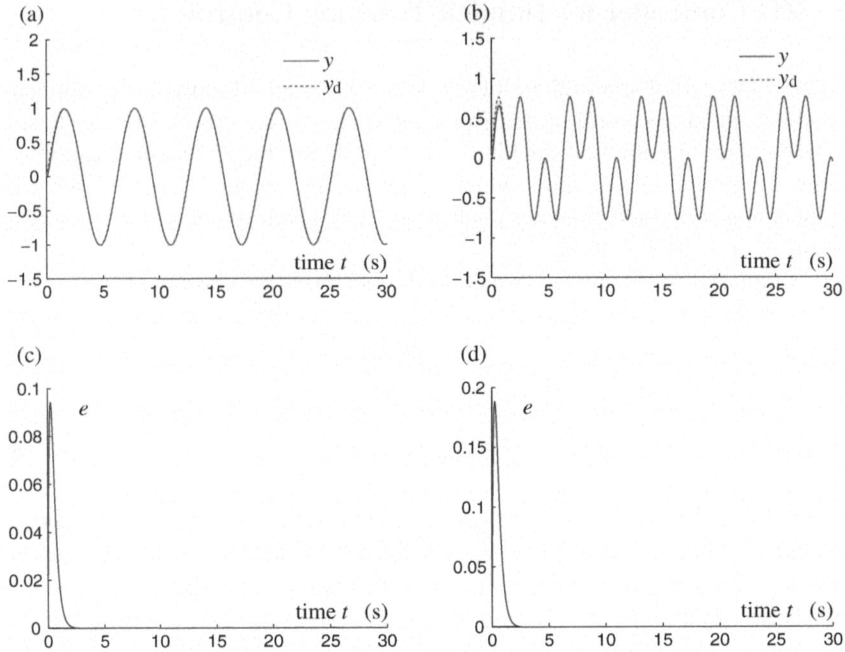

Fig. 8.2 Tracking performance of simple pendulum system (8.1) equipped with z2g0 controller (8.2) for explicit tracking control with desired trajectories (8.3) and (8.4), respectively. (**a**) Output trajectory and desired trajectory (8.3). (**b**) Output trajectory and desired trajectory (8.4). (**c**) Tracking error with desired trajectory (8.3). (**d**) Tracking error with desired trajectory (8.4)

$\gamma = 1000$, and the values of other parameters are given in Table 8.1. Specifically, Fig. 8.3 displays the simulation results of z1g0 controller (8.5) applied to the implicit tracking control of the simple pendulum system (8.1) with $y = x_1x_2+x_2$ and desired trajectory (8.3). As seen from Fig. 8.3, when the state x_1 approaches -1, the control input u becomes extremely large, which leads to system crash.

In contrast, by further using the GD method [8, 9, 11], an EF can be defined as $\epsilon = \phi_1^2/2$. Then, for the implicit tracking control of simple pendulum system (8.1) with $y = x_1x_2 + x_2$, a z1g1 controller in the form of $\dot{u}$ is designed as

$$\dot{u} = -\gamma \frac{\partial \epsilon}{\partial u} = -\gamma(x_1 + 1)\phi_1, \tag{8.6}$$

where the design parameter $\gamma \in \mathbb{R}^+$ is used to scale the convergence rate of the GD solution. By using the same initial states, parameters shown in Table 8.1, and $u(0) = 0$, the simulation results of z1g1 controller (8.6) applied to the simple pendulum system (8.1) with desired trajectories (8.3) and (8.4) are shown in Fig. 8.4. As seen from the figure, we can observe that the system equipped with z1g1 controller (8.6) gets through the DBZ point of $x_1 = -1$ successfully.

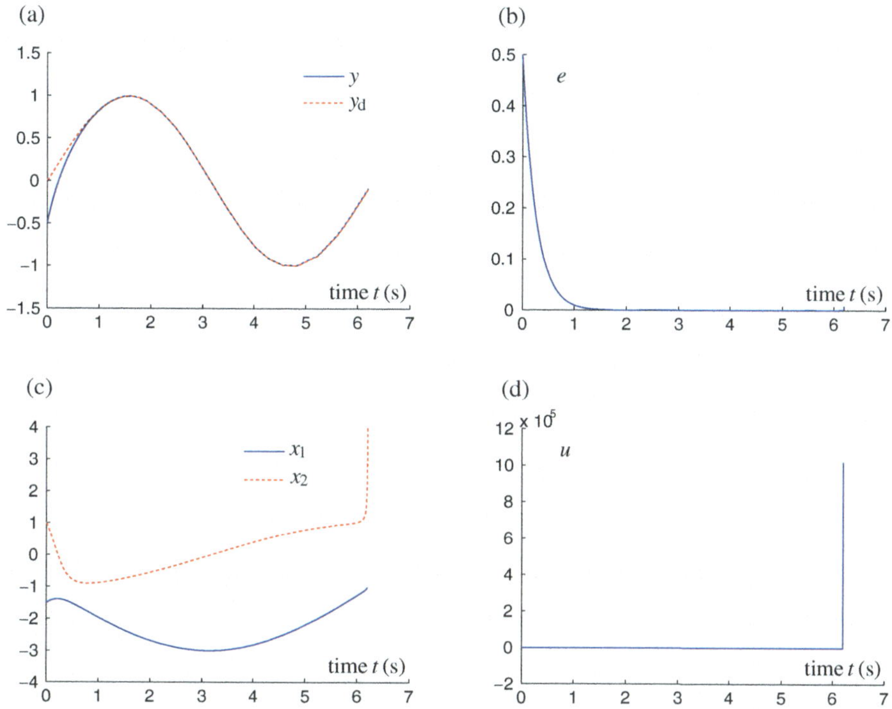

Fig. 8.3 Tracking performance and crash of simple pendulum system (8.1) equipped with z1g0 controller (8.5) for DBZ-containing implicit tracking control with desired trajectory (8.3). (**a**) Output trajectory and desired trajectory. (**b**) Tracking error. (**c**) System states. (**d**) Control input

Example 8.2 As another example of tracking desired trajectory y_d, a different output is set as $y = \sin x_1 \cos x_2 + \sin x_2$. Similarly, we can define

$$\phi_2 = ml^2(x_2 \cos x_1 \cos x_2 - \dot{y}_d) + \alpha_1(u - mgl \sin x_1 - \sigma l^2 x_2) + \lambda ml^2(y - y_d),$$

where $\alpha_1 = \cos x_2 - \sin x_1 \sin x_2$. Thus, a z1g1 controller in the form of $\dot{u}$ for the implicit tracking control of simple pendulum system (8.1) with $y = \sin x_1 \cos x_2 + \sin x_2$ is designed as

$$\dot{u} = -\gamma \alpha_1 \phi_2, \tag{8.7}$$

which can conquer the DBZ problem of $\alpha_1 = 0$.

By using the parameters shown in Table 8.1 and initial values $x_1(0) = 2$, $x_2(0) = 1$ and $u(0) = 0$, the corresponding simulation results of z1g1 controller (8.7) applied to the implicit tracking control of simple pendulum system (8.1) with $y = \sin x_1 \cos x_2 + \sin x_2$ for desired trajectory (8.3) are shown in Fig. 8.5. To illustrate the DBZ problem more clearly, Fig. 8.5d displays the trajectory of $\alpha_1 =$

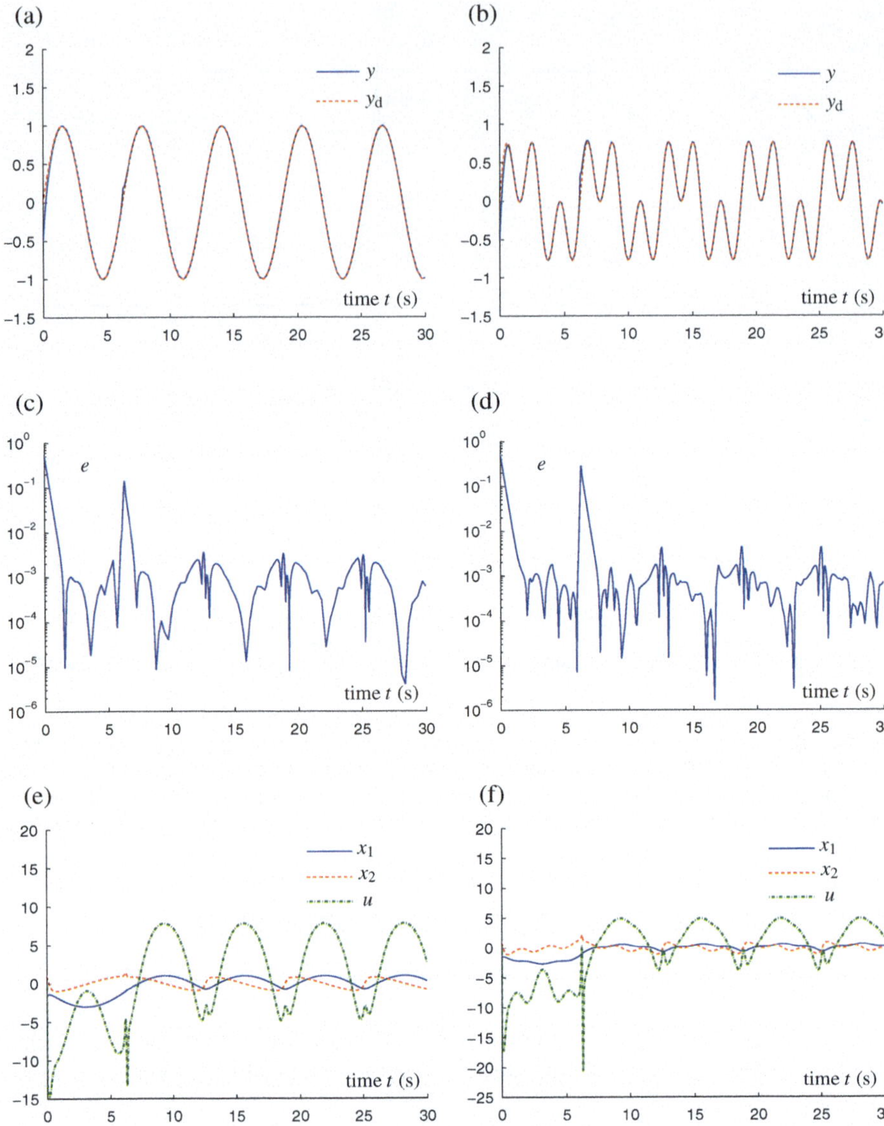

Fig. 8.4 Tracking performance of simple pendulum system (8.1) equipped with z1g1 controller (8.6) for DBZ-containing implicit tracking control with desired trajectories (8.3) and (8.4), respectively. (**a**) Output trajectory and desired trajectory (8.3). (**b**) Output trajectory and desired trajectory (8.4). (**c**) Tracking error with desired trajectory (8.3). (**d**) Tracking error with desired trajectory (8.4). (**e**) System states and control input with desired trajectory (8.3). (**f**) System states and control input with desired trajectory (8.4)

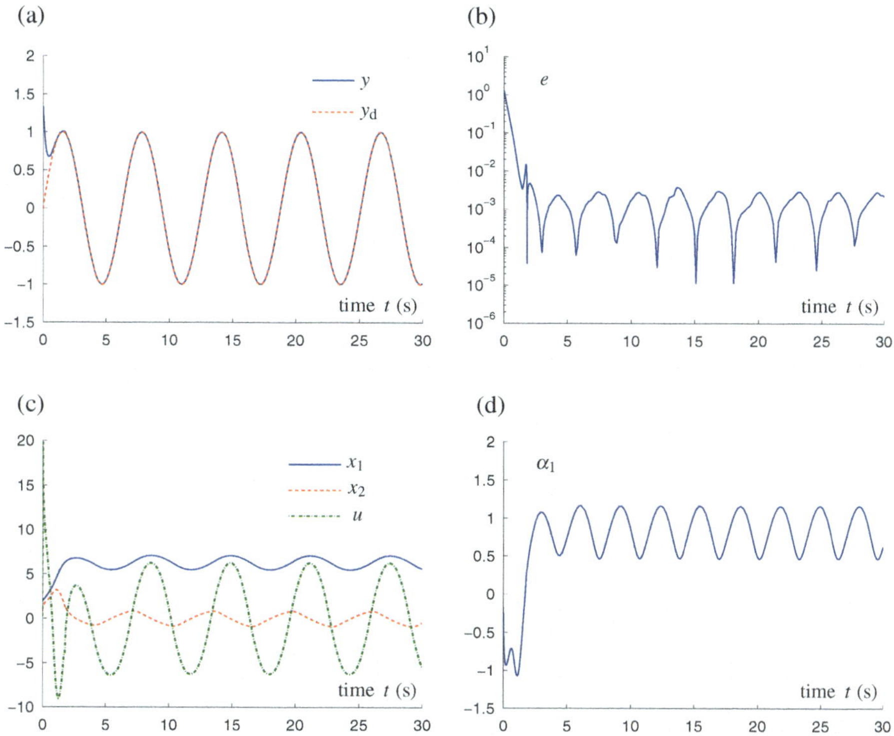

Fig. 8.5 Tracking performance of simple pendulum system (8.1) equipped with z1g1 controller (8.7) for DBZ-containing implicit tracking control with desired trajectory (8.3), which gets through DBZ point of $\alpha_1 = 0$ successfully. (**a**) Output trajectory and desired trajectory. (**b**) Tracking error. (**c**) System states and control input. (**d**) Trajectory of α_1

$\cos x_2 - \sin x_1 \sin x_2$, which evidently passes zero. It can be readily observed that, in the first 2 s, z1g1 controller (8.7) successfully gets through the DBZ point of $\alpha_1 = 0$. In other words, the ZG controller derived from the ZG method conquers the DBZ-containing implicit tracking control problem.

8.4 Chapter Summary

Based on the simple pendulum system (8.1), this chapter has investigated the explicit and implicit tracking control by using the ZG method. Specifically, we have shown the efficacy of the ZD controller for the explicit tracking control of simple pendulum system (8.1) at first. Then, we have investigated how the DBZ problem exists in the more general implicit tracking control and how it leads to system crash. Finally, by using the ZG controllers, the DBZ problem existing in the implicit tracking control has been solved. The comparative simulation results have further substantiated that the ZG controllers possess superior performance for achieving the DBZ-containing implicit tracking control of simple pendulum system (8.1).

References

1. Zhang C, Tang G, Han S (2009) Approximate design of optimal tracking controller for systems with delayed state and control. In: Proceedings of international conference on control and automation, pp 1168–1172
2. Tsourdos A, White B (2005) Adaptive flight control design for nonlinear missile. Control Eng Pract 13(3):373–382
3. Wang Z, Guo W, Guo Y, Zhang J (2010) Output tracking controller design for networked control systems. In: Proceedings of international conference on modelling, identification and control, pp 650–652
4. Ha IJ, Gilbert EG (1985) Robust tracking in nonlinear systems and its applications to robotics. In: Proceedings of the 24th conference on decision and control, pp 1009–1017
5. Trinh H, Aldeen M (1996) Output tracking for linear uncertain time-delay systems. IEE Proc Control Theory Appl 143(6):481–488
6. Khalil HK, Grizzle JW (1996) Nonlinear systems. Prentice Hall, New Jersey
7. Zhang Y, Peng C, Yu X, Yin, Y, Ling Y (2013) ZD and ZG controllers for explicit and implicit tracking of pendulum with singularity finally conquered. In: Proceedings of international conference on machine learning and cybernetics, pp 777–782
8. Zhang Y, Ke Z, Xu P, Yi C (2010) Time-varying square roots finding via Zhang dynamics versus gradient dynamics and the former's link and new explanation to Newton-Raphson iteration. Inf Process Lett 110(24):1103–1109
9. Cai B, Zhang Y (2012) Different-level redundancy-resolution and its equivalent relationship analysis for robot manipulators using gradient-descent and Zhang et al.'s neural-dynamic methods. IEEE Trans Ind Electron 59(8):3146–3155
10. Zhang Y, Yang Y, Tang N, Cai B (2011) Zhang neural network solving for time-varying full-rank matrix Moore-Penrose inverse. Computing 92(2):97–121
11. Zhang Y, Yi C, Guo D, Zheng J (2011) Comparison on Zhang neural dynamics and gradient-based neural dynamics for online solution of nonlinear time-varying equation. Neural Comput Appl 20(1):1–7

Chapter 9
Cart Path Tracking Control of IPC System

Abstract With wider investigations and applications of autonomous robotics and intelligent vehicles, the inverted-pendulum-on-a-cart (IPC) system has become more attractive for numerous researchers due to its concise and representative structure. In this chapter, the cart path tracking control of the IPC system is considered and investigated. Based on the ZG method, the controllers of z2g0 and z2g1 types are designed to achieve the tracking-control purpose. Besides, theoretical results and analyses are presented to guarantee the global and exponential convergence performance of both z2g0 and z2g1 controllers. Computer simulations are further performed to illustrate the feasibility and efficacy of both z2g0 and z2g1 controllers. More importantly, comparative simulation results indicate that z2g1 controllers can effectively conquer the DBZ problem.

9.1 Introduction

Mechatronic systems are becoming more and more popular with numerous researchers of robotics and control theory since the fusion of electronics and mechanics technologies is required by industrial applications [1–3]. As a typical mechatronic system and control model, the inverted pendulum system has attracted the attention of both researchers and educators in the last decades due to its inherently nonlinear and unstable characteristics [4–6]. The basic model of the inverted pendulum system is an inverted pendulum on a cart that can be moved horizontally, which is termed the inverted-pendulum-on-a-cart (IPC) system. In general, the control objective of the IPC system is to keep the pendulum at the upper unstable equilibrium position by moving the cart on the horizontal plane [4, 6, 7]. Recently, with wider investigations and applications of autonomous robotics and intelligent vehicles, the IPC system has also been considered and investigated for achieving the tracking-control purpose (i.e., tracking desired trajectories) [5, 8, 9].

As known, the tracking-control problem of nonlinear systems has been widely encountered in various applications, such as flight control, robot control and motion planning [8–11]. For the purpose of tracking control, we need to design a controller

Y. Zhang et al., *Zhang-Gradient Control*,
https://doi.org/10.1007/978-981-15-8257-8_9

in terms of the input u for nonlinear systems such that the actual output y can track the desired trajectory y_{d}. In the tracking control application of the IPC system, the output y can be defined in terms of specific needs, such as the position variable of the cart or the angle variable of the pendulum. In this chapter [12], we mainly consider and investigate the tracking-control problem of the IPC system in terms of the position variable of the cart.

For solving the tracking-control problem of nonlinear systems (e.g., the IPC system), a number of methods have been investigated. Isidori [13] presented a traditional method available for the design of a nonlinear controller, which is the exact feedback linearization technique based on geometric control theory. El-Farra et al. [14] adopted multiple Lyapunov functions for output feedback control of switched nonlinear systems. Based on neural-network approximation, a control scheme was developed for tracking control of a spherical inverted pendulum [15]. Recently, sliding-mode control has been used and investigated for control of inverted pendulum systems [4, 6]. However, most of the aforementioned control methods are relatively complex for their design procedures of controllers and practical implementations. Thus, it is meaningful and reasonable to find a simple and effective control method (i.e., the design method of controllers).

Differing from GD, a special class of neural dynamics, which is called ZD, has recently been proposed by Zhang et al. for online solution of various problems [16]. The design of ZD is based on an indefinite matrix-/vector-valued ZF, while the design of GD is based on a scalar-valued nonnegative EF. In the previous studies [17, 18], the ZD and GD methods have shown the efficacy for tracking control. In this chapter [12], by combining the ZD and GD methods together, a kind of tracking controllers, i.e., ZG controllers, are designed and investigated for the cart path tracking control of the IPC system. Specifically, according to the number of times of using ZD and GD methods, controllers of z2g0 and z2g1 types are developed for the cart path tracking control of the IPC system. The z2g0 controllers are designed by adopting the ZD method twice and without using the GD method, while the z2g1 controllers are designed by adopting the ZD method twice and the GD method once. Furthermore, the DBZ problem is also discussed and investigated in this chapter [12], which has been rarely studied among the traditional investigations and is a difficult problem in the conventional tracking controller design. Note that, in the conventional controller design [13], the divisor of a controller is simply assumed to be nonzero at any time instant, which often leads to contradictions between theoretical investigations and practical applications.

Through the theoretical analyses, both z2g0 and z2g1 controllers possess the global and exponential convergence performance. For the purpose of illustration and comparison, different desired trajectories are tested for the cart path tracking control of the IPC system equipped with both z2g0 and z2g1 controllers. Computer simulations illustrate the feasibility and efficacy of both z2g0 and z2g1 controllers for the cart path tracking control of the IPC system. More importantly, the superiority of z2g1 controllers in conquering the DBZ problem is also substantiated by the comparative simulation results.

9.2 Mathematical Model of IPC System

The control of pendulum tracking (including swinging up) is studied on the basis of the mathematical model of the IPC system shown in Fig. 9.1. The standard assumptions are considered, i.e., massless rod and point masses. Let (m_c, p) be the mass and position of the cart, which can move freely on the horizontal plane. In addition, m_p is the mass of the pendulum, concentrated in the ball, θ is the angle between the vertical line and the pendulum (positive clockwise), l is the length of the pendulum, and I is the moment of inertia. By using the Newton's second law or Euler-Lagrange formulation, the dynamics of the complete system can be obtained as

$$\begin{bmatrix} m_p + m_c & m_p l \cos\theta \\ m_p l \cos\theta & I + m_p l^2 \end{bmatrix} \begin{bmatrix} \ddot{p} \\ \ddot{\theta} \end{bmatrix} = \begin{bmatrix} m_p l \dot{\theta}^2 \sin\theta - \sigma \dot{p} \\ m_p g l \sin\theta \end{bmatrix} + \begin{bmatrix} u \\ 0 \end{bmatrix}, \tag{9.1}$$

where g is the gravitational acceleration constant, σ is the coefficient of viscous friction for motion of the cart, and u is the control input of the IPC system, corresponding to the horizontal force applied to the cart. Note that, for the convenience of further research, it is assumed that the moment of inertia of the IPC

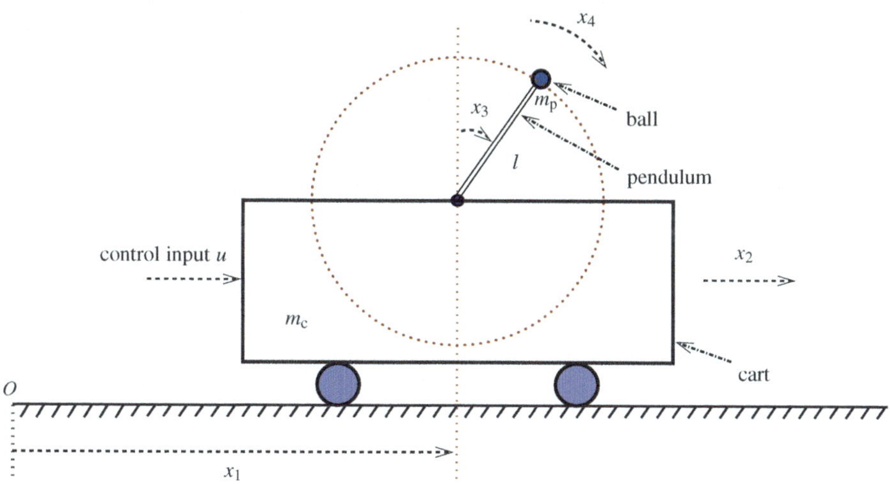

Fig. 9.1 Schematic of IPC system

system is negligible, as done in [19–21]. According to Eq. (9.1), the state equations of the IPC system can be expressed as [12, 22]:

$$\begin{cases} \dot{x}_1 = f_1(x_2) = x_2, \\ \dot{x}_2 = f_2(x_2, x_3, x_4, u) = \dfrac{u - \sigma x_2 + m_\mathrm{p}(l x_4^2 - g\cos x_3)\sin x_3}{m_\mathrm{c} + m_\mathrm{p}\sin^2 x_3}, \\ \dot{x}_3 = f_3(x_4) = x_4, \\ \dot{x}_4 = f_4(x_2, x_3, x_4, u) = \dfrac{(m_\mathrm{c} + m_\mathrm{p}) g \sin x_3 - (u - \sigma x_2 + m_\mathrm{p} l x_4^2 \sin x_3)\cos x_3}{l(m_\mathrm{c} + m_\mathrm{p}\sin^2 x_3)}, \end{cases} \tag{9.2}$$

where $x_1 = p$, $x_2 = \dot{p}$, $x_3 = \theta$, and $x_4 = \dot{\theta}$ are selected as state variables. Evidently, x_2 and x_4 correspond to the velocity of the cart and the angular velocity of the pendulum, respectively. For the purpose of simulation, the parameter values of the IPC system are set as shown in Table 9.1.

Remark 9.1 It is worth pointing out here that, with the consideration of robustness, the model of IPC system (9.2) can be generalized and then applied in the situation that the mass of the rod for the pendulum is m_r (with $m_\mathrm{r} > 0$). In the center of mass system for the pendulum, the total mass is $m_\mathrm{t} = m_\mathrm{r} + m_\mathrm{p}$. Besides, for the rigid and regular rod of the pendulum with uniform mass distribution, the mass point of the rod locates at the center (i.e., $l_\mathrm{r} = l/2$) and the mass point of the ball locates at the end of the pendulum (i.e., $l_\mathrm{b} = l$) with the assumption that the radius of the ball is small enough. Then, according to the geometric relationship of the center of mass system for the pendulum, it is readily derived that the mass point of the whole pendulum is $l_\mathrm{w} = m_\mathrm{p} l/(2m_\mathrm{t}) + l/2$. Note that, if $m_\mathrm{p} \gg m_\mathrm{r}$, then $l_\mathrm{w} \approx l$, which is similar to the situation that the rod of pendulum is assumed as zero mass. Thus, replacing m_p and l with m_t and l_w in IPC system (9.2), respectively, we have the more practical and general model of IPC system as

$$\begin{cases} \dot{x}_1 = f_1(x_2) = x_2, \\ \dot{x}_2 = f_2(x_2, x_3, x_4, u) = \dfrac{u + m_\mathrm{t}(l_\mathrm{w} x_4^2 - g\cos x_3)\sin x_3 - \sigma x_2}{m_\mathrm{c} + m_\mathrm{t}\sin^2 x_3}, \\ \dot{x}_3 = f_3(x_4) = x_4, \\ \dot{x}_4 = f_4(x_2, x_3, x_4, u) = \dfrac{(m_\mathrm{c} + m_\mathrm{t}) g \sin x_3 + (\sigma x_2 - u - m_\mathrm{t} l_\mathrm{w} x_4^2 \sin x_3)\cos x_3}{l_\mathrm{w}(m_\mathrm{c} + m_\mathrm{t}\sin^2 x_3)}. \end{cases}$$

Table 9.1 Parameter values of IPC system

Parameter	m_c	m_p	l	σ	g
Value	1.378 kg	0.051 kg	0.325 m	12.98 N s/m	9.81 m/s^2

9.3 Design of Controllers

In this section, we apply the ZG method to design z2g0 and z2g controllers for the cart path tracking control of IPC system (9.2).

9.3.1 Design of ZD Controllers

For the design of z2g0 controllers, we adopt the ZD method twice without using the GD method. In consideration of completeness, the tracking-control problems of explicit and implicit types are both investigated.

9.3.1.1 Explicit Tracking Control

Without loss of generality, the output of IPC system (9.2) can be set as $y = x_1$. Controllers of z2g0 type can be obtained by using the ZD method twice and without using the GD method, which are constructed by the following three steps.

Step 1 Based on the ZD method [16, 17], the first ZF is defined as

$$z_1 = y - y_{\mathrm{d}} = x_1 - y_{\mathrm{d}},$$

where y_{d} is the desired trajectory. Then, the following ZD design formula is adopted:

$$\dot{z}_1 = -\lambda_1 z_1, \tag{9.3}$$

where the design parameter $\lambda_1 \in \mathbb{R}^+$ is used to scale the convergence rate of the ZD solution.

Step 2 From the first ZF and (9.3), we further have

$$\dot{x}_1 - \dot{y}_{\mathrm{d}} = -\lambda_1(x_1 - y_{\mathrm{d}}).$$

Thus, in light of IPC system (9.2), the second ZF is defined as

$$z_2 = \dot{x}_1 - \dot{y}_{\mathrm{d}} + \lambda_1(x_1 - y_{\mathrm{d}}).$$

Using the ZD method again (i.e., the second time), we have the second ZD design formula:

$$\dot{z}_2 = -\lambda_2 z_2,$$

where the design parameter $\lambda_2 \in \mathbb{R}^+$ is used to scale the convergence rate of the ZD solution. Thus, we have

$$\dot{x}_2 - \ddot{y}_\mathrm{d} + \lambda_1(x_2 - \dot{y}_\mathrm{d}) = -\lambda_2\left(x_2 - \dot{y}_\mathrm{d} + \lambda_1(x_1 - y_\mathrm{d})\right). \tag{9.4}$$

Step 3 Based on Eq. (9.4) and IPC system (9.2), a z2g0 controller in the form of u is derived as

$$u = \alpha_1\alpha_2 - \alpha_3, \tag{9.5}$$

where $\alpha_1 = \lambda_1\lambda_2(y_\mathrm{d} - x_1) + (\lambda_1 + \lambda_2)(\dot{y}_\mathrm{d} - x_2) + \ddot{y}_\mathrm{d}$, $\alpha_2 = m_\mathrm{c} + m_\mathrm{p}\sin^2 x_3$ and $\alpha_3 = m_\mathrm{p}(lx_4^2 - g\cos x_3)\sin x_3 - \sigma x_2$, with $\dot{y}_\mathrm{d}$ and $\ddot{y}_\mathrm{d}$ denoting the first and second time derivatives of y_d, respectively.

From the above three steps, z2g0 controller (9.5) is obtained for the cart path tracking control of IPC system (9.2). Note that z2g0 controller (9.5) is simple and concise, as compared with the traditional design methods [6, 13–15, 19, 23].

9.3.1.2 Implicit Tracking Control

For completeness of investigation on the cart path tracking control of IPC system (9.2), it is necessary and meaningful to consider the output in a complex form (with high-order/nonlinear term of x_1), such as $y = x_1^2 + x_1$ to track y_d. With a complex form of the output, the cart path tracking control of IPC system (9.2) is implicit.

Similarly, by adopting the ZD method twice and without using the GD method, the following z2g0 controller can be obtained with the output of IPC system (9.2) being $y = x_1^2 + x_1$:

$$u = \frac{\alpha_2\alpha_4}{\alpha_5} - \alpha_3, \tag{9.6}$$

where $\alpha_4 = \lambda_1\lambda_2(y_\mathrm{d} - x_1^2 - x_1) + (\lambda_1 + \lambda_2)(\dot{y}_\mathrm{d} - 2x_1x_2 - x_2) - 2x_2^2 + \ddot{y}_\mathrm{d}$ and $\alpha_5 = 1 + 2x_1$. Note that z2g0 controller (9.6) also has the simple and concise characteristics. However, if the divisor of z2g0 controller (9.6) is zero (i.e., $\alpha_5 = 0$), the value of u is infinite. In other words, the DBZ problem may lead to crash of physical systems in practical applications.

9.3.2 Design of ZG Controllers

In this subsection, the z2g1 controllers are further designed to conquer the DBZ problem encountered in the cart path tracking control of IPC system (9.2).

9.3.2.1 Explicit Tracking Control

For the case of $y = x_1$, from z2g0 controller (9.5), we can define $\phi_1 = u - \alpha_1\alpha_2 + \alpha_3$, and ϕ_1 should theoretically be zero. Then, a corresponding EF ϵ_1 with the form of $\epsilon_1 = \phi_1^2/2$ can be defined. According to the GD method [16, 17], the following GD design formula can be adopted:

$$\dot{u} = -\gamma \frac{\partial \epsilon_1}{\partial u},$$

where the design parameter $\gamma \in \mathbb{R}^+$ is used to scale the convergence rate of the GD solution. Thus, with the output of IPC system (9.2) being $y = x_1$, a z2g1 controller is obtained as

$$\dot{u} = -\gamma\phi_1 = -\gamma(u - \alpha_1\alpha_2 + \alpha_3). \tag{9.7}$$

9.3.2.2 Implicit Tracking Control

For the case of $y = x_1^2 + x_1$, by defining $\phi_2 = \alpha_5(u + \alpha_3) - \alpha_2\alpha_4$ and $\epsilon_2 = \phi_2^2/2$, the following z2g1 controller can similarly be obtained:

$$\dot{u} = -\gamma\phi_2\alpha_5 = -\gamma\alpha_5\left(\alpha_5(u + \alpha_3) - \alpha_2\alpha_4\right). \tag{9.8}$$

Note that z2g1 controllers (9.7) and (9.8) both are in the form of $\dot{u}$. More importantly, z2g1 controller (9.8) can conquer the DBZ problem encountered in the cart path tracking control, since it has no division operation. Thus, z2g1 controllers can be termed DBZ-conquering ZG controllers. For reading and comparison convenience, the z2g0 and z2g1 controllers developed and investigated in this chapter [12] are shown in Table 9.2.

Remark 9.2 When the rod of pendulum is nonzero mass (i.e., the model of IPC system shown in Remark 9.1), by adopting the above similar procedures for the design of ZG controllers, the design results of z2g0 and z2g1 controllers are obtained and presented in Table 9.3, where $\bar{\alpha}_1 = \lambda_1\lambda_2(y_d - x_1) + (\lambda_1 + \lambda_2)(\dot{y}_d - x_2) + \ddot{y}_d$, $\bar{\alpha}_2 = m_c + m_t \sin^2 x_3$, $\bar{\alpha}_3 = m_t(l_w x_4^2 - g\cos x_3)\sin x_3 - \sigma x_2$, $\bar{\alpha}_4 = \lambda_1\lambda_2(y_d - x_1^2 - x_1) + (\lambda_1 + \lambda_2)(\dot{y}_d - 2x_1x_2 - x_2) - 2x_2^2 + \ddot{y}_d$ and $\bar{\alpha}_5 = 1 + 2x_1$. Note that, compared with the design results shown in Table 9.2, it can be readily found

Table 9.2 Controllers of z2g0 and z2g1 types for explicit and implicit tracking control of IPC system with pendulum rod being zero mass

Type	$y = x_1$ (explicit)	$y = x_1^2 + x_1$ (implicit)
z2g0	$u = \alpha_1\alpha_2 - \alpha_3$	$u = \alpha_2\alpha_4/\alpha_5 - \alpha_3$
z2g1	$\dot{u} = -\gamma(u - \alpha_1\alpha_2 + \alpha_3)$	$\dot{u} = -\gamma\alpha_5(\alpha_5(u + \alpha_3) - \alpha_2\alpha_4)$

Table 9.3 Controllers of z2g0 and z2g1 types for explicit and implicit tracking control of IPC system with pendulum rod being nonzero mass

Type	$y = x_1$ (explicit)	$y = x_1^2 + x_1$ (implicit)
z2g0	$u = \bar{\alpha}_1\bar{\alpha}_2 - \bar{\alpha}_3$	$u = \bar{\alpha}_2\bar{\alpha}_4/\bar{\alpha}_5 - \bar{\alpha}_3$
z2g1	$\dot{u} = -\gamma(u - \bar{\alpha}_1\bar{\alpha}_2 + \bar{\alpha}_3)$	$\dot{u} = -\gamma\bar{\alpha}_5(\bar{\alpha}_5(u + \bar{\alpha}_3) - \bar{\alpha}_2\bar{\alpha}_4)$

that $\bar{\alpha}_1 = \alpha_1, \bar{\alpha}_4 = \alpha_4$ and $\bar{\alpha}_5 = \alpha_5$. Besides, replacing m_p and l with m_t and l_w in the expressions of α_2 and α_3, respectively, we then have the expressions of $\bar{\alpha}_2$ and $\bar{\alpha}_3$. These results show the robustness of the derived results as well as the feasibility and efficacy of the ZG design method.

9.3.3 Discussion on Controller Implementation

As seen from the design results of z2g0 and z2g1 controllers shown in Tables 9.2 and 9.3, z2g0 controllers contain derivative signals of $\dot{y}_\text{d}$ and $\ddot{y}_\text{d}$, while z2g1 controllers contain derivative signals of $\dot{y}_\text{d}$, $\ddot{y}_\text{d}$ and $\dot{u}$.

Without loss of generality, let us consider the derivative signals of $\dot{y}_\text{d}$ and $\ddot{y}_\text{d}$ first. Generally speaking, for the cart path tracking control of the IPC system, y_d can usually be expressed as a function with respect to t. For example, if we select $y_\text{d} = \sin(t)$ for the cart path tracking control, it is readily derived that $\dot{y}_\text{d} = \cos(t)$ and $\ddot{y}_\text{d} = -\sin(t)$. On the other hand, for the complicated situation that y_d cannot be expressed as an explicit function or y_d has sudden changes because of tracking tasks, e.g., the tracking control of the flight trajectory, it is reasonable and effective to obtain the derivative signals by estimation [24]. Specifically, if Euler numerical differentiation formula is applied on the estimation, we have $\dot{y}_\text{d}(t_k) = (y_\text{d}(t_k) - y_\text{d}(t_{k-1}))/\tau$, where $y_\text{d}(t_k)$ and $\dot{y}_\text{d}(t_k)$ denote the desired trajectory and its derivative signal at time instant t_k (with $k = 1, 2, \ldots$), respectively, and $\tau > 0$ denotes the sampling time interval. Similarly, $\ddot{y}_\text{d}$ can also be estimated by using Euler numerical differentiation formula. Note that the algorithms for the estimation of derivative signals have been widely investigated, such as Euler numerical differentiation formula [25].

Besides, for the z2g1 controllers containing derivative signal of $\dot{u}$, it can be well understood and realized from a viewpoint of circuit implementation. Specifically, a block diagram of the circuit implementation can be designed on the basis of Eq. (9.7), as shown in Fig. 9.2. Thus, the circuit implementation of z2g1 controller (9.7) consists of summators, integrators and weighted connections. It is worth pointing out here that the controllers for implicit tracking, e.g., z2g1 controller (9.8), can also be designed in a similar manner. Therefore, it is feasible and effective to apply the ZG controllers to the cart path tracking control of IPC system.

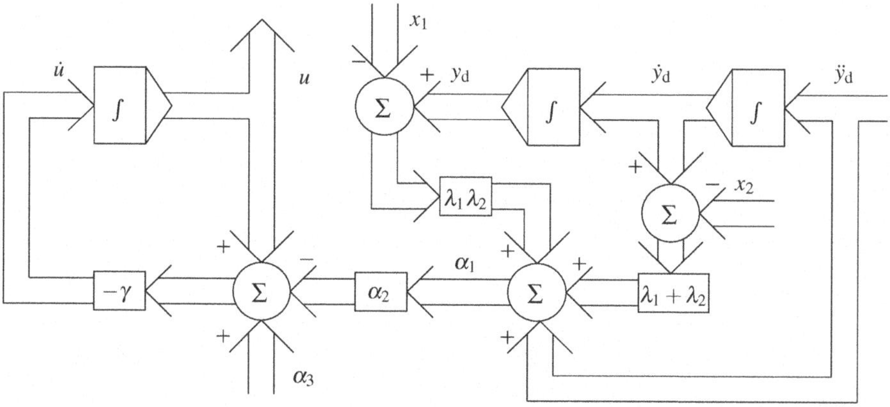

Fig. 9.2 Block diagram of circuit implementation for z2g1 controller (9.7)

9.4 Convergence Performance Analyses

This section presents the convergence performance analyses on the z2g0 and z2g1 controllers shown in Table 9.2, respectively.

9.4.1 Analyses on ZD Controllers

Based on the above design procedures, the analyses on z2g0 controllers, i.e., controllers (9.5) and (9.6), can also be divided into two types: the explicit tracking control and the implicit tracking control.

9.4.1.1 Analysis on Explicit Tracking Control

For the explicit tracking control of IPC system (9.2), the tracking error is defined as $e = y - y_d$ (with $y = x_1$). Then, the convergence performance analysis on z2g0 controller (9.5) is given as the following theorem.

Theorem 9.1 *For smooth and bounded desired trajectory y_d, starting with bounded initial state $\mathbf{x}(0) = [x_1(0), x_2(0), x_3(0), x_4(0)]^T \in \mathbb{R}^4$, the tracking error of IPC system (9.2) equipped with z2g0 controller (9.5) exponentially converges to zero on a large scale.*

Proof According to the ZD design formulas about z_1 and z_2, we readily obtain the following equation by substituting $z_2 = \dot{z}_1 + \lambda_1 z_1$ into $\dot{z}_2 = -\lambda_2 z_2$:

$$\ddot{z}_1 + (\lambda_1 + \lambda_2)\dot{z}_1 + \lambda_1\lambda_2 z_1 = 0.$$

If $\lambda_1 \neq \lambda_2$, we have

$$z_1 = c_1 \exp(-\lambda_1 t) + c_2 \exp(-\lambda_2 t),$$

where c_1 and c_2 are constants, and $t \in [0, +\infty)$. Otherwise, we have

$$z_1 = c_3 \exp(-\lambda_1 t) + c_4 t \exp(-\lambda_1 t),$$

where c_3 and c_4 are constants. In view of $\lambda_1 > 0$ and $\lambda_2 > 0$, there exist $\bar{c} > 0$ and $\bar{\lambda} > 0$ such that $z_1 \leq \bar{c} \exp(-\bar{\lambda} t)$ [26].

According to the above analysis, we have the conclusion that, starting with bounded initial state $\mathbf{x}(0)$, the tracking error of IPC system (9.2) equipped with z2g0 controller (9.5) exponentially converges to zero on a large scale. The proof is thus completed. □

9.4.1.2 Analysis on Implicit Tracking Control

In the case of the implicit tracking control of IPC system (9.2), the tracking error is thus defined as $e = y - y_{\mathrm{d}}$ (with $y = x_1^2 - x_1$). Then, the convergence performance analysis on z2g0 controller (9.6) is given as follows.

Theorem 9.2 *For smooth and bounded desired trajectory y_{d}, starting with bounded initial state $\mathbf{x}(0) \in \mathbb{R}^4$, the tracking error of IPC system (9.2) equipped with z2g0 controller (9.6) exponentially converges to zero on a large scale, provided that $\alpha_5 \neq 0$, $\forall t \in [0, +\infty)$.*

Proof It can be generalized from the proof of Theorem 9.1. □

9.4.2 Analyses on ZG Controllers

In this subsection, the analyses on z2g1 controllers (9.7) and (9.8) are presented in terms of the explicit tracking control and the implicit tracking control, respectively.

9.4.2.1 Analyses on Explicit Tracking Control

For the explicit tracking control of IPC system (9.2) equipped with z2g1 controller (9.7), the tracking error is defined as $e = y - y_{\mathrm{d}}$ (with $y = x_1$). Based on z2g0 controllers (9.5) and (9.7), it is readily found that the desired time-varying solution of z2g1 controller (9.7) is $u^* = \alpha_1\alpha_2 - \alpha_3$. Then, the convergence performance analysis on z2g1 controller (9.7) is given as the following theorems.

Theorem 9.3 *Consider IPC system (9.2) equipped with z2g1 controller (9.7) for smooth and bounded desired trajectory* y_d*. Starting with bounded initial state* $\mathbf{x}(0) \in \mathbb{R}^4$ *and control input* $u(0) \in \mathbb{R}$*, provided that* $|\dot{u}^*| = |\mathrm{d}u^*/\mathrm{d}t| \leq \varrho$, $\exists 0 \leq \varrho < +\infty$*, the steady-state tracking error of the system is upper bounded tightly as*

$$\limsup_{t\to+\infty} |e| \leq \frac{\varrho}{\lambda_1\lambda_2\gamma m_\mathrm{c}}.$$

Proof Before analyzing the convergence performance of the tracking error, we firstly present the convergence performance of solution error of z2g1 controller (9.7). Specifically, a solution error is defined as $\psi = u - u^* = u - \alpha_1\alpha_2 + \alpha_3$ with $\dot{u} = -\gamma(u - \alpha_1\alpha_2 + \alpha_3)$. Then, we have $u = \psi + u^*$ and its time derivative $\dot{u} = \dot{\psi} + \dot{u}^*$. In view of $\dot{u} = -\gamma\psi$, we have $\dot{\psi} = -\gamma\psi - \dot{u}^*$.

For further analysis, a Lyapunov function candidate is defined as $L = \psi^2/2 \geq 0$. Then, the time derivative of L is obtained as

$$\dot{L} = \psi\dot{\psi} = -\gamma\psi^2 - \psi\dot{u}^*. \tag{9.9}$$

Evidently, there are two terms in the expression of $\dot{L}$, i.e., $-\gamma\psi^2$ and $-\psi\dot{u}^*$. For the first term, $-\gamma\psi^2 \leq 0$ always holds true. For the second term, making use of Cauchy inequality [27], we have

$$-\psi\dot{u}^* \leq |\psi||\dot{u}^*| \leq \varrho|\psi|.$$

Then, (9.9) can be derived as

$$\dot{L} = -\gamma\psi^2 - \psi\dot{u}^* \leq -\gamma\psi^2 + \varrho|\psi| = -|\psi|(\gamma|\psi| - \varrho). \tag{9.10}$$

During the time evolution of solution error ψ, (9.10) falls into one of the following three situations.

- In the first situation (i.e., $\gamma|\psi| - \varrho > 0$), we have $\gamma|\psi| > \varrho$ and then $\dot{L} < 0$, which implies that ψ approaches zero (i.e., u approaches u^*) as time evolves.
- In the second situation (i.e., $\gamma|\psi| - \varrho = 0$), we have $\gamma|\psi| = \varrho$ and then $\dot{L} \leq 0$, which implies that ψ approaches zero (i.e., u approaches u^*) or ψ stays on the ball surface with $|\psi| = |u - u^*| = \varrho/\gamma$.
- In the third situation (i.e., $\gamma|\psi| - \varrho < 0$), we have $\gamma|\psi| < \varrho$ and thus $\dot{L}$ is less than a positive constant, which consists of sub-situations $\dot{L} \leq 0$ and $0 < \dot{L} \leq -|\psi|(\gamma|\psi| - \varrho)$. On the one hand, if $\dot{L} \leq 0$ holds true, then it returns to the same case of the second situation. On the other hand, if $0 < \dot{L} \leq -|\psi|(\gamma|\psi| - \varrho)$ holds true, then L and ψ would increase, which implies that $\gamma|\psi| - \varrho$ increases as well. Thus, since $\gamma|\psi| - \varrho < 0$, it must exist a certain time instant t_1 such that $\gamma|\psi| - \varrho = 0$, which would return to the second situation, i.e., $\dot{L} \leq 0$.

Summarizing the above three situations, we can conclude that the solution error of z2g1 controller (9.7) is upper bounded by ϱ/γ, i.e.,

$$\limsup_{t\to+\infty}|\psi| = \limsup_{t\to+\infty}|u-u^*| \leq \frac{\varrho}{\gamma}. \tag{9.11}$$

From the design procedure of z2g1 controller (9.7), we can derive that $\dot{z}_2 + \lambda_2 z_2 = (u+\alpha_3)/\alpha_2 - \alpha_1 = \psi/\alpha_2$. Besides, we have $\dot{z}_2 + \lambda_2 z_2 = \ddot{z}_1 + (\lambda_1 + \lambda_2)\dot{z}_1 + \lambda_1\lambda_2 z_1$, which can be further derived as $\ddot{e} + (\lambda_1+\lambda_2)\dot{e} + \lambda_1\lambda_2 e = \psi/\alpha_2$ (in view of $e = z_1$). From (9.11), we have

$$\limsup_{t\to+\infty}|\,(\ddot{e} + (\lambda_1+\lambda_2)\dot{e} + \lambda_1\lambda_2 e)\,\alpha_2| \leq \frac{\varrho}{\gamma}.$$

Note that $0 < m_c \leq \alpha_2 = m_c + m_p \sin^2 x_3 \leq m_c + m_p$ always holds true. Thus, we have

$$\limsup_{t\to+\infty}|(\ddot{e}+(\lambda_1+\lambda_2)\dot{e}+\lambda_1\lambda_2 e)m_c| \leq \limsup_{t\to+\infty}|(\ddot{e}+(\lambda_1+\lambda_2)\dot{e}+\lambda_1\lambda_2 e)\alpha_2| \leq \frac{\varrho}{\gamma},$$

which can be further derived as

$$\limsup_{t\to+\infty}|(\ddot{e} + (\lambda_1+\lambda_2)\dot{e} + \lambda_1\lambda_2 e)| \leq \frac{\varrho}{\gamma m_c}.$$

For an enough large time instant t_e, we further have $|(\ddot{e}(t_e) + (\lambda_1+\lambda_2)\dot{e}(t_e) + \lambda_1\lambda_2 e(t_e)| \leq \varrho/(\gamma m_c)$. Based on Gronwall inequality [28], it can be reformulated as $|e(t_e)| \leq \varrho/(\lambda_1\lambda_2\gamma m_c)$. In view of $t_e \to +\infty$, we can obtain

$$\limsup_{t\to+\infty}|e| \leq \frac{\varrho}{\lambda_1\lambda_2\gamma m_c}.$$

The proof is thus completed. □

Theorem 9.3 indicates that the tracking error converges toward the ball of $\varrho/(\lambda_1\lambda_2\gamma m_c)$, which implies that, when γ tends to infinite, the tracking error bound approaches zero. However, the nature of Theorem 9.3 is asymptotic convergence, which may not be good enough in practice as it may require an infinitely long time period to accomplish the cart path tracking control. Via further investigation, the tracking error is proven to be exponentially convergent toward a relatively loose tracking-error bound, which is presented as the following theorem.

Theorem 9.4 *Consider IPC system (9.2) equipped with z2g1 controller (9.7) for smooth and bounded desired trajectory* y_d. *Starting with bounded initial state* $\mathbf{x}(0) \in \mathbb{R}^4$ *and control input* $u(0) \in \mathbb{R}$, *provided that* $|\dot{u}^*| \leq \varrho$, $\exists 0 \leq \varrho < +\infty$, *the tracking error of the system exponentially converges or stays within the error bound* $\varrho/(\omega\lambda_1\lambda_2\gamma m_c)$ *on a large scale, where* $\omega \in (0, 1)$ *is a loosening parameter.*

Proof As shown in Theorem 9.3, the solution error ψ of z2g1 controller (9.7) asymptotically converges toward the ball of ϱ/γ. Then, from (9.10), we have

$$\dot{L} \leq -\gamma\psi^2 + \varrho|\psi| = -(1-\omega)\gamma\psi^2 + (-\omega\gamma\psi^2 + \varrho|\psi|). \tag{9.12}$$

On the one hand, for the right side of (9.12), the first term $-(1-\omega)\gamma\psi^2 \leq 0$ always holds true. On the other hand, the second term on the right side of (9.12) consists of two situations, i.e., $-\omega\gamma\psi^2 + \varrho|\psi| \leq 0$ and $-\omega\gamma\psi^2 + \varrho|\psi| > 0$.

For the first situation, we have $|\psi| \geq \varrho/(\omega\gamma)$ and further obtain

$$\dot{L} \leq -(1-\omega)\gamma\psi^2 = -2(1-\omega)\gamma L. \tag{9.13}$$

Then, the analytic solution of (9.13) is $L \leq \exp(-2(1-\omega)\gamma t)L(0)$, which can be further formulated as

$$|\psi| \leq |\psi(0)|\exp(-(1-\omega)\gamma t), \forall t \in [0, t_c], \tag{9.14}$$

where the convergence time $t_c = \ln(\omega\gamma|\psi(0)|/\varrho)/((1-\omega)\gamma)$ in view of $\varrho/(\omega\gamma) = |\psi(0)|\exp(-(1-\omega)\gamma t_c)$ and the exponential convergence rate is $(1-\omega)\gamma$.

For the second situation, such ψ can never leave the new ball of $\varrho/(\omega\gamma)$. The analysis is similar to that of (9.10) in the third situation.

Furthermore, from the above analysis, we have the following facts.

- If $|\psi(0)| \geq \varrho/(\omega\gamma)$, then

$$|\psi| \begin{cases} \leq |\psi(0)|\exp(-(1-\omega)\gamma t), \forall t \in [0, t_c], \\ \leq \varrho/(\omega\gamma), \forall t \in [t_c, +\infty). \end{cases}$$

- If $|\psi(0)| \leq \varrho/(\omega\gamma)$, then $|\psi| \leq \varrho/(\omega\gamma), \forall t \in [0, +\infty)$.

Evidently, even in the worst case, the exponential convergence rate is $(1-\omega)\gamma$. Besides, for any $t \geq t_c$, $|\psi| \leq \varrho/(\omega\gamma)$. Similar to the proof of Theorem 9.3, the following inequality can be obtained:

$$|\ddot{e} + (\lambda_1 + \lambda_2)\dot{e} + \lambda_1\lambda_2 e| \leq \frac{\varrho}{\omega\gamma m_c}, \forall t \geq t_c.$$

It can be derived from Gronwall inequality [28] that $|e|$ exponentially converges toward the new error bound $\varrho/(\omega\lambda_1\lambda_2\gamma m_c)$ when $t \geq t_c$.

By the above analysis, the tracking error of IPC system (9.2) equipped with z2g1 controller (9.7) exponentially converges toward or stays within the error bound $\varrho/(\omega\lambda_1\lambda_2\gamma m_c)$ on a large scale. The proof is thus completed. □

9.4.2.2 Analyses on Implicit Tracking Control

For the implicit tracking control of IPC system (9.2) equipped with z2g1 (9.8), the convergence performance analysis on z2g1 controller (9.8) is presented below, of which the tracking error is defined as $e = y - y_{\mathrm{d}}$ (with $y = x_1^2 - x_1$).

Theorem 9.5 *Consider IPC system (9.2) equipped with z2g1 controller (9.8) for smooth and bounded desired trajectory y_{d}. Starting with bounded initial state $\mathbf{x}(0) \in \mathbb{R}^4$ and control input $u(0) \in \mathbb{R}$, the following results are achieved on a large scale for the cart path tracking control of the system.*

- *For the case of $\alpha_5 \neq 0$ (i.e., the non-DBZ case), the tracking error of the system converges toward or stays within the error bound $\varrho\sqrt{\eta_2}/(\lambda_1\lambda_2\gamma\eta_1 m_{\mathrm{c}})$, provided that (i) $\eta_1 \le \alpha_5^2 \le \eta_2$, $\exists 0 < \eta_1 \le \eta_2 < +\infty$, and (ii) $|\dot{u}^*| \le \varrho$, $\exists 0 \le \varrho < +\infty$.*
- *For the case of $\alpha_5 = 0$ (i.e., the DBZ case), the tracking error of the system is bounded.*

Proof For the case of $\alpha_5 \neq 0$ (i.e., the non-DBZ case), based on controllers (9.6) and (9.8), it is readily found that the optimal solution of z2g1 controller (9.8) is $u^* = \alpha_2\alpha_4/\alpha_5 - \alpha_3$. Similar to the proof of Theorem 9.3, a solution error is defined as $\psi = u - u^* = u + \alpha_3 - \alpha_2\alpha_4/\alpha_5$ with $\dot{u} = -\gamma\alpha_5(\alpha_5(u + \alpha_3) - \alpha_2\alpha_4)$. Then, we have $u = \psi + u^*$ and its time derivative $\dot{u} = \dot{\psi} + \dot{u}^*$. In view of $\dot{u} = -\gamma\alpha_5^2\psi$, we have $\dot{\psi} = -\gamma\alpha_5^2\psi - \dot{u}^*$. For further analysis, a Lyapunov function candidate is defined as $L = \psi^2/2 \ge 0$. Then, the time derivative of L is obtained as

$$\dot{L} = \psi\dot{\psi} = -\gamma\alpha_5^2\psi^2 - \psi\dot{u}^*. \tag{9.15}$$

In view of $\eta_1 \le \alpha_5^2$, $-\gamma\alpha_5^2\psi^2 \le -\gamma\eta_1\psi^2$ always holds true for the first term on the right side of (9.15). Besides, making use of Cauchy inequality, we have $-\psi\dot{u}^* \le |\psi||\dot{u}^*| \le \varrho|\psi|$ for the second term on the right side of (9.15).

Evidently, (9.15) can be derived as

$$\dot{L} = -\gamma\alpha_5^2\psi^2 - \psi\dot{u}^* \le -\gamma\eta_1\psi^2 + \varrho|\psi| = -|\psi|(\gamma\eta_1|\psi| - \varrho). \tag{9.16}$$

Similar to the analysis of (9.10), from (9.16), we have

$$\limsup_{t\to+\infty} |\psi| = \limsup_{t\to+\infty} |u - u^*| \le \frac{\varrho}{\gamma\eta_1}. \tag{9.17}$$

From the design procedure of z2g1 controller (9.8), we can derive $\dot{z}_2 + \lambda_2 z_2 = \alpha_5(u + \alpha_3)/\alpha_2 - \alpha_4 = \alpha_5\psi/\alpha_2$. Besides, we have $\dot{z}_2 + \lambda_2 z_2 = \ddot{z}_1 + (\lambda_1 + \lambda_2)\dot{z}_1 + \lambda_1\lambda_2 z_1$, which can be further derived as $\ddot{e} + (\lambda_1 + \lambda_2)\dot{e} + \lambda_1\lambda_2 e = \alpha_5\psi/\alpha_2$ (in view of $e = z_1$). Similarly, from the proof of Theorem 9.3, we have

$$\limsup_{t\to+\infty} |e| \le \frac{\varrho\sqrt{\eta_2}}{\lambda_1\lambda_2\gamma\eta_1 m_{\mathrm{c}}}.$$

For the case of $\alpha_5 = 0$ (i.e., the DBZ case), it can be derived that $\lim_{t \to t_s} \dot{u} = \lim_{\alpha_5 \to 0} \dot{u} = 0$ in view of $\dot{u} = -\gamma\alpha_5(\alpha_5(u + \alpha_3) - \alpha_2\alpha_4)$. The control input at the DBZ time instant t_s is the same as that at the previous time instant t_{s^-}, which implies that $u(t_s) = u(t_{s^-})$. Similarly, at time instant t_{s^+} (which is after the DBZ time instant), $u(t_s) = u(t_{s^+})$. Note that, at the time instant t_{s^-}, the control input $u(t_{s^-})$ is bounded and the tracking error converges toward or stays within an error bound. Then, we have the result that $u(t_{s^-}) = u(t_s) = u(t_{s^+})$ is bounded. For a bounded control input, the output of system (9.2) is bounded. Since the desired trajectory y_d is bounded, the tracking error e is thus bounded at the time instants t_{s^-}, t_s and t_{s^+}. After getting through the DBZ point (i.e., for time instants after t_{s^+}), the tracking error converges toward an error bound again, which implies that IPC system (9.2) equipped with z2g1 controller (9.8) finally conquers the DBZ problem.

By the above analyses, the proof is thus completed. □

Theorem 9.6 *Consider IPC system (9.2) equipped with z2g1 controller (9.8) for smooth and bounded desired trajectory* y_d. *Starting with bounded initial state* $\mathbf{x}(0) \in \mathbb{R}^4$ *and control input* $u(0) \in \mathbb{R}$, *the following results are achieved on a large scale for the cart path tracking control of the system.*

- *For the case of* $\alpha_5 \neq 0$ *(i.e., the non-DBZ case), provided that (i)* $\eta_1 \leq \alpha_5^2 \leq \eta_2$, $\exists 0 < \eta_1 \leq \eta_2 < +\infty$, *and (ii)* $|\dot{u}^*| \leq \varrho$, $\exists 0 \leq \varrho < +\infty$, *the tracking error of the system exponentially converges toward or stays within the error bound* $\varrho\sqrt{\eta_2}/(\omega\lambda_1\lambda_2\gamma\eta_1 m_c)$ *with* $\omega \in (0, 1)$.
- *For the case of* $\alpha_5 = 0$ *(i.e., the DBZ case), the tracking error of the system is bounded.*

Proof It can be generalized from the proofs of Theorems 9.3, 9.4, and 9.5. □

Remark 9.3 As seen from the above descriptions and proofs of Theorems 9.3, 9.4, 9.5, and 9.6, the boundedness of tracking error has been discussed and investigated adequately, instead of the limit of tracking error. In addition, the computer simulations will be conducted in the ensuing section to verify the corresponding theoretical results (i.e., the boundedness of tracking error in the cart path tracking control of the IPC system). It is worth pointing out that, when the mass of the pendulum rod is nonzero, we have the same analyses of convergence performance for the cart path tracking control of the IPC system equipped with controllers of z2g0 and z2g1 types shown in Table 9.3, respectively. Moreover, the corresponding simulation results of cart path tracking control for the IPC system (where the rod is nonzero mass) equipped with controllers of z2g0 and z2g1 types, respectively, are shown in the Appendix, which further substantiates the robustness of the presented design method as well as the derived results.

9.5 Simulation, Verification and Comparison

In this section, three different desired trajectories are presented and then tracked by IPC system (9.2) equipped with controllers of z2g0 and z2g1 types, respectively, shown in Table 9.2. Simulation results substantiate the feasibility and efficacy of z2g0 and z2g1 controllers in the cart path tracking control, and the superiority of z2g1 controllers in conquering the DBZ problem. In these simulations, the running time $T = 50$ s. Besides, the initial states $x_1(0) = 0.1$, $x_2(0) = x_3(0) = x_4(0) = 0$, and the initial control value of z2g1 controllers $u(0) = 0$.

Example 9.1 This example is investigated with the desired trajectory $y_d = \cos(\pi t/10)$ and the output of IPC system (9.2) $y = x_1$. To achieve the purpose of tracking y_d, z2g0 controller (9.5) and z2g1 controller (9.7) are used in the cart path tracking control of IPC system (9.2), respectively. For simulations, related design parameters are set as $\lambda_1 = \lambda_2 = 8$ and $\gamma = 10$.

The simulation results of this example are shown in Figs. 9.3 and 9.4. For illustration and comparison, z2g0 controller (9.5) and z2g1 controller (9.7) are both exploited for the cart path tracking control of IPC system (9.2) with desired trajectory $y_d = \cos(\pi t/10)$. Specifically, Fig. 9.3 illustrates output trajectories and control inputs of IPC system (9.2) equipped with z2g0 controller (9.5) and z2g1 controller (9.7), respectively. On the one hand, the tracking processes by using controllers (9.5) and (9.7) are respectively shown in Fig. 9.3a, b, from which we can find that z2g0 and z2g1 controllers both achieve the tracking-control purpose within 1 s. On the other hand, Fig. 9.3c, d illustrate the control inputs for IPC system (9.2) by using controllers (9.5) and (9.7). It is worth pointing out that the control input of z2g1 controller (9.7) is more reasonable and acceptable in practical applications, because control values of z2g1 controller (9.7) are smaller than those of z2g0 controller (9.5) and have no peaks during the tracking process. Furthermore, Fig. 9.4 shows the tracking errors of IPC system (9.2) equipped with z2g0 controller (9.5) and z2g1 controller (9.7), respectively, with tracking error $e = y - y_d$. From Fig. 9.4a, b, it is readily found that the tracking errors of z2g0 and z2g1 controllers both rapidly approach zero. For further illustration, Fig. 9.4c, d show the tracking errors of z2g0 controller (9.5) and z2g1 controller (9.7) in the logarithmic scale, respectively. From a viewpoint of more accurate control, it seems that z2g0 controller (9.5) is a better choice. Note that the tracking error of z2g1 controller (9.7) is small enough for general practical applications with the magnitude of steady-state tracking errors being 10^{-3}, as shown in Fig. 9.4d. In summary, z2g0 controller (9.5) and z2g1 controller (9.7) are both feasible and effective for the cart path tracking control of IPC system (9.2) with desired trajectory $y_d = \cos(\pi t/10)$ and output $y = x_1$.

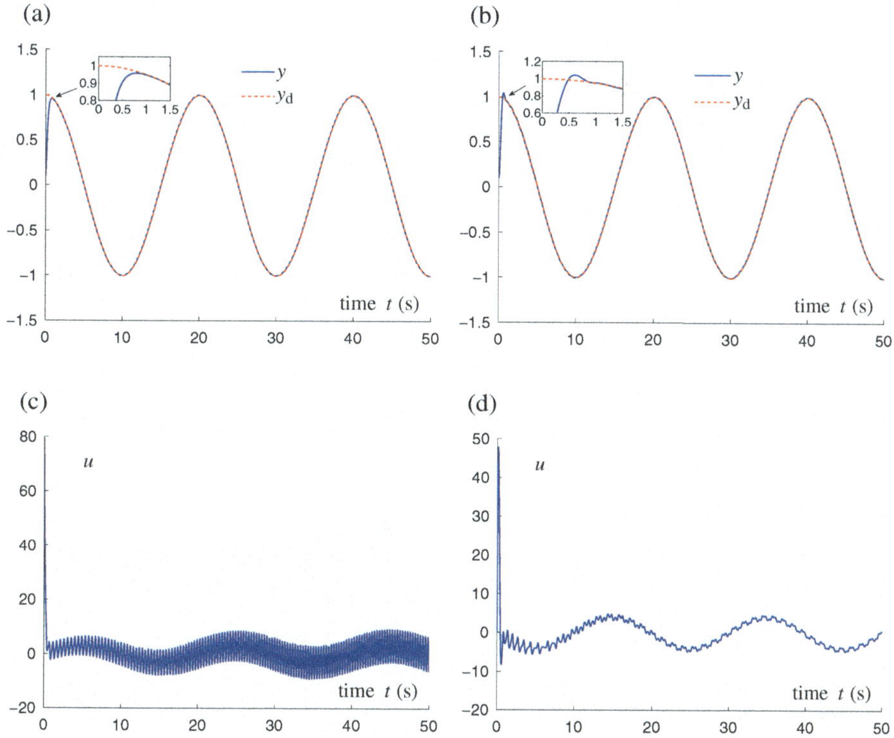

Fig. 9.3 Output trajectories and control inputs of IPC system (9.2) equipped with z2g0 controller (9.5) and z2g1 controller (9.7), respectively, for desired trajectory $y_\mathrm{d} = \cos(\pi t/10)$. (**a**) Output trajectory with z2g0 controller (9.5) and desired trajectory. (**b**) Output trajectory with z2g1 controller (9.7) and desired trajectory. (**c**) Control input with z2g0 controller (9.5). (**d**) Control input with z2g1 controller (9.7)

Example 9.2 For this example, we consider and investigate a complex output of IPC system (9.2), i.e., $y = x_1^2 + x_1$. Without loss of generality, the desired trajectory is set as $y_\mathrm{d} = \sin(t)\exp(-t/5) + 0.12$. In this simulation example, related design parameters are set as $\lambda_1 = \lambda_2 = 10$ and $\gamma = 50$. Then, z2g0 controller (9.6) and z2g1 controller (9.8) are exploited for the cart path tracking control of IPC system (9.2), respectively.

The simulation results of this example are illustrated in Figs. 9.5 and 9.6. Specifically, Fig. 9.5 shows output trajectories and tracking errors of IPC system (9.2) equipped with z2g0 controller (9.6) and z2g1 controller (9.8), respectively. Note that the tracking-control process stops when IPC system (9.2) equipped with z2g0 controller (9.6) encounters the zero-crossing DBZ problem, as shown in Fig. 9.5a. By contrast, the tracking-control process runs uninterruptedly in Fig. 9.5b for IPC system (9.2) equipped with z2g1 controller (9.8). As seen from the tracking error shown in Fig. 9.5d, IPC system (9.2) equipped with z2g1 controller (9.8) achieves the tracking-control purpose, with the magnitude of steady-state tracking error being

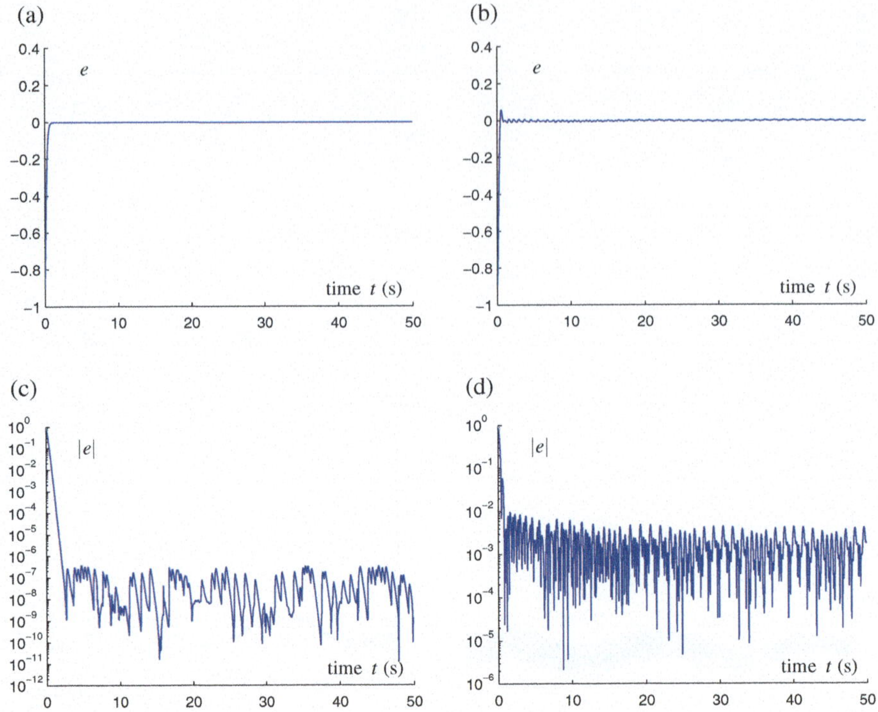

Fig. 9.4 Tracking errors of IPC system (9.2) equipped with z2g0 controller (9.5) and z2g1 controller (9.7), respectively, for desired trajectory $y_d = \cos(\pi t/10)$. (**a**) Tracking error with z2g0 controller (9.5). (**b**) Tracking error with z2g1 controller (9.7). (**c**) Order of $|e|$ with z2g0 controller (9.5). (**d**) Order of $|e|$ with z2g1 controller (9.7)

10^{-4}. For further illustration, Fig. 9.6 shows how z2g0 controller (9.6) fails and how z2g1 controller (9.8) works. Specifically, from the formulations of these two controllers (i.e., controllers (9.6) and (9.8) listed in Table 9.2), when $\alpha_5 = 0$ holds true, IPC system (9.2) equipped with z2g0 controller (9.6) encounters the DBZ problem. That is, the control value of z2g0 controller (9.6) is infinite, which leads to crash of IPC system (9.2) in practice. However, in the case of using z2g1 controller (9.8), the DBZ problem is successfully solved, as shown in Fig. 9.6b. Furthermore, as seen from Fig. 9.6c, d, IPC system (9.2) equipped with z2g0 controller (9.6) fails even in the first DBZ point, i.e., the DBZ point denoted by DBZ1 in Fig. 9.6c, while IPC system (9.2) equipped with z2g1 controller (9.8) conquers two DBZ points, i.e., the DBZ points denoted by DBZ1 and DBZ2, respectively, in Fig. 9.6d. The comparative simulation results validate the efficacy of z2g1 controller (9.8) in conquering the DBZ problem for tracking control of IPC system (9.2).

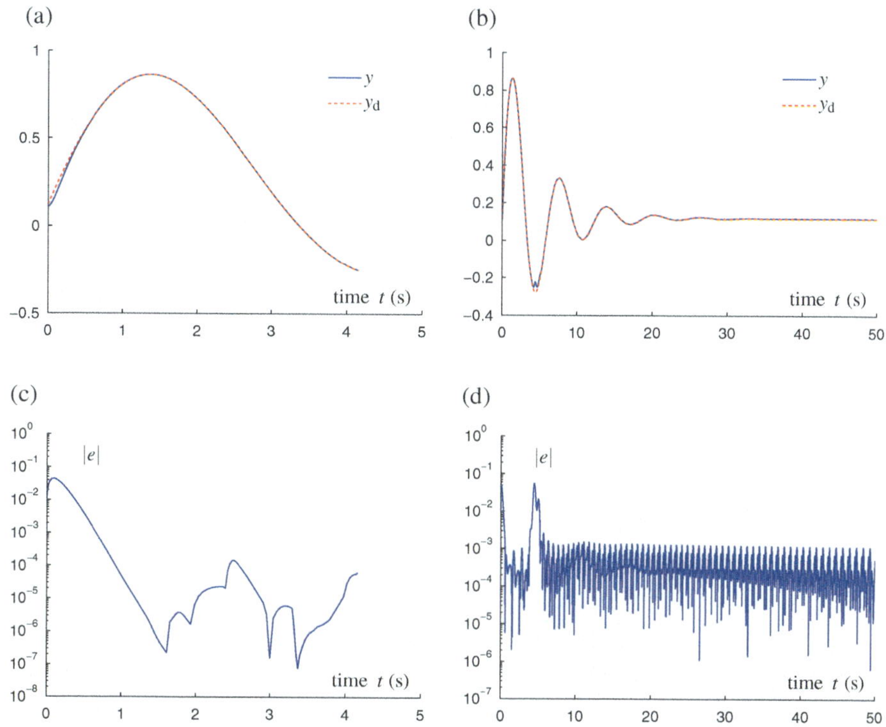

Fig. 9.5 Output trajectories and absolute tracking errors of IPC system (9.2) equipped with z2g0 controller (9.6) and z2g1 controller (9.8), respectively, for desired trajectory $y_d = \sin(t)\exp(-t/5) + 0.12$. (**a**) Output trajectory with z2g0 controller (9.6) and desired trajectory. (**b**) Output trajectory with z2g1 controller (9.8) and desired trajectory. (**c**) Order of $|e|$ with z2g0 controller (9.6). (**d**) Order of $|e|$ with z2g1 controller (9.8)

Example 9.3 For further verification of DBZ-conquering z2g1 controllers, it is worth investigating the efficacy of z2g1 controllers for the cart path tracking control with other different desired trajectories. This example further illustrates the cart path tracking control of IPC system (9.2) equipped with z2g1 controller (9.8) in conquering the DBZ problem, where the desired trajectory is $y_d = \sin(t)\cos(t) + 0.25$. For comparison, z2g0 controller (9.6) is also used for the cart path tracking control of IPC system (9.2). In addition, related design parameters are set as $\lambda_1 = \lambda_2 = 15$ and $\gamma = 150$.

As shown in Figs. 9.7 and 9.8, the efficacy of z2g1 controller (9.8) is further verified for the cart path tracking control in conquering the DBZ problem. On the one hand, as seen from Fig. 9.7, in the case of using z2g0 controller (9.6), the tracking-control process is terminated as time t approaches around 2.3554 s in Fig. 9.7a, c. By contrast, as shown in Fig. 9.7b, d, the tracking-control process is successful and smooth by using z2g1 controller (9.8), with the magnitude of steady-state tracking error being 10^{-4}. On the other hand, from Fig. 9.8, it is readily found that, when α_5 approaches zero, the control value of z2g0 controller (9.6) is too

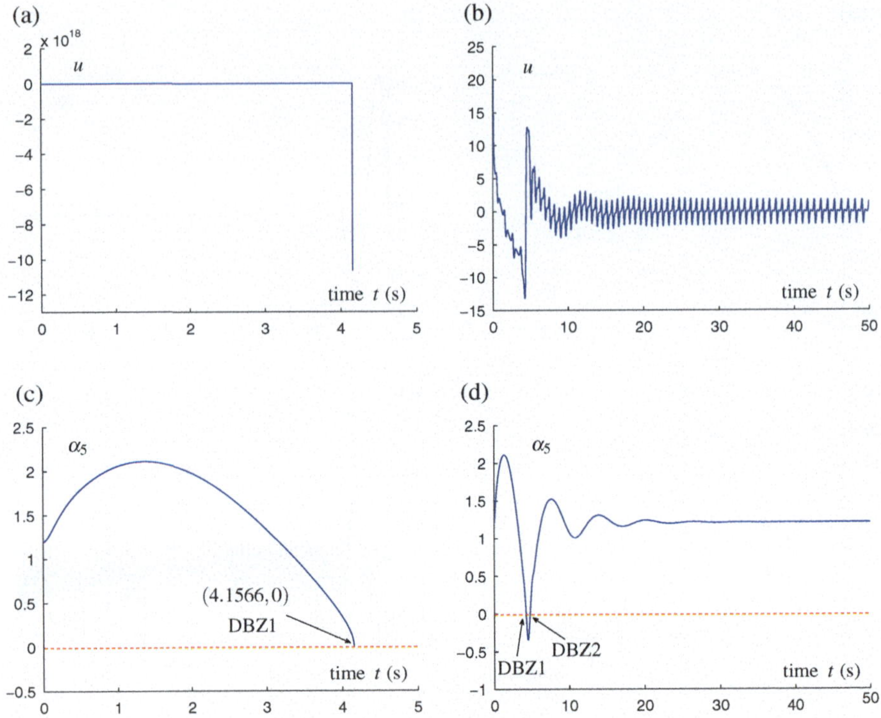

Fig. 9.6 Control inputs and trajectories of denominator α_5 of IPC system (9.2) equipped with z2g0 controller (9.6) and z2g1 controller (9.8), respectively, for desired trajectory $y_\mathrm{d} = \sin(t)\exp(-t/5) + 0.12$. (**a**) Control input with z2g0 controller (9.6). (**b**) Control input with z2g1 controller (9.8). (**c**) Trajectory of α_5 with z2g0 controller (9.6). (**d**) Trajectory of α_5 with z2g1 controller (9.8)

large to be implemented. Thus, IPC system (9.2) equipped with z2g0 controller (9.6) fails to track the desired trajectory even in the first DBZ point, i.e., the DBZ point denoted by DBZ1 in Fig. 9.8c. However, IPC system (9.2) equipped with z2g1 controller (9.8) successfully tracks the desired trajectory by conquering all DBZ points (including the DBZ point DBZ1), as shown in Fig. 9.8d. It is worth pointing out that, for other different desired trajectories, simulation results also indicate the same conclusion that z2g1 controllers can conquer the DBZ problem, which are omitted here due to the results' similarity. In summary, the efficacy of DBZ-conquering z2g1 controllers has been well verified.

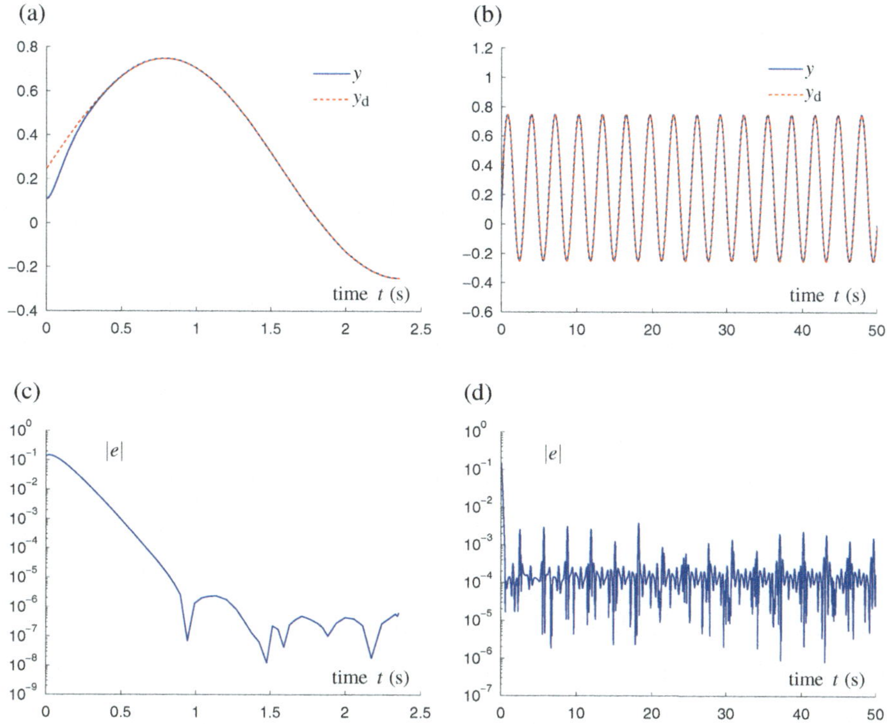

Fig. 9.7 Output trajectories and absolute tracking errors of IPC system (9.2) equipped with z2g0 controller (9.6) and z2g1 controller (9.8), respectively, for desired trajectory $y_{\mathrm{d}} = \sin(t)\cos(t) + 0.25$. (**a**) Output trajectory with z2g0 controller (9.6) and desired trajectory. (**b**) Output trajectory with z2g1 controller (9.8) and desired trajectory. (**c**) Order of $|e|$ with z2g0 controller (9.6). (**d**) Order of $|e|$ with z2g1 controller (9.8)

9.6 Chapter Summary

The cart path tracking control of the IPC system is considered and investigated in this chapter. As known, the IPC system is a typical control model due to its inherently unstable and nonlinear characteristics. The investigations on the cart path tracking control of the IPC system can promote the development and study of autonomous robotics and intelligent vehicles. For achieving the cart path tracking control of IPC system (9.2), based on the ZG method, the controllers of z2g0 and z2g1 types have been designed and investigated. Furthermore, both the z2g0 and z2g1 controllers can fulfill the tracking-control task, and the z2g1 controllers can conquer the DBZ problem. Theoretical results and analyses have been presented to guarantee the global and exponential convergence performance of both the z2g0 and z2g1 controllers. The corresponding computer simulations

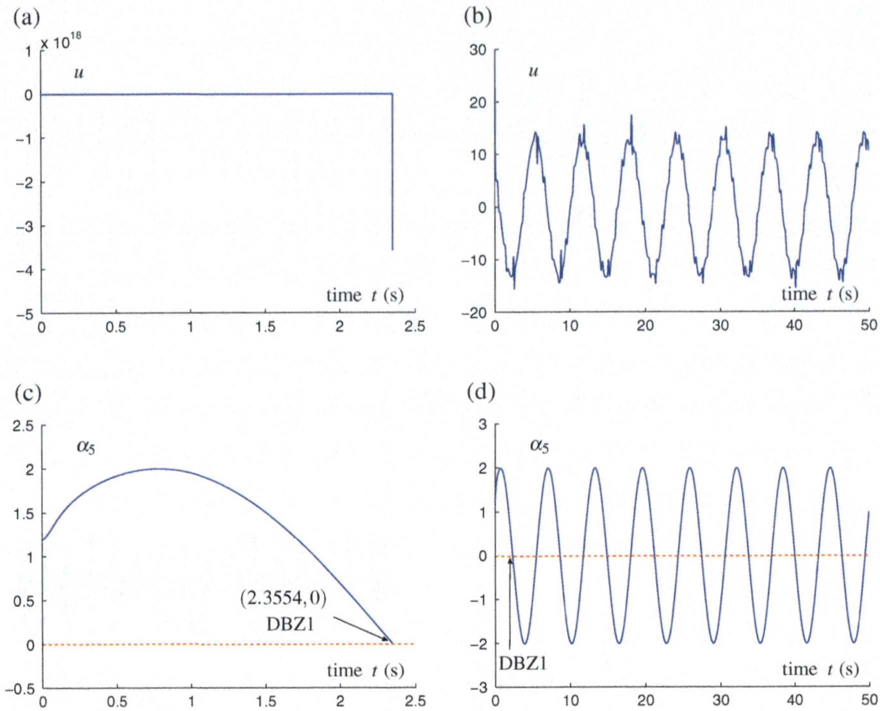

Fig. 9.8 Control inputs and trajectories of denominator α_5 for IPC system (9.2) equipped with z2g0 controller (9.6) and z2g1 controller (9.8), respectively, for desired trajectory $y_d = \sin(t)\cos(t) + 0.25$. (**a**) Control input with z2g0 controller (9.6). (**b**) Control input with z2g1 controller (9.8). (**c**) Trajectory of α_5 with z2g0 controller (9.6). (**d**) Trajectory of α_5 with z2g1 controller (9.8)

have further substantiated the feasibility and efficacy of both the z2g0 and z2g1 controllers for the cart path tracking control of IPC system (9.2). More importantly, the comparative simulation results have illustrated the superiority of the z2g1 controllers for conquering the DBZ problem.

Appendix

In the Appendix, simulation results are further presented to show the robustness of the ZG design method. Without loss of generality, this example is investigated with the desired trajectory $y_d = \cos(\pi t/10)$ and the output of the IPC system $y = x_1$ (note that the IPC system expressed in Remark 9.1 is used here). To achieve the purpose of tracking y_d, z2g0 controller and z2g1 controller for explicit tracking control (shown in Table 9.3) are applied to the cart path tracking control of the IPC system, respectively. For simulations, the related design parameters are set as $\lambda_1 = \lambda_2 = 8$ and $\gamma = 10$. In addition, the mass of pendulum rod $m_r = 0.1$kg, the initial states $x_1(0) = 0.1$ and $x_2(0) = x_3(0) = x_4(0) = 0$, the initial control input of z2g1 controllers $u(0) = 0$, as well as the running time $T = 50$ s.

The simulation results of this example are shown in Fig. 9.9. As seen from the figure, the desired trajectory is well tracked by respectively using z2g0 controller or z2g1 controller for explicit tracking control shown in Table 9.3, when the rod of the pendulum has nonzero mass. Besides, the tracking errors are small enough for general practical applications, as shown in Fig. 9.9e, f. Note that, compared with the simulation results shown in Example 9.1, the robustness of the ZG controllers (as well as the ZG design method) is well substantiated.

Moreover, the tracking errors with different values of design parameters (i.e., different values of λ_1, λ_2 and γ) for the cart path tracking control of the IPC system are shown in Fig. 9.10. From this figure, we can draw the conclusion that the boundary of tracking error is smaller when the values of design parameters become larger, which coincides well with the theoretical analyses in Sect. 9.4.

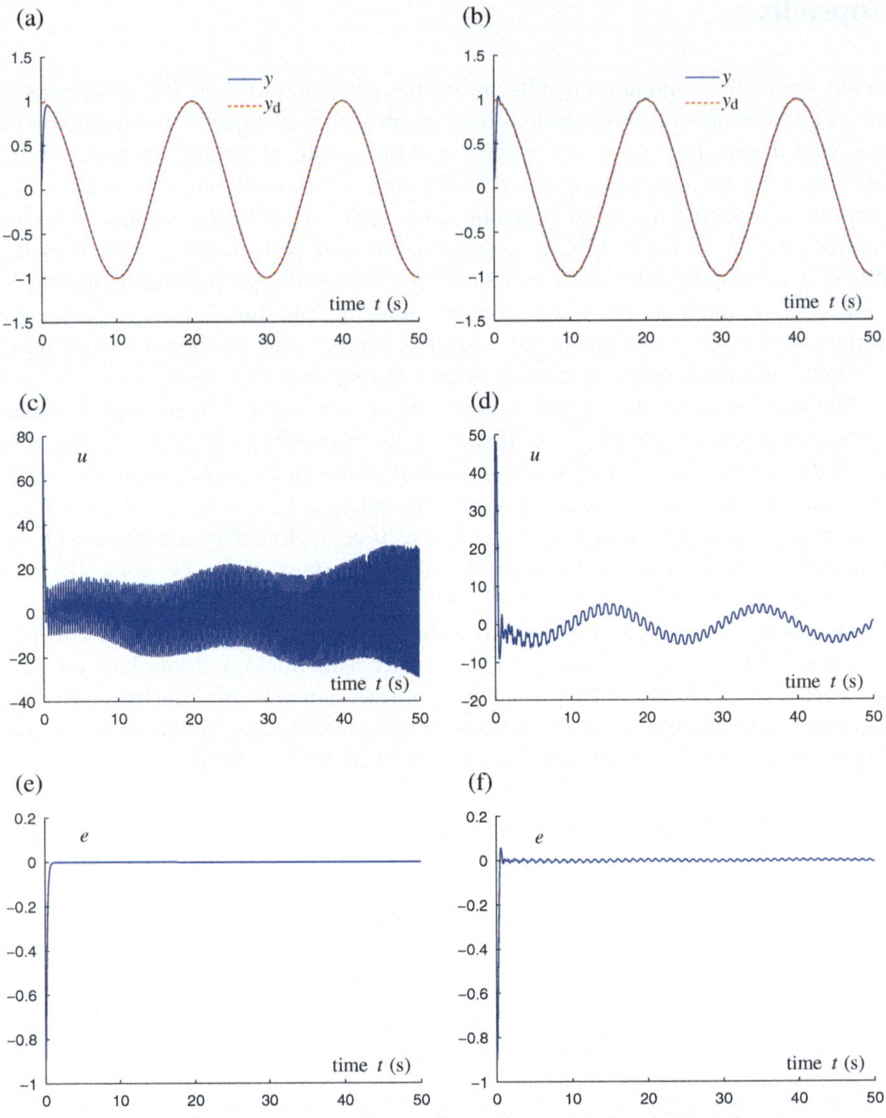

Fig. 9.9 Output trajectories, control inputs and tracking errors of IPC system in Remark 9.1 respectively equipped with z2g0 controller and z2g1 controller for explicit tracking control, shown in Table 9.3, for desired trajectory $y_\mathrm{d} = \cos(\pi t/10)$. (**a**) Output trajectory with z2g0 controller and desired trajectory. (**b**) Output trajectory with z2g1 controller and desired trajectory. (**c**) Control input with z2g0 controller. (**d**) Control input with z2g1 controller. (**e**) Tracking error with z2g0 controller. (**f**) Tracking error with z2g1 controller

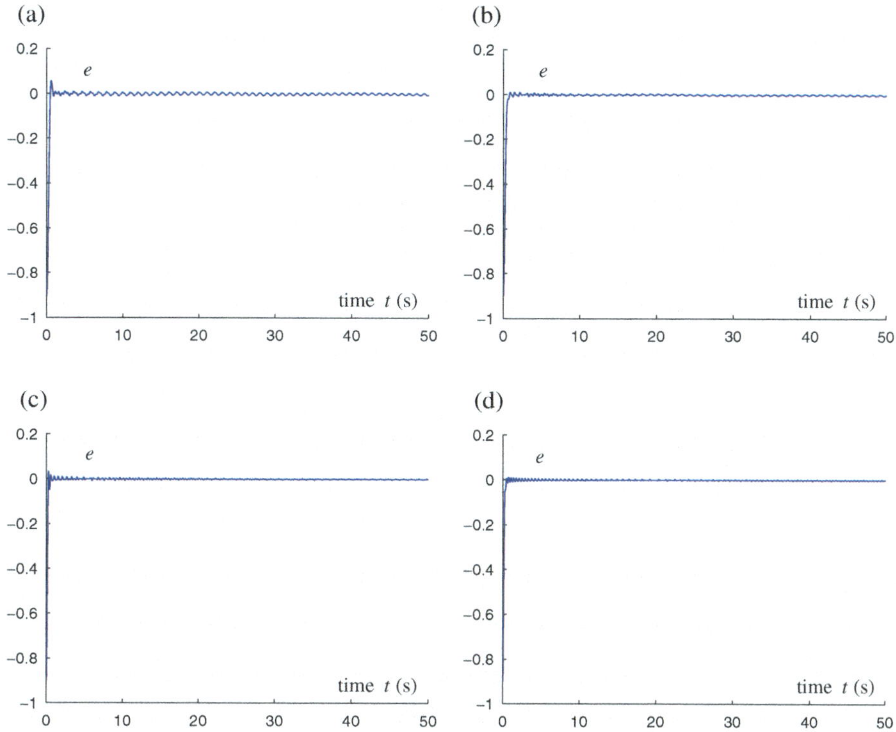

Fig. 9.10 Tracking errors with different values of design parameters for IPC system in Remark 9.1 equipped with z2g1 controller for explicit tracking control, shown in Table 9.3, for desired trajectory $y_d = \cos(\pi t/10)$. (**a**) With $\lambda_1 = \lambda_2 = 8$ and $\gamma = 10$. (**b**) With $\lambda_1 = \lambda_2 = 8$ and $\gamma = 20$. (**c**) With $\lambda_1 = \lambda_2 = 15$ and $\gamma = 20$. (**d**) With $\lambda_1 = \lambda_2 = 15$ and $\gamma = 40$

References

1. Luo R, Chang C (2010) Multisensor fusion and integration aspects of mechatronics. IEEE Ind Electron Mag 4(2):20–27
2. Meng D, Jia Y, Du J, Yu F (2011) Data-driven control for relative degree systems via iterative learning. IEEE Trans Neural Netw 22(12):2213–2225
3. Zhang Y, Yang Y, Zhao Y, Wen G (2013) Distributed finite-time tracking control for nonlinear multi-agent systems subject to external disturbances. Int J Control 86(1):29–40
4. Chen C, Chen W (1998) Robust adaptive sliding-mode control using fuzzy modeling for an inverted-pendulum system. IEEE Trans Ind Electron 45(2):297–306
5. Consolini L, Tosques M (2009) On the exact tracking of the spherical inverted pendulum via an homotopy method. Syst Control Lett 58(1):1–6
6. Huang J, Guan Z, Matsuno T, Fukuda T, Sekiyama K (2010) Sliding-mode velocity control of mobile-wheeled inverted-pendulum systems. IEEE Trans Robot 26(4):750–758
7. Lozano R, Fantoni I, Block D (2000) Stabilization of the inverted pendulum around its homoclinic orbit. Syst Control Lett 40(3):197–204
8. Ilchmann A, Ryan E, Townsend P (2006) Tracking control with prescribed transient behaviour for systems of known relative degree. Syst Control Lett 55(5):396–406

9. Meng D, Jia Y, Du J, Yu F (2012) Tracking control over a finite interval for multi-agent systems with a time-varying reference trajectory. Syst Control Lett 61(7):807–818
10. Do KK (2013) Global tracking control of underactuated ODINs in three-dimensional space. Int J Control 86(2):183–196
11. Meng D, Jia Y, Du J, Yu F (2013) Tracking algorithms for multiagent systems. IEEE Trans Neural Netw Learn Syst 24(10):1660–1676
12. Zhang Y, Yu X, Yin Y, Peng C, Fan Z (2014) Singularity-conquering ZG controllers of z2g1 type for tracking control of the IPC system. Int J Control 87(9):1729–1746
13. Isidori A (1989) Nonlinear control systems: an introduction, 2nd edn. Springer, New York
14. El-Farra N, Mhaskar P, Christofides P (2005) Output feedback control of switched nonlinear systems using multiple Lyapunov functions. Syst Control Lett 54(12):1163–1182
15. Ping Z (2013) Tracking problems of a spherical inverted pendulum via neural network enhanced design. Neurocomputing 106(6):137–147
16. Zhang Y, Xiao L, Xiao Z, Mao M (2015) Zeroing dynamics, gradient dynamics, and Newton iterations. CRC Press, Boca Raton
17. Zhang Y, Yin Y, Wu H, Guo D (2012) Zhang dynamics and gradient dynamics with tracking-control application. In: Proceedings of the 5th international symposium on computational intelligence and design, pp 235–238
18. Zhang Y, Liu J, Yin Y, Guo D, Luo F (2013) Zhang-gradient tracking controllers of Z1G0 and Z1G1 types for time-invariant linear systems. In: Proceedings of the 2nd international conference on computer science and network technology, pp 146–150
19. Zhang Y, Wang J (2001) Recurrent neural networks for nonlinear output regulation. Automatica 37(8):1161–1173
20. Wang JJ (2011) Simulation studies of inverted pendulum based on PID controllers. Simul Model Pract Theory 19(1):440–449
21. Prasad LB, Tyagi B, Gupta HO (2014) Optimal control of nonlinear inverted pendulum system using PID controller and LQR: performance analysis without and with disturbance input. Int J Autom Comput 11(6):661–670
22. Zhang Y, Qiu B, Liao B, Yang Z (2017) Control of pendulum tracking (including swinging up) of IPC system using zeroing-gradient method. Nonlinear Dyn 89(1):1–25
23. Duan C, Wu F (2012) Output-feedback control for switched linear systems subject to actuator saturation. Int J Control 85(10):1532–1545
24. Dorleansa P, Massieua JF, Ahmed-Alia T (2011) High-gain observer design with sampled measurements: application to inverted pendulum. Int J Control 84(4):801–807
25. Zhang Y, Mu B, Zheng H (2013) Link between and comparison and combination of Zhang neural network and quasi-Newton BFGS method for time-varying quadratic minimization. IEEE Trans Cybern 43(2):490–503
26. Zhang Z, Zhang Y (2013) Design and experimentation of accelerationlevel drift-free scheme aided by two recurrent neural networks. IET Control Theory Appl 7(1):25–42
27. Abramowitz M, Stegun IA (1972) Handbook of mathematical functions with formulas, graphs, and mathematical tables. Dover Publications, New York
28. Chu S, Metcalf F (1967) On Gronwall's inequality. Proc Amer Math Soc 18(3):439–440

Chapter 10
Pendulum Tracking Control of IPC System

Abstract The pendulum control of the IPC system is one of the most important issues in nonlinear control theory and has been widely investigated. Nevertheless, the control of pendulum tracking and swinging up has often been addressed separately. In this chapter, by applying the ZG method, two different types of tracking controllers (termed z2g0 controller and z2g1 controller, respectively) are designed for the IPC system. Importantly, the z2g1 controller not only realizes the simultaneous control of pendulum swinging up and pendulum angle tracking, but also conquers the DBZ problem elegantly without using any switching strategy. Besides, corresponding theoretical analyses on the convergence performance of z2g0 and z2g1 controllers are provided. Moreover, the boundedness of both control input u and its derivative $\dot{u}$ of the z2g1 controller is investigated and proven. Computer simulations with three illustrative examples are conducted to show the efficacy of z2g0 and z2g1 controllers for the pendulum tracking control of the IPC system. In particular, comparative simulation results substantiate the superiority of the z2g1 controller for the control of pendulum tracking (including swinging up) of the IPC system in conquering the DBZ problem.

10.1 Introduction

In recent years, the topic of nonlinear control has attracted increasing attention and research enthusiasm. This is because: on the one hand, the advent of microprocessors with high performance has made the implementation of nonlinear controllers become relatively simple; on the other hand, modern technologies (such as high-accuracy robots and high-performance aircrafts) are demanding control systems with more stringent design specifications [1, 2]. Many researchers have devoted much effort to the development and applications of nonlinear control methods [1–3]. Because of the inherently nonlinear, unstable, and underactuated characteristics, various types of inverted pendulum models have been widely developed and investigated in both academia and industry [4–7]. Being a classic control example, the IPC system has been used as a benchmark to test nonlinear control methods

Y. Zhang et al., *Zhang-Gradient Control*,
https://doi.org/10.1007/978-981-15-8257-8_10

[4–6]. Therefore, the research of control methods for the IPC system has important theoretical and practical significance.

According to the control purposes of inverted pendulum, the control of the IPC system can be generally divided into three aspects, i.e., swing-up control [8, 9], stabilization control [10, 11], and tracking control [12, 13]. Specifically, the swing-up control is basically used to swing up the pendulum from the stable pendant position toward the unstable upward position; the stabilization (or say, balance) control is to maintain the pendulum at its upright position; and the tracking control is to achieve the purpose that the cart or the pendulum can track a desired trajectory, which is often more difficult to realize than the balance control. Owing to the important roles of the IPC system, many control methods have been put forward by researchers. For example, the dynamics of an inverted pendulum with delayed feedback control has been studied in [4]. In [5], a Lyapunov-based controller has been developed to stabilize the inverted pendulum cart system. In [14], fuzzy controllers have been presented for the stabilization control of inverted pendulum systems. In addition, Mazenc and Praly [15] have presented a control law based on the technique consisting of adding integrators to handle the control problems of the IPC system. It is worth pointing out that, for almost all of the aforementioned methods, the initial pendulum angles of the IPC system are all assumed to be above the horizontal position or even located near the upright position with a small angle deviation from the vertical line, which means that the swing-up part has not been included in those control schemes. That is, the problem of getting into the vertically upward region, i.e., the swinging up, has typically been considered and investigated separately [8, 9, 16]. On the other hand, a few literatures [6, 10, 11] have considered such a combined problem, but they mainly focus on addressing the swing-up control and the stabilization control of the IPC system, and often require complicated switching between swinging up and control around the upright position. To the best of the authors' knowledge, the investigation on the simultaneous control of pendulum swinging up and pendulum angle tracking has rarely been studied before, which is exactly the main research motivation of this work.

In this chapter [7], by applying the ZG method, two tracking controllers are then designed for the pendulum control of the IPC system. According to the numbers of times of utilizing ZD and GD methods, such two controllers are referred to as z2g0 controller and z2g1 controller, respectively. By employing the ZD method twice and without using the GD method, the z2g0 controller is thus obtained. By contrast, the z2g1 controller is developed by using the ZD method twice and the GD method once. Besides, the control objective of this chapter [7] is to swing up the pendulum from the stable pendant position to the unstable upright position, and then let the pendulum track a desired trajectory effectively. As we know, there usually exists a DBZ problem when the pendulum is horizontal. This DBZ problem would directly lead to the failure of swinging up for the conventional controllers, which thus makes the control process complicated to realize. To address the DBZ problem, a global stabilization strategy for an inverted pendulum has been presented in [11], which uses actuator saturation to handle the DBZ point, and switches the reference position to realize the global stabilization of the IPC system.

However, though the IPC system with DBZ problem can be controlled by using the stabilization control, that approach may cost much in terms of implementation and complicate the stability analysis. Compared with the stabilization control strategy, the DBZ-conquering ZG controller presented in this chapter [7] not only realizes the simultaneous control of pendulum swinging up and pendulum angle tracking, but also conquers the DBZ problem elegantly in a unified form and without using any switching strategy. Besides, it is worth pointing out here that this chapter [7] mainly focuses on designing and investigating a ZG controller to achieve the control of pendulum tracking (including swinging up) instead of the stabilization control.

10.2 Design of Controllers

As presented in the previous chapter, the corresponding mathematical model of IPC system has been provided. This chapter [7] aims at developing a control law that can combine swing-up control and tracking control of the pendulum for IPC system (9.2). The output of the IPC system is selected as $y = \theta = x_3$. In the section, based on the ZD and GD methods, z2g0 and z2g1 controllers are designed for the pendulum tracking control of IPC system (9.2).

10.2.1 Design of ZD Controller

In order to construct the z2g0 controller, the ZD method is exploited two times, while the GD method is not used. Specifically, the following three steps are adopted to develop the z2g0 controller.

In the first step, by following ZD method [17], the first ZF is defined as

$$z_1 = y - y_{\mathrm{d}} = x_3 - y_{\mathrm{d}}. \tag{10.1}$$

Then, the ZD design formula is employed:

$$\dot{z}_1 = -\lambda_1 z_1, \tag{10.2}$$

where the design parameter $\lambda_1 \in \mathbb{R}^+$ is used to scale the convergence rate of the ZD solution. Substituting (10.1) into (10.2), we have $\dot{x}_3 - \dot{y}_{\mathrm{d}} = -\lambda_1(x_3 - y_{\mathrm{d}})$.

In the second step, to generate a direct relationship between the output y and the input u, the second ZF is defined as $z_2 = \dot{x}_3 - \dot{y}_{\mathrm{d}} + \lambda_1(x_3 - y_{\mathrm{d}})$. Afterwards, applying the ZD design formula (i.e., $\dot{z}_2 = -\lambda_2 z_2$) once again, we obtain

$$\dot{x}_4 - \ddot{y}_{\mathrm{d}} + \lambda_1(x_4 - \dot{y}_{\mathrm{d}}) = -\lambda_2\left(x_4 - \dot{y}_{\mathrm{d}} + \lambda_1(x_3 - y_{\mathrm{d}})\right). \tag{10.3}$$

Finally, defining $\alpha_1 = \ddot{y}_\mathrm{d} + (\lambda_1 + \lambda_2)(\dot{y}_\mathrm{d} - x_4) + \lambda_1\lambda_2(y_\mathrm{d} - x_3)$, which is a function of state variables, and combining Eqs. (9.2) and (10.3), we obtain a tracking controller in the form of u:

$$\begin{cases} \alpha_1 = \ddot{y}_\mathrm{d} + (\lambda_1 + \lambda_2)(\dot{y}_\mathrm{d} - x_4) + \lambda_1\lambda_2(y_\mathrm{d} - x_3), \\ \alpha_2 = l(m_\mathrm{c} + m_\mathrm{p}\sin^2 x_3), \\ \alpha_3 = (m_\mathrm{c} + m_\mathrm{p})g\sin x_3, \\ \alpha_4 = \sigma x_2 - m_\mathrm{p} l x_4^2 \sin x_3, \\ u = \alpha_4 - \dfrac{\alpha_1\alpha_2 - \alpha_3}{\cos x_3}. \end{cases} \tag{10.4}$$

For presentation convenience, the above controller can be termed z2g0 controller, in view of the fact that the ZD method is exploited twice and without using the GD method during the controller design procedure.

Based on the above three steps, a concise z2g0 controller is obtained for the tracking control of IPC system (9.2). This controller design strategy can also be applied to many other nonlinear systems. It is noted that, similar to other conventional controllers, the z2g0 controller has a fundamental drawback, i.e., (10.4) has DBZ points at $x_3 = \theta = (i + 1/2)\pi$ with $i = 0, \pm 1, \pm 2, \cdots$. In other words, when the pendulum is horizontal, controller (10.4) will collapse. This means that z2g0 controller (10.4) cannot achieve the swing-up control.

10.2.2 Design of ZG Controller

To remedy the basic drawback (i.e., the DBZ problem) of z2g0 controller (10.4), we can adopt the ZD and GD methods in a unified manner and then present the z2g1 controller for the pendulum control of IPC system (9.2).

Specifically, given z2g0 controller (10.4), we first define $\phi = \cos x_3(u - \alpha_4) + (\alpha_1\alpha_2 - \alpha_3)$. Evidently, in order to realize the pendulum control, ϕ should theoretically be zero. Subsequently, EF $\epsilon = \phi^2/2$ is constructed accordingly. Finally, employing the GD design formula, i.e., $\dot{u} = -\gamma\partial\epsilon/\partial u$, with the design parameter $\gamma \in \mathbb{R}^+$ used to scale the convergence rate of the GD solution, we obtain a tracking controller in the form of $\dot{u}$:

$$\begin{cases} \alpha_1 = \ddot{y}_\mathrm{d} + (\lambda_1 + \lambda_2)(\dot{y}_\mathrm{d} - x_4) + \lambda_1\lambda_2(y_\mathrm{d} - x_3), \\ \alpha_2 = l(m_\mathrm{c} + m_\mathrm{p}\sin^2 x_3), \\ \alpha_3 = (m_\mathrm{c} + m_\mathrm{p})g\sin x_3, \\ \alpha_4 = \sigma x_2 - m_\mathrm{p} l x_4^2 \sin x_3, \\ \dot{u} = -\gamma\cos x_3\phi = -\gamma\cos x_3\left(\cos x_3(u - \alpha_4) + (\alpha_1\alpha_2 - \alpha_3)\right), \end{cases} \tag{10.5}$$

for the pendulum control of IPC system (9.2). Evidently, such a controller is designed by combining the ZD and GD methods, i.e., containing two procedures. Specifically, the ZD method is used twice and the GD method is used once. Thus, controller (10.5) can be termed z2g1 controller for comparative purposes. Intuitively, z2g1 controller (10.5) has no division operation, and thus has no DBZ point. That is, the z2g1 controller can conquer the DBZ problem, which means that the pendulum can pass the horizontal position. As a result, z2g1 controller (10.5) can achieve the pendulum tracking control and swing-up control simultaneously.

Remark 10.1 As mentioned above, coefficients λ_1, λ_2 and γ are used as the design parameters. It is evident that the ZG controllers are designed with only one restriction, i.e., design parameters being larger than zero. Basically speaking, the control of a ZG controller can be effective, provided that such a restriction is satisfied. Moreover, the basic principle on the choice of design parameters can be outlined as follows. In general, if the values of design parameters become larger, the convergence rate of control process is faster and the tracking error is smaller. Thus, the control effectiveness is directly proportional to the values of design parameters; in other words, the tracking error is inversely proportional to the values of design parameters. Thus, the values of design parameters need to be set sufficiently large or selected appropriately for simulative purposes [17, 18]. In the numerical tests for a specific example, we can try different values of the design parameters to show their effects on the tracking performance and then to determine their optimal values or intervals for usage. Besides, in the previous work [18] as well as in this chapter [7], the DBZ-conquering property of the GD method has been theoretically analyzed and numerically substantiated. Theoretically speaking, when the value of the GD design parameter tends to infinity, the upper bound of the tracking error of the GD-aided controller would converge toward zero. On the other hand, from Theorem 10.1 in the ensuing Sect. 10.3, we know that the tracking error of the IPC system equipped with the z2g0 controller exponentially converges to zero. Therefore, it can be concluded that, even though the control (10.4) (i.e., the z2g0 controller) is modified as Eq. (10.5) by using the GD method, the characteristics of control (10.4) can still be kept up, with the GD design parameter being sufficiently large.

10.3 Convergence Performance Analyses

This section provides the theoretical analyses on the convergence performance of z2g0 controller (10.4) and z2g1 controller (10.5) for IPC system (9.2). Specifically, the following theoretical result on the convergence performance of the z2g0 controller is firstly presented. In what follows, $e = z_1 = y - y_\mathrm{d} = x_3 - y_\mathrm{d} = \theta - y_\mathrm{d}$ represents the tracking error.

Theorem 10.1 *For smooth and bounded desired trajectory* y_d, *starting with bounded initial state* $\mathbf{x}(0) = [x_1(0), x_2(0), x_3(0), x_4(0)]^\mathrm{T} \in \mathbb{R}^4$, *the tracking error* *of IPC system (9.2) equipped with z2g0 controller (10.4) exponentially converges to zero on a large scale, provided that* $\cos x_3 \neq 0$, $\forall t \in [0, +\infty)$.

Proof It can be generalized from the proof of Theorem 9.1. □

Next, the convergence performance of z2g1 controller (10.5) for the pendulum control of IPC system (9.2) is provided. Moreover, based on controllers (10.4) and (10.5), it can be found that the desired time-varying solution of z2g1 controller (10.5) is $u^* = \alpha_4 - (\alpha_1\alpha_2 - \alpha_3)/\cos x_3$. Then, for z2g1 controller (10.5), the following theoretical results on its convergence performance are obtained.

Theorem 10.2 *Consider IPC system (9.2) equipped with z2g1 controller (10.5) for smooth and bounded desired trajectory* y_d. *Starting with bounded initial state* $\mathbf{x}(0) \in \mathbb{R}^4$ *and control input* $u(0) \in \mathbb{R}$, *the following results are achieved on a large scale for the pendulum tracking control of the system.*

- *For the case of* $\cos x_3 \neq 0$ *(i.e., the non-DBZ case), the tracking error of the system converges toward or stays within the error bound* $\varrho\sqrt{\eta_2}/(\lambda_1\lambda_2\gamma\eta_1 l m_\mathrm{c})$, *provided that (i)* $\eta_1 \leq \cos^2 x_3 \leq \eta_2$, $\exists 0 < \eta_1 \leq \eta_2 \leq 1$, *and (ii)* $|\dot{u}^*| = |\mathrm{d}u^*/\mathrm{d}t| \leq \varrho$, $\exists 0 \leq \varrho < +\infty$.
- *For the case of* $\cos x_3 = 0$ *(i.e., the DBZ case), the tracking error of the system is bounded.*

Proof For the case of $\cos x_3 \neq 0$ (i.e., the non-DBZ case), we define a solution error of z2g1 controller (10.5) as $\psi = u - u^* = u - \alpha_4 + (\alpha_1\alpha_2 - \alpha_3)/\cos x_3$. Evidently, it can be further derived that $\dot{u} = -\gamma \cos x_3 (\cos x_3(u - \alpha_4) + (\alpha_1\alpha_2 - \alpha_3)) = -\gamma \cos^2 x_3 \psi$. Then, we have the derivative of ψ as $\dot{\psi} = \dot{u} - \dot{u}^* = -\gamma \cos^2 x_3 \psi - \dot{u}^*$. Defining a Lyapunov function candidate $L = \psi^2/2 \geq 0$, we can obtain its time derivative as

$$\dot{L} = \psi\dot{\psi} = -\gamma \cos^2 x_3 \psi^2 - \psi\dot{u}^*. \tag{10.6}$$

It follows from Eq. (10.6) that $\dot{L}$ has two terms, i.e., $-\gamma \cos^2 x_3 \psi^2$ and $-\psi\dot{u}^*$. For the first term, given $\cos^2 x_3 \geq \eta_1$, $-\gamma \cos^2 x_3 \psi^2 \leq -\gamma\eta_1\psi^2$ holds true. For the second term, according to Cauchy inequality [19], we obtain $-\psi\dot{u}^* \leq |\psi||\dot{u}^*| \leq \varrho|\psi|$. Then, we have

$$\dot{L} = -\gamma \cos^2 x_3 \psi^2 - \psi\dot{u}^* \leq -\gamma\eta_1\psi^2 + \varrho|\psi| = -|\psi|(\gamma\eta_1|\psi| - \varrho). \tag{10.7}$$

During the time evolution of solution error ψ, there exist three situations for (10.7): (i) $\gamma\eta_1|\psi| - \varrho > 0$; (ii) $\gamma\eta_1|\psi| - \varrho = 0$; (iii) $\gamma\eta_1|\psi| - \varrho < 0$. The detailed analyses are presented as below.

- In the first situation (i.e., $\gamma\eta_1|\psi| - \varrho > 0$), $\dot{L} < 0$, which means that ψ tends to zero (i.e., u tends to u^*) with time.
- In the second situation (i.e., $\gamma\eta_1|\psi| - \varrho = 0$, a so-called ball surface), $\dot{L} \leq 0$, which means that ψ tends to zero (i.e., u tends to u^*) or stays on the surface with

$|\psi| = |u - u^*| = \varrho/(\gamma\eta_1)$. In other words, ψ would not go outside the ball of $\varrho/(\gamma\eta_1)$ in this situation.

- In the third situation (i.e., $\gamma\eta_1|\psi| - \varrho < 0$, inside the ball of $\varrho/(\gamma\eta_1)$), $\dot{L}$ is less than a positive constant, comprising sub-situations $\dot{L} \leq 0$ and $0 < \dot{L} \leq -|\psi|(\gamma\eta_1|\psi| - \varrho)$. If $\dot{L} \leq 0$ holds, then it returns to the second situation. Then, let us analyze the worst situation, i.e., $0 < \dot{L} \leq -|\psi|(\gamma\eta_1|\psi| - \varrho)$. In this situation, $L = \psi^2/2$ and $|\psi|$ would increase with time, which means that $\gamma\eta_1|\psi| - \varrho$ increases as well. As a result of the worst situation, there exists a certain time instant such that $\gamma\eta_1|\psi| - \varrho = 0$, which returns to the second situation, i.e., $\dot{L} \leq 0$. In brief, ψ would not go outside the ball of $\varrho/(\gamma\eta_1)$ in any situation.

Summarizing the above three situations, we conclude that the solution error ψ of z2g1 controller (10.5) is upper bounded by $\varrho/(\gamma\eta_1)$ when the solving process enters the steady state, i.e.,

$$\limsup_{t\to+\infty} |\psi| = \limsup_{t\to+\infty} |u - u^*| \leq \frac{\varrho}{\gamma\eta_1}. \tag{10.8}$$

By following state equation (9.2) and the design procedure of z2g1 controller (10.5), it can be obtained that

$$\dot{z}_2 + \lambda_2 z_2 = \dot{x}_4 - \alpha_1 = \frac{\alpha_3 + (\alpha_4 - u)\cos x_3}{\alpha_2} - \alpha_1 = -\frac{\cos x_3}{\alpha_2}\psi. \tag{10.9}$$

In addition, we have $\dot{z}_2+\lambda_2 z_2 = \ddot{\sigma}_1+(\lambda_1+\lambda_2)\dot{z}_1+\lambda_1\lambda_2 z_1$. Considering $e = z_1$ and Eq. (10.9), we further have $(\ddot{e} + (\lambda_1 + \lambda_2)\dot{e} + \lambda_1\lambda_2 e)\alpha_2/\cos x_3 = -\psi$. Therefore, it follows from (10.8) that

$$\limsup_{t\to+\infty} \left|(\ddot{e} + (\lambda_1 + \lambda_2)\dot{e} + \lambda_1\lambda_2 e)\frac{\alpha_2}{\cos x_3}\right| = \limsup_{t\to+\infty} |\psi| \leq \frac{\varrho}{\gamma\eta_1}.$$

Taking into account that $0 < lm_{\mathrm{c}} \leq \alpha_2 = l(m_{\mathrm{c}} + m_{\mathrm{p}}\sin^2 x_3) \leq l(m_{\mathrm{c}} + m_{\mathrm{p}})$ and $0 < |\cos x_3| \leq \sqrt{\eta_2}$, we obtain

$$\begin{aligned}&\limsup_{t\to+\infty} \left|(\ddot{e} + (\lambda_1 + \lambda_2)\dot{e} + \lambda_1\lambda_2 e)\frac{lm_{\mathrm{c}}}{\sqrt{\eta_2}}\right| \\ &\leq \limsup_{t\to+\infty} \left|(\ddot{e} + (\lambda_1 + \lambda_2)\dot{e} + \lambda_1\lambda_2 e)\frac{\alpha_2}{\cos x_3}\right| \leq \frac{\varrho}{\gamma\eta_1},\end{aligned}$$

which leads to

$$\limsup_{t\to+\infty} |(\ddot{e} + (\lambda_1 + \lambda_2)\dot{e} + \lambda_1\lambda_2 e)| \leq \frac{\varrho\sqrt{\eta_2}}{\gamma\eta_1 lm_{\mathrm{c}}}. \tag{10.10}$$

For an enough large time instant t_e, (10.10) reduces to

$$|(\ddot{e}(t_e) + (\lambda_1 + \lambda_2)\dot{e}(t_e) + \lambda_1\lambda_2 e(t_e))| \leq \frac{\varrho\sqrt{\eta_2}}{\gamma\eta_1 l m_c}.$$

According to Gronwall inequality [20], the above inequality can be reformulated as $|e(t_e)| \leq \varrho\sqrt{\eta_2}/(\lambda_1\lambda_2\gamma\eta_1 l m_c)$, i.e.,

$$\limsup_{t\to+\infty} |e| \leq \frac{\varrho\sqrt{\eta_2}}{\lambda_1\lambda_2\gamma\eta_1 l m_c},$$

by considering $t_e \to +\infty$.

For the case of $\cos x_3 = 0$ (i.e., the DBZ case), we can obtain $\lim_{\cos x_3 \to 0} \dot{u} = 0$, since $\dot{u} = -\gamma \cos x_3 (\cos x_3 (u - \alpha_4) + (\alpha_1\alpha_2 - \alpha_3))$. As a result, the control input at the DBZ time instant t_s is equal to that at the previous time instant t_{s-}, i.e., $u(t_s) = u(t_{s-})$. Similarly, we have $u(t_s) = u(t_{s+})$ with t_{s+} being after the DBZ time instant t_s. Note that, at the time instant t_{s-}, the control input $u(t_{s-})$ is bounded, which implies that $u(t_{s+}) = u(t_s) = u(t_{s-})$ is bounded. When the control input is bounded, the output y of IPC system (9.2) is also bounded. Then, for the bounded desired trajectory y_d, the tracking error $e = y - y_d$ is bounded at the time instants t_{s-}, t_s and t_{s+}. In other words, by means of the z2g1 controller, the DBZ problem is conquered successfully. This implies that the pendulum can pass the horizontal position and achieve the swing-up control. After getting through the DBZ point, the tracking error converges toward an error bound again, which implies that IPC system (9.2) equipped with z2g1 controller (10.5) finally conquers the DBZ problem.

By the above analyses, the proof is thus completed. □

According to Theorem 10.2, we know that the tracking error converges toward the ball of $\varrho\sqrt{\eta_2}/(\lambda_1\lambda_2\gamma\eta_1 l m_c)$ on a large scale. This means that, when the values of design parameters λ_1, λ_2 and γ are chosen to be infinite, the tracking error bound diminishes to zero. However, Theorem 10.2 shows that the tracking error is just asymptotically convergent, which may be less acceptable in applications since it requires an infinitely long time period to achieve the tracking control. Actually, by choosing a relatively loose upper bound, the tracking error can be proven to be exponentially convergent, which is presented as the following theorem.

Theorem 10.3 *Consider IPC system (9.2) equipped with z2g1 controller (10.5) for smooth and bounded desired trajectory* y_d. *Starting with bounded initial state* $\mathbf{x}(0) \in \mathbb{R}^4$ *and control input* $u(0) \in \mathbb{R}$, *the following results are achieved on a large scale for the pendulum tracking control of the system.*

- *For the case of* $\cos x_3 \neq 0$ *(i.e., the non-DBZ case), provided that (i)* $\eta_1 \leq \cos^2 x_3 \leq \eta_2$, $\exists 0 < \eta_1 \leq \eta_2 \leq 1$, *and (ii)* $|\dot{u}^*| \leq \varrho$, $\exists 0 \leq \varrho < +\infty$, *the tracking error of the system exponentially converges toward or stays within the error bound* $\varrho\sqrt{\eta_2}/(\omega\lambda_1\lambda_2\gamma\eta_1 l m_c)$, *where* $\omega \in (0, 1)$ *is a loosening parameter.*

- *For the case of* $\cos x_3 = 0$ *(i.e., the DBZ case), the tracking error of the system is bounded.*

Proof For the case of $\cos x_3 \neq 0$ (i.e., the non-DBZ case), from (10.7), we can obtain

$$\dot{L} \leq -\gamma\eta_1\psi^2 + \varrho|\psi| = -(1-\omega)\gamma\eta_1\psi^2 + (-\omega\gamma\eta_1\psi^2 + \varrho|\psi|), \tag{10.11}$$

with the loosening parameter $\omega \in (0, 1)$. Evidently, on the right side of (10.11), the first term $-(1-\omega)\gamma\eta_1\psi^2 \leq 0$ holds all the time. Besides, for the second term on the right side of (10.11), there exist two sub-situations, i.e., $-\omega\gamma\eta_1\psi^2 + \varrho|\psi| \leq 0$ and $-\omega\gamma\eta_1\psi^2 + \varrho|\psi| > 0$.

When $-\omega\gamma\eta_1\psi^2 + \varrho|\psi| \leq 0$, i.e., $|\psi| \geq \varrho/(\omega\gamma\eta_1)$, we have

$$\dot{L} \leq -(1-\omega)\gamma\eta_1\psi^2 = -2(1-\omega)\gamma\eta_1 L.$$

Solving the above inequality, we further obtain

$$\frac{1}{2}\psi^2 = L \leq L(0)\exp\left(-2(1-\omega)\gamma\eta_1 t\right) = \frac{1}{2}\psi^2(0)\exp\left(-2(1-\omega)\gamma\eta_1 t\right),$$

which leads to

$$\frac{\varrho}{\omega\gamma\eta_1} \leq |\psi| \leq |\psi(0)|\exp\left(-(1-\omega)\gamma\eta_1 t\right), \ \forall t \in [0, t_c], \tag{10.12}$$

where the convergence time $t_c = \ln\left(\omega\gamma\eta_1|\psi(0)|/\varrho\right) / \left((1-\omega)\gamma\eta_1\right)$. This means that $|\psi|$ exponentially converges toward the new ball of $\varrho/(\omega\gamma\eta_1)$.

When $-\omega\gamma\eta_1\psi^2 + \varrho|\psi| > 0$, i.e., $|\psi| < \varrho/(\omega\gamma\eta_1)$, ψ stays within the new ball of $\varrho/(\omega\gamma\eta_1)$. The analysis is similar to that of the third situation of (10.7).

In light of the above analysis, the results are summarized as follows.

- If $|\psi(0)| \geq \varrho/(\omega\gamma\eta_1)$, then

$$|\psi| \begin{cases} \leq |\psi(0)|\exp\left(-(1-\omega)\gamma\eta_1 t\right), \ \forall t \in [0, t_c], \\ \leq \varrho/(\omega\gamma\eta_1), \ \forall t \in [t_c, +\infty). \end{cases}$$

- If $|\psi(0)| \leq \varrho/(\omega\gamma\eta_1)$, then $|\psi| \leq \varrho/(\omega\gamma\eta_1)$, $\forall t \in [0, +\infty)$.

Evidently, the exponential convergence rate of the solution error is $(1-\omega)\gamma\eta_1$. Moreover, when $t \geq t_c$, $|\psi| \leq \varrho/(\omega\gamma\eta_1)$ always holds. Similar to the proof given in Theorem 10.2, the following inequality can be obtained:

$$|(\ddot{e} + (\lambda_1 + \lambda_2)\dot{e} + \lambda_1\lambda_2 e)| \leq \frac{\varrho\sqrt{\eta_2}}{\omega\gamma\eta_1 l m_c}, \ \forall t \geq t_c.$$

Similarly, based on Gronwall inequality [20], it can be concluded that $|e|$ exponentially converges toward the new error bound $\varrho\sqrt{\eta_2}/(\omega\lambda_1\lambda_2\gamma\eta_1 l m_c)$ when $t \geq t_c$.

In summary, for the case of $\cos x_3 \neq 0$, the tracking error of IPC system (9.2) equipped with z2g1 controller (10.5) exponentially converges toward or stays within the error bound $\varrho\sqrt{\eta_2}/(\omega\lambda_1\lambda_2\gamma\eta_1 l m_c)$ on a large scale.

For the case of $\cos x_3 = 0$ (i.e., the DBZ case), similar to Theorem 10.2, it can be proven that the tracking error of IPC system (9.2) equipped with z2g1 controller (10.5) is bounded. That is, the z2g1 controller can conquer the DBZ problem, thereby implying that the pendulum can pass the horizontal position and then accomplish the process of swinging up.

By the above analyses, the proof is thus completed. □

Moreover, it is worth investigating the boundedness of both control input u and its derivative $\dot{u}$ in the pendulum tracking control of IPC system (9.2) equipped with z2g1 controller (10.5). Thus, we have the following theorem.

Theorem 10.4 *Consider IPC system (9.2) equipped with z2g1 controller (10.5) for smooth and bounded desired trajectory y_d. Starting with bounded initial state $\mathbf{x}(0) \in \mathbb{R}^4$ and control input $u(0) \in \mathbb{R}$, the following results are achieved on a large scale for the pendulum tracking control of the system.*

- *For the case of $\cos x_3 \neq 0$ (i.e., the non-DBZ case), the control input u and its derivative $\dot{u}$ of the system are upper bounded, respectively, as*

$$\begin{cases} \limsup_{t\to+\infty} |\dot{u}| \leq \varrho\eta_2/\eta_1, \\ \limsup_{t\to+\infty} |u| \leq \varrho/(\gamma\eta_1) + \zeta, \end{cases}$$

  provided that (i) $\eta_1 \leq \cos^2 x_3 \leq \eta_2$, $\exists 0 < \eta_1 \leq \eta_2 \leq 1$, and (ii) $|u^*| \leq \zeta$, $\exists 0 \leq \zeta < +\infty$, $|\dot{u}^*| \leq \varrho$, $\exists 0 \leq \varrho < +\infty$.
- *For the case of $\cos x_3 = 0$ (i.e., the DBZ case), both the control input u and its derivative $\dot{u}$ of the system are bounded.*

Proof For the case of $\cos x_3 \neq 0$ (i.e., the non-DBZ case), a solution error of controller (10.5) is defined as $\psi = u - u^* = u - \alpha_4 + (\alpha_1\alpha_2 - \alpha_3)/\cos x_3$. Then, we have $e = \cos x_3\psi$, and $\dot{u} = -\gamma\cos^2 x_3\psi = -\gamma\cos x_3 e$. In view of $0 < |\cos x_3| \leq \sqrt{\eta_2}$, we have $|e| = |\cos x_3||\psi| \leq \sqrt{\eta_2}|\psi|$, and further obtain

$$|\dot{u}| = |-\gamma\cos x_3 e| = \gamma|\cos x_3||e| \leq \gamma\eta_2|\psi|.$$

In addition, from (10.8), we know that the solution error ψ of z2g1 controller (10.5) is upper bounded by $\varrho/(\gamma\eta_1)$, i.e., $\lim_{t\to+\infty}\sup|\psi| = \lim_{t\to+\infty}\sup|u - u^*| \leq \varrho/(\gamma\eta_1)$. Thus, we have

$$\limsup_{t\to+\infty} |\dot{u}| \leq \gamma\eta_2 \limsup_{t\to+\infty} |\psi| \leq \frac{\varrho\eta_2}{\eta_1}.$$

That is, the derivative of u (i.e., $\dot{u}$) is upper bounded by $\varrho\eta_2/\eta_1$.

The boundedness of the control input u is proven as below. Taking $\psi = u - u^*$ into account and according to the triangle inequality [21], we can obtain $|u|-|u^*| \leq |u - u^*| = |\psi|$. Considering that $\lim_{t\to+\infty} \sup |\psi| \leq \varrho/(\gamma\eta_1)$ and $|u^*| \leq \zeta$, we have

$$\lim_{t\to+\infty} \sup |u| \leq \frac{\varrho}{\gamma\eta_1} + \zeta.$$

From the above inequality, we can readily know that the control input u is upper bounded by $\varrho/(\gamma\eta_1) + \zeta$.

For the case of $\cos x_3 = 0$ (i.e., the DBZ case), we have $\lim_{\cos x_3\to 0} \dot{u} = 0$, in view of $\dot{u} = -\gamma \cos x_3 (\cos x_3(u - \alpha_4) + (\alpha_1\alpha_2 - \alpha_3))$. Evidently, $\dot{u}$ is bounded and tends to zero, as time t evolves to the DBZ time instant t_s. Besides, it can be readily known that the control input at the DBZ time instant t_s is equal to that at the previous time instant t_{s^-}, i.e., $u(t_s) = u(t_{s^-})$. Analogously, we have $u(t_s) = u(t_{s^+})$ with t_{s^+} being the time instant after the DBZ point. Note that, at the time instant t_{s^-}, the control input $u(t_{s^-})$ is bounded, which implies that $u(t_{s^+}) = u(t_s) = u(t_{s^-})$ is bounded. In other words, for the case of $\cos x_3 = 0$ (i.e., the DBZ case), the control input u of the IPC system is bounded. After getting through the DBZ point, the control input u and its derivative $\dot{u}$ converge toward respective bounds again (which are given in the non-DBZ case).

By the above analyses, the proof is thus completed. □

10.4 Simulation, Verification and Comparison

As presented above, for comparative purposes, two tracking controllers, i.e., z2g0 controller (10.4) and z2g1 controller (10.5), have been designed and analyzed for the pendulum control of the IPC system. In this section, three illustrative simulation examples are provided and performed. Without loss of generality, the parameter values of the IPC system are set as $m_c = 0.378$ kg, $m_p = 0.037$ kg, $l = 0.125$ m, $g = 9.81$ m/s^2, and $\sigma = 0.001$ N s/m.

Example 10.1 A sinusoidal desired trajectory $y_d = \sin(0.1\pi t)\cos(0.2\pi t)$ is considered in this example. Both z2g0 controller (10.4) and z2g1 controller (10.5) are applied in the pendulum tracking control of IPC system (9.2). The initial states are set as $x_3(0) = 0.5$ and $x_1(0) = x_2(0) = x_4(0) = 0$, as well as the initial input of z2g1 controller $u(0) = 0$. Besides, the corresponding design parameters are chosen as $\lambda_1 = \lambda_2 = 15$ and $\gamma = 1000$. Note that, in this example, $x_3 = \theta \in (-\pi/2, \pi/2)$; that is to say, the pendulum is above the horizontal position.

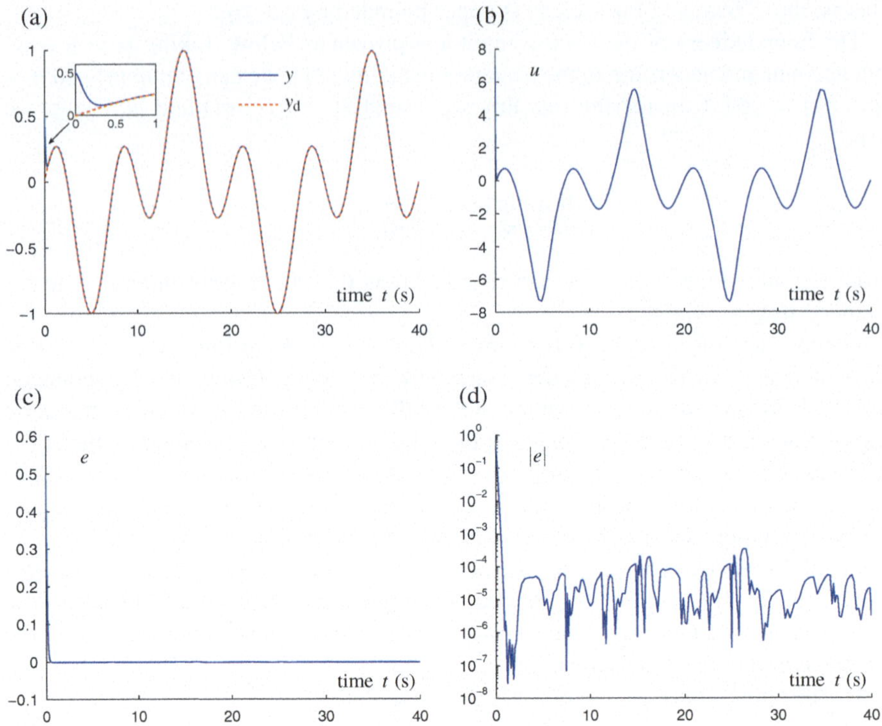

Fig. 10.1 Output trajectory, control input and tracking error of IPC system (9.2) equipped with z2g0 controller (10.4) for desired trajectory $y_d = \sin(0.1\pi t)\cos(0.2\pi t)$. (**a**) Output trajectory and desired trajectory. (**b**) Control input. (**c**) Tracking error. (**d**) Order of $|e|$

The corresponding simulation results are displayed in Figs. 10.1 and 10.2. Specifically, Fig. 10.1 shows the output trajectory, the control input, and the tracking error of IPC system (9.2) equipped with z2g0 controller (10.4). As illustrated in Fig. 10.1a, the output y (i.e., pendulum angle θ) converges to the desired trajectory y_d within a short time. Besides, it can be seen from Fig. 10.1c that the tracking error e decreases to zero. From Fig. 10.1d, we can observe that the maximal steady-state tracking error is of order 10^{-4}. Furthermore, as shown in Fig. 10.1b, the control input u (i.e., control force) is smooth and has not undergone abrupt changes, which is suitable for engineering applications. Additionally, Fig. 10.2 shows the output trajectory, the control input, and the tracking error of IPC system (9.2) using z2g1 controller (10.5). From Fig. 10.2, the similar conclusion can be obtained. That is, the presented z2g1 controller (10.5) works well for the pendulum tracking control of IPC system (9.2). It is worth pointing out here that the maximal steady-state tracking error with z2g1 controller (10.5) is slightly larger than that with z2g0 controller (10.4) but still small enough for general practical applications. Thus, the feasibility and efficacy of z2g0 controller (10.4) and z2g1 controller (10.5) for the pendulum tracking control is substantiated, when $x_3 = \theta \in (-\pi/2, \pi/2)$.

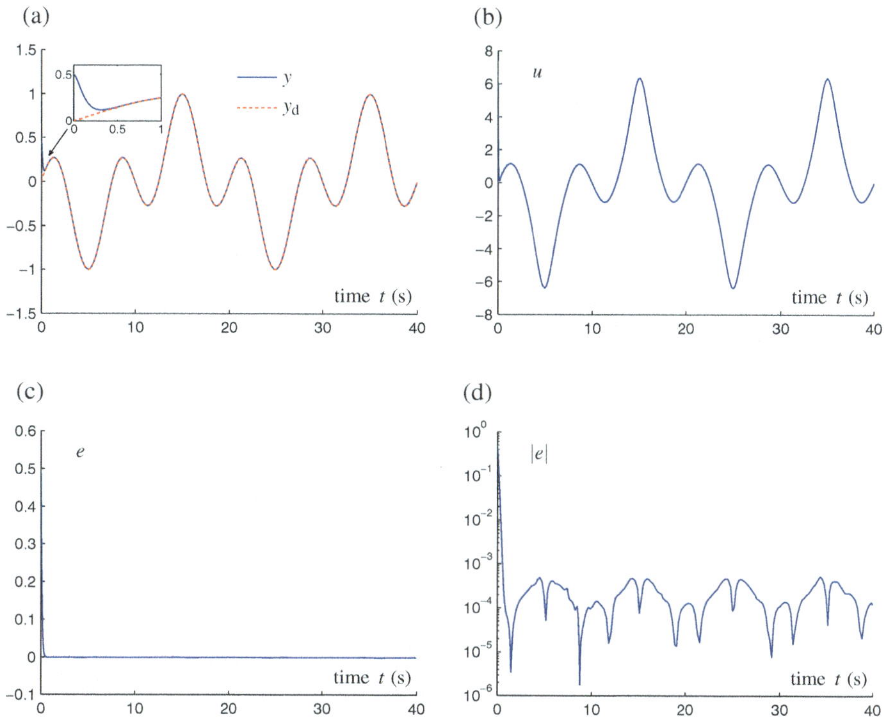

Fig. 10.2 Output trajectory, control input and tracking error of IPC system (9.2) equipped with z2g1 controller (10.5) for desired trajectory $y_{\mathrm{d}} = \sin(0.1\pi t)\cos(0.2\pi t)$. (**a**) Output trajectory and desired trajectory. (**b**) Control input. (**c**) Tracking error. (**d**) Order of $|e|$

Example 10.2 In this example, let us consider an eventually constant desired trajectory as $y_{\mathrm{d}} = 0.3\pi\cos(0.5t)\exp(-0.2t)$. The initial state $x_3(0)$, i.e., $\theta(0)$, is chosen as π. That is, the pendulum is vertical down at the initial time. Hence, the task also includes the swing-up control. The other initial states are chosen as $x_1(0) = x_2(0) = x_4(0) = 0$ for both z2g0 and z2g1 controllers, and the initial input of z2g1 controller $u(0) = 0$. In addition, the design parameters are set as $\lambda_1 = \lambda_2 = 15$ and $\gamma = 100$.

The simulation results are illustrated in Figs. 10.3 and 10.4. Specifically, in Fig. 10.3, the output trajectories and tracking errors of IPC system (9.2) equipped with z2g0 controller (10.4) and z2g1 controller (10.5) are shown. As seen from Fig. 10.3a, b, the tracking-control process of IPC system (9.2) equipped with z2g0 controller (10.4) stops at around 0.3 s. In contrast, the tracking-control process of IPC system (9.2) equipped with z2g1 controller (10.5) runs uninterruptedly, which is illustrated in Fig. 10.3c, d. Meanwhile, it can be seen from Fig. 10.3d that the tracking error with z2g1 controller (10.5) goes to zero quickly, which means that the tracking-control purpose is achieved by means of z2g1 controller (10.5). Considering that the initial value of θ is π, we can readily know that the swing-up

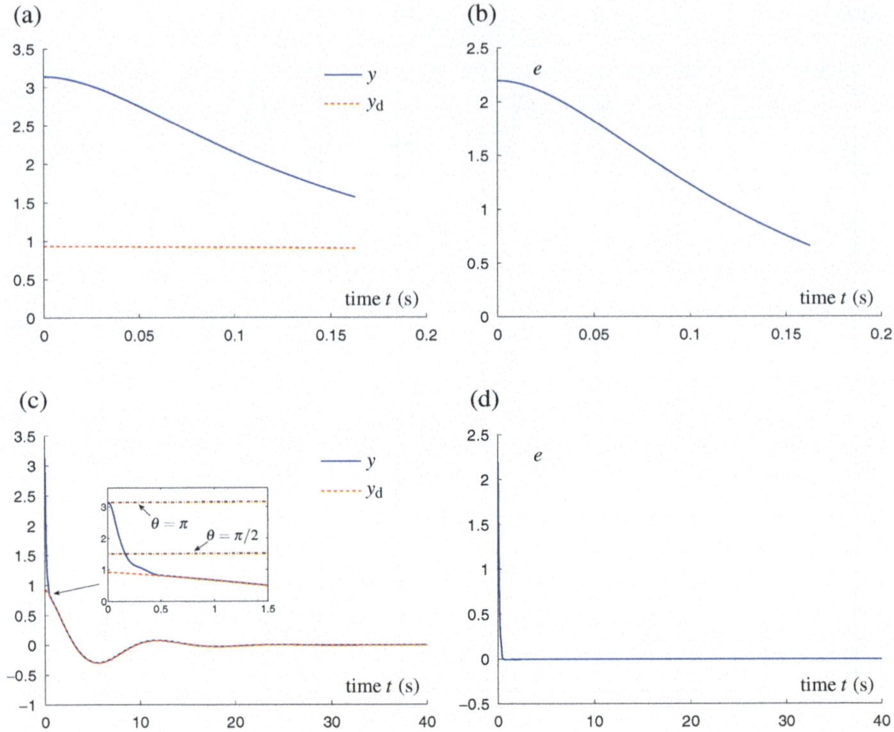

Fig. 10.3 Output trajectories and tracking errors of IPC system (9.2) equipped with z2g0 controller (10.4) and z2g1 controller (10.5), respectively, for desired trajectory $y_d = 0.3\pi \cos(0.5t) \exp(-0.2t)$. (**a**) Output trajectory with z2g0 controller (10.4) and desired trajectory. (**b**) Tracking error with z2g0 controller (10.4). (**c**) Output trajectory with z2g1 controller (10.5) and desired trajectory. (**d**) Tracking error with z2g1 controller (10.5)

control is achieved simultaneously, as visualized in Fig. 10.3c. Besides, Fig. 10.4 further reveals the reason why z2g0 controller (10.4) fails but z2g1 controller (10.5) works well. As shown in Fig. 10.4a, the value of control input tends to infinity at around 0.16 s, which leads to the crash of z2g0 controller (10.4). Indeed, when time t is around 0.16 s, $\cos x_3 = 0$ holds true such that z2g0 controller (10.4) encounters the DBZ problem, just as displayed in Fig. 10.4b. Therefore, it follows from (10.4) that the control input u tends to infinity theoretically, resulting in the failure of z2g0 controller (10.4). In contrast, as displayed in Fig. 10.4c, d, even when $\cos x_3 = 0$ holds true, the control input of z2g1 controller (10.5) is bounded and acceptable, which implies that this controller runs well. In other words, by making use of z2g1 controller (10.5), the DBZ problem is conquered successfully. Evidently, the above comparative simulation results show well the superiority of z2g1 controller (10.5) in conquering the DBZ problem for the control of pendulum tracking (including swinging up) of IPC system (9.2).

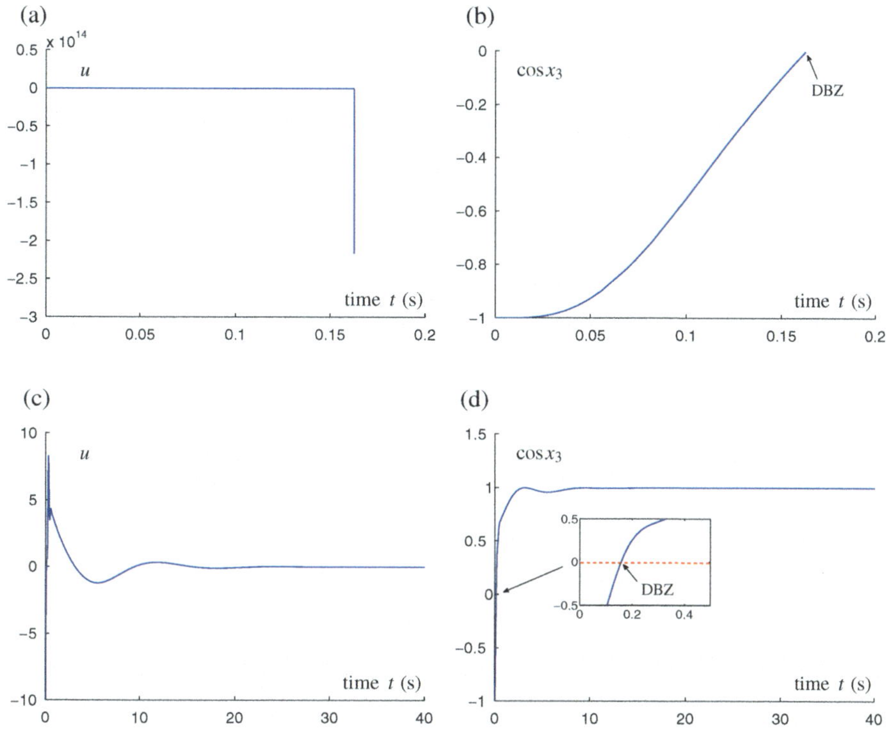

Fig. 10.4 Control inputs and trajectories of denominator $\cos x_3$ of IPC system (9.2) equipped with z2g0 controller (10.4) and z2g1 controller (10.5), respectively, for desired trajectory $y_d = 0.3\pi \cos(0.5t)\exp(-0.2t)$. (**a**) Control input with z2g0 controller (10.4). (**b**) Trajectory of $\cos x_3$ with z2g0 controller (10.4). (**c**) Control input with z2g1 controller (10.5). (**d**) Trajectory of $\cos x_3$ with z2g1 controller (10.5)

Example 10.3 In this example, the efficacy of z2g1 controller (10.5) and the important effect of design parameters λ_1, λ_2 and γ are further investigated. The desired trajectory $y_d = 0.5(\sin(t) + \cos(0.5\pi t))$ is considered. The initial values are chosen to be the same as those in Example 10.2. Note that, in this example, the initial state $x_3(0)$, i.e., $\theta(0)$, is also selected as π, which means that the pendulum needs to be swung up from the pendant position during the tracking-control process of IPC system (9.2).

For comparison, both z2g0 controller (10.4) and z2g1 controller (10.5) are employed. The design parameters are set as $\lambda_1 = \lambda_2 = 16$ and $\gamma = 1000$. The simulation results are shown in Fig. 10.5. As seen from Fig. 10.5a, b, when z2g0 controller (10.4) is used, the control input u tends to infinity and the tracking-control process stops at around 0.13 s. In contrast, as indicated in Fig. 10.5c, d, the desired trajectory is tracked well and the control input is continuous and smooth. Evidently, the pendulum tracking control is achieved by means of z2g1 controller (10.5). Besides, from Fig. 10.5c, we can find that this tracking-control process achieves

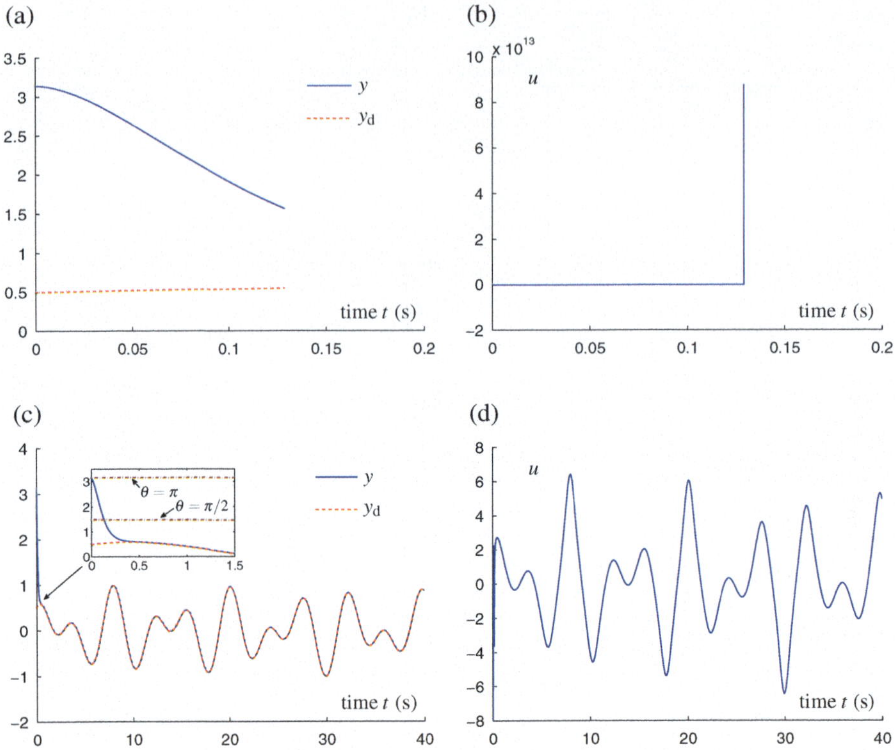

Fig. 10.5 Output trajectories and control inputs of IPC system (9.2) equipped with z2g0 controller (10.4) and z2g1 controller (10.5), respectively, for desired trajectory $y_d = 0.5(\sin(t)+\cos(0.5\pi t))$. (**a**) Output trajectory with z2g0 controller (10.4) and desired trajectory. (**b**) Control input with z2g0 controller (10.4). (**c**) Output trajectory with z2g1 controller (10.5) and desired trajectory. (**d**) Control input with z2g1 controller (10.5)

the swinging up successfully. In brief, the efficacy of z2g1 controller (10.5) in conquering the DBZ problem is verified once more for the control of pendulum tracking (including swinging up) of IPC system (9.2).

Next, in order to further investigate the simulation results in the situation of relatively small values of design parameters (i.e., λ_1, λ_2 and γ) as well as their effect on the tracking performance of z2g1 controller (10.5), different settings of design parameters with small values are tested in the simulations accordingly. Specifically, Fig. 10.6 shows the absolute tracking errors of IPC system (9.2) equipped with z2g1 controller (10.5) by using different small values of λ_1, λ_2 and γ. From this figure, we can observe that, although relatively small values of λ_1, λ_2 and γ are used, the absolute tracking errors all converge toward zero rapidly and remain with an acceptably small level, which means that the z2g1 controller as well as the ZG design method is still effective for the pendulum tracking control of IPC system (9.2). Besides, it can be found that the tracking error bound becomes smaller

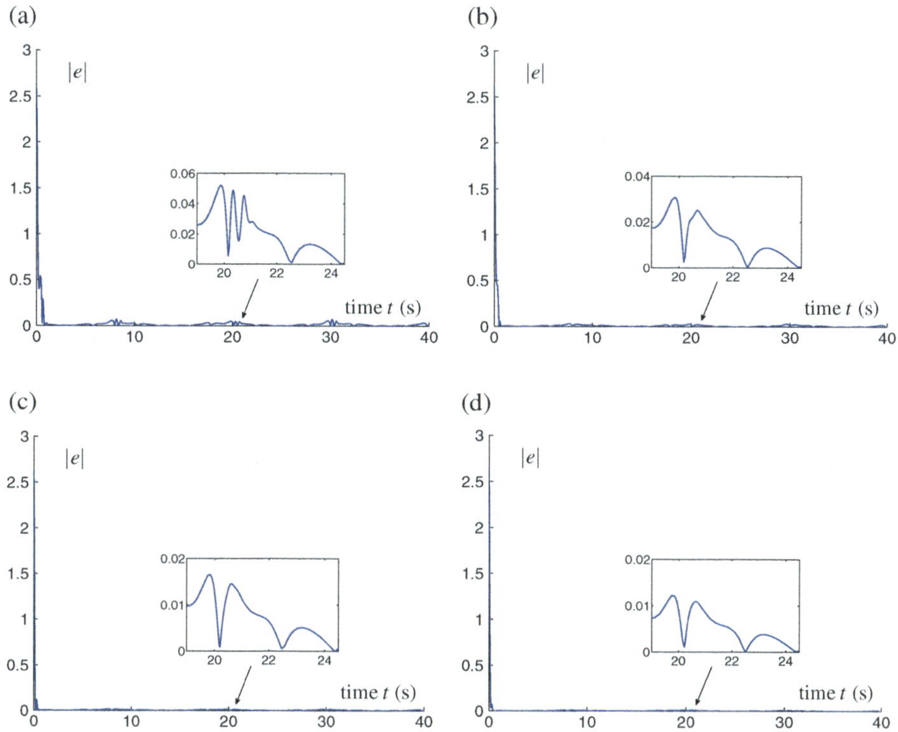

Fig. 10.6 Absolute tracking errors with different small values of design parameters for IPC system (9.2) equipped with z2g1 controller (10.5) for desired trajectory $y_d = 0.5(\sin(t) + \cos(0.5\pi t))$. (**a**) With $\lambda_1 = \lambda_2 = 12$ and $\gamma = 30$. (**b**) With $\lambda_1 = \lambda_2 = 12$ and $\gamma = 45$. (**c**) With $\lambda_1 = \lambda_2 = 16$ and $\gamma = 45$. (**d**) With $\lambda_1 = \lambda_2 = 16$ and $\gamma = 60$

with larger values of design parameters used, which coincides with the theoretical analyses provided in Sect. 10.3. In practical applications, the practitioners can choose the appropriate values of λ_1, λ_2 and γ in accordance with specific needs.

In summary, from the above three illustrative examples, the efficacy and superiority of the DBZ-conquering z2g1 controller (10.5) for the control of pendulum tracking (including swinging up) of IPC system (9.2) in conquering the DBZ problem have been substantiated. Besides, the suitability for the situation of relatively small values of design parameters (i.e., λ_1, λ_2 and γ) and their effect on the tracking performance of z2g1 controller (10.5) have also been illustrated.

10.5 Chapter Summary

In this chapter, two tracking controllers, i.e., z2g0 controller (10.4) and z2g1 controller (10.5), have been designed for the pendulum control of IPC system (9.2). Between them, z2g1 controller (10.5) not only achieves the control of pendulum tracking (including swinging up) of IPC system (9.2) with satisfactory tracking accuracy, but also successfully conquers the DBZ problem. In addition, the theoretical results and analyses on the convergence performance of both z2g0 and z2g1 controllers have been presented. Besides, the boundedness of both control input and its derivative has been investigated and proven. The comparative simulation results have substantiated the efficacy and superiority of z2g1 controller (10.5) in conquering the DBZ problem for the control of pendulum tracking (including swinging up) of IPC system (9.2). In contrast, just like other conventional controllers, z2g0 controller (10.4) fails to achieve the pendulum control of the IPC system when encountering the DBZ problem.

References

1. Khalil HK (2014) Nonlinear control. Prentice Hall, New Jersey
2. Zak SH (2003) Systems and control. Oxford University Press, New York
3. Wang X, Zhang X, Ma C (2012) Modified projective synchronization of fractional-order chaotic systems via active sliding mode control. Nonlinear Dyn 69(1–2):511–517
4. Yang R, Peng Y, Song Y (2013) Stability and Hopf bifurcation in an inverted pendulum with delayed feedback control. Nonlinear Dyn 73(1–2):737–749
5. Ibanez CA, Frias OG, Castanon MS (2005) Lyapunov-based controller for the inverted pendulum cart system. Nonlinear Dyn 40(4):367–374
6. Adhikary N, Mahanta C (2013) Integral backstepping sliding mode control for underactuated systems: swing-up and stabilization of the Cart-Pendulum System. ISA Trans 52(6):870–880
7. Zhang Y, Qiu B, Liao B, Yang Z (2017) Control of pendulum tracking (including swinging up) of IPC system using zeroing-gradient method. Nonlinear Dyn 89(1):1–25
8. Yang JH, Shim SY, Seo JH, Lee YS (2009) Swing-up control for an inverted pendulum with restricted cart rail length. Int J Control Autom Syst 7(4):674–680
9. Kassem AH (2005) Swing-up control of inverted pendulum. J Eng Appl Sci 52(6):1163–1178
10. Angeli D (2001) Almost global stabilization of the inverted pendulum via continuous state feedback. Automatica 37(7):1103–1108
11. Srinivasan B, Huguenin P, Bonvin D (2009) Global stabilization of an inverted pendulum-control strategy and experimental verification. Automatica 45(1):265–269
12. Wei E, Li T, Li J, Hu Y, Li Q (2014) Neural network-based adaptive dynamic surface control for inverted pendulum system. Adv Intell Syst Comput 215:695–704
13. Dusek F, Honc D, Sharma KR, Havlicek L (2016) Inverted pendulum optimal control based on first principle model. Adv Intell Syst Comput 466:63–74
14. Yi J, Yubazaki N (2000) Stabilization fuzzy control of inverted pendulum systems. Artif Intell Eng 14(2):153–163
15. Mazenc F, Praly L (1996) Adding integrations, saturated controls, and stabilisation of feedforward systems. IEEE Trans Autom Control 41(11):1559–1578
16. Astrom KJ, Furuta K (2000) Swinging up a pendulum by energy control. Automatica 36(2):287–295

17. Zhang Y, Xiao L, Xiao Z, Mao M (2015) Zeroing dynamics, gradient dynamics, and Newton iterations. CRC Press, Boca Raton
18. Zhang Y, Chen D, Jin L, Zhang Y, Yin Y (2016) GD-aided IOL (input-output linearisation) controller for handling affine-form nonlinear system with loose condition on relative degree. Int J Control 89(4):757–769
19. Strang G (2006) Linear algebra and its applications, 4th edn. Cengage Learning, Stamford
20. Dragomir SS (2003) Some Gronwall type inequalities and applications. Nova Science Publishers, New York
21. Anton H, Rorres C (2010) Elementary linear algebra, 10th edn. Wiley, Hoboken

Part IV
AFN Systems Using ZG Control

Chapter 11
GD-Aided IOL Tracking Control of AFN System

Abstract IOL may encounter the knotty DBZ problem when applied to the tracking control of affine-form nonlinear (AFN) system, which may not have a well-defined relative degree. In this chapter, we incorporate the GD into IOL, which leads to the GD-aided IOL method for conquering the DBZ problem encountered in the AFN system, with the proposition of the loose condition on relative degree. Corresponding theoretical analyses on tracking-error bound and convergence performance of the GD-aided IOL controller are provided. Moreover, comparative simulation results further substantiate that the GD-aided IOL controller is capable of fulfilling the tracking-control task with the DBZ problem conquered.

11.1 Introduction

The tracking control of nonlinear systems is often encountered in practical applications [1–4]. In general, the objective of tracking control is to design an appropriate controller such that the output of the system tracks a desired trajectory and the tracking error is kept at an acceptable tolerance level. Studies have shown that IOL is a widely used design method for nonlinear tracking control and has been applied to the tracking control of different types of nonlinear systems [5–7]. The affine-form nonlinear (AFN) system is one of the commonly used nonlinear systems, which has gained increasing interest from researchers in fuzzy control, numerical analysis, and robot control [8–10]. Besides, the AFN system has extensive applications [11, 12]. For example, the necessary and sufficient condition for global controllability of planar affine nonlinear systems was investigated by Sun [9]. For the optimal control of nonlinear systems, the reinforcement learning controller was designed by Yang and Jagannathan [10] for affine nonlinear discrete-time systems. In [8], a stable adaptive fuzzy sliding-mode control was presented for affine nonlinear systems. In [13], the neural network-based finite-horizon optimal control of uncertain affine nonlinear discrete-time systems was investigated.

Note that the IOL method is no longer valid when the nonlinear system encounters DBZ points, at which the nonlinear system fails to have a well-defined

Y. Zhang et al., *Zhang-Gradient Control*,
https://doi.org/10.1007/978-981-15-8257-8_11

relative degree. In these DBZ situations, the controller designed by the IOL method encounters the DBZ problem (especially for the AFN system), which generates an infinitely large control input and leads to system crash [14]. This harsh problem has become the main restriction for IOL applications to nonlinear tracking control. It has also attracted significant attention from researches [14–16]. Naturally, different approaches, as well as their improvements, have been presented to tackle the DBZ problem [17–20]. For example, a method of approximate IOL for the ball-and-beam system was presented in [17], which fails to have a well-defined relative degree in a neighborhood of the origin. The method of approximate IOL constructs a tracking controller by neglecting or modifying some nonlinear terms that generate DBZ points, and then gets through the DBZ points. As mentioned in [18], because of the approximation error generated by the neglected/modified terms, the method of approximate IOL may not work well when the system is away from the DBZ points [17]. A scheme of switching between approximate and exact IOL was presented in [19], so as to avoid the DBZ points and improve the tracking performance. However, this approach requires a high cost of implementation because it requires two or more controllers to solve a single problem.

Zhang et al. [20] proposed a simple and effective controller-design method, i.e., aiding the conventional controller with GD, and yielding a GD-aided controller in the framework of the ZG method. For further research, this chapter [4] applies it to the conventional IOL method for the tracking control of AFN system, yields the GD-aided IOL controller, and provides the corresponding convergence performance analyses of the presented controller. Note that Zhang et al. [20] used GD to conquer the DBZ problem of conventional controller for the output tracking of a class of nonlinear system, providing a new direction to tackle DBZ points, but the rigorous theoretical analysis was not provided. Besides, it is worth pointing out that GD is intrinsically designed for solving time-invariant problems and has recently been generalized to solve time-varying problems [15, 21, 22].

In this chapter [4], the GD-aided IOL method for the tracking control of AFN system is presented to conquer the DBZ problem, with the proposition of the loose condition on relative degree. Note that the DBZ problem has existed for twelve centuries [23]. However, much effort has been spent in studying the problem under a time-invariant premise, i.e., studying the division operation with fixed operands at a certain time instant. In contrast, this chapter [4] investigates the DBZ problem from the perspective of temporal evolution. By employing the GD, the division operation is transformed into a generalized version that contains no division, specifically, a time-varying minimization problem. With the GD incorporated in solving the time-varying minimization problem, the simple and effective method presented in this chapter [4] is capable of designing a division-free controller. That is, the controller gets rid of the potential possibility of encountering the DBZ problem, and thus remains valid at the DBZ points encountered during the tracking-control process of AFN system.

11.2 AFN System and Problem Description

In this section, the general AFN system is firstly formulated. Then, we present the problem description for such an AFN system, i.e., the DBZ problem occurring in the IOL controller of AFN system, which fails to have a well-defined relative degree.

11.2.1 AFN System

Note that the AFN system has many derivations and thus can be applied in specific applications as discussed in Sect. 11.1. Let us consider the general AFN system as below:

$$\begin{cases} \dot{\mathbf{x}} = \mathbf{f}(\mathbf{x}, u) = \mathbf{g}(\mathbf{x}) + \mathbf{h}(\mathbf{x})u, \\ y = y(\mathbf{x}), \end{cases} \tag{11.1}$$

where $\mathbf{x} = [x_1,\ x_2,\ \cdots,\ x_n]^{\mathrm{T}} \in \mathbb{R}^n$ is the state vector, $u \in \mathbb{R}$ is the control input, $y \in \mathbb{R}$ is the output, and the functions $\mathbf{g}(\mathbf{x}) : \mathbb{R}^n \to \mathbb{R}^n$, $\mathbf{h}(\mathbf{x}) : \mathbb{R}^n \to \mathbb{R}^n$, and $y(\mathbf{x}) : \mathbb{R}^n \to \mathbb{R}$ are smooth (as differentiable as needed), which also satisfy the condition that $\mathbf{g}(\mathbf{x})$, $\mathbf{h}(\mathbf{x})$ and $y(\mathbf{x})$ remain bounded for bounded $\mathbf{x}$. With a smooth and bounded desired trajectory $y_{\mathrm{d}} \in \mathbb{R}$, the objective is to design a controller for AFN system (11.1) such that the output y tracks the desired trajectory y_{d}, with the tracking error $e = y - y_{\mathrm{d}}$ asymptotically approaching zero (or near zero in practice).

Moreover, in order to focus on the DBZ problem, AFN system (11.1) should satisfy the uniformly bounded-input bounded-state (UBIBS) stability property [24], where, for each bounded initial pair $(t_0, \mathbf{x}(0))$ and each admissible input u, the corresponding solution $\mathbf{x}$ of AFN system (11.1) remains bounded for all time $t > t_0$.

11.2.2 Problem Description

The IOL method for the tracking control of AFN system (11.1) is summarized in [7]. When the IOL method is applied, the relative degree needs to be obtained firstly, and the controller is then designed on the basis of the relative degree. For the convenience of further analysis, let $L_g^i y(\mathbf{x})$ (with integer $i \geq 1$) denote the ith Lie derivative of $y(\mathbf{x})$ with respect to $\mathbf{g}(\mathbf{x})$ [5, 7]. For $i = 0$, $L_g^i y(\mathbf{x}) = L_g^0 y(\mathbf{x}) = y(\mathbf{x})$; for $i = 1$, $L_g^i y(\mathbf{x})$ is defined as

$$L_g^1 y(\mathbf{x}) = L_g y(\mathbf{x}) = \frac{\partial y(\mathbf{x})}{\partial \mathbf{x}} \mathbf{g}(\mathbf{x}).$$

In addition, for $i \geq 2$, $L_g^i y(\mathbf{x})$ is recursively defined as

$$L_g^i y(\mathbf{x}) = \frac{\partial L_g^{i-1} y(\mathbf{x})}{\partial \mathbf{x}} \mathbf{g}(\mathbf{x}).$$

Moreover, with $i \geq 1$, it follows from the above that $L_h L_g^i y(\mathbf{x})$ is defined as

$$L_h L_g^i y(\mathbf{x}) = \frac{\partial L_g^i y(\mathbf{x})}{\partial \mathbf{x}} \mathbf{h}(\mathbf{x}).$$

For further discussion, let $\mathbb{W}$ denote the region of interest (that is, the possible region that state vector $\mathbf{x}$ can reach) for AFN system (11.1), which has a well-defined relative degree. For completeness and readability, the IOL method for the tracking control of AFN system (11.1), which requires a well-defined relative degree in $\mathbb{W}$, can be summarized as the following definition and lemma [7, 25].

Definition 11.1 AFN system (11.1) is said to have a well-defined relative degree r_{w} in the region of interest $\mathbb{W}$ if the following two properties hold true:

- $\forall \mathbf{x} \in \mathbb{W}$, $L_h L_g^i y(\mathbf{x}) = 0$, for $0 \leq i < r_{\mathrm{w}} - 1$;
- $\forall \mathbf{x} \in \mathbb{W}$, $L_h L_g^{r_{\mathrm{w}}-1} y(\mathbf{x}) \neq 0$.

Lemma 11.1 *Let AFN system (11.1) have a well-defined relative degree* $r_{\mathrm{w}} = n$ *in the region of interest denoted by* $\mathbb{W}$ *and satisfy the UBIBS stability property; let* $\tilde{\mathbf{y}} = [y,\ \dot{y},\ \cdots,\ y^{(r_{\mathrm{w}}-1)}]^{\mathrm{T}} \in \mathbb{R}^{r_{\mathrm{w}}}$ *and* $\tilde{\mathbf{y}}_{\mathrm{d}} = [y_{\mathrm{d}},\ \dot{y}_{\mathrm{d}},\ \cdots,\ y_{\mathrm{d}}^{(r_{\mathrm{w}}-1)}]^{\mathrm{T}} \in \mathbb{R}^{r_{\mathrm{w}}}$. *Besides, the tracking-error vector is defined as* $\tilde{\mathbf{e}} = [e,\ \dot{e},\ \cdots,\ e^{(r_{\mathrm{w}}-1)}]^{\mathrm{T}} = \tilde{\mathbf{y}} - \tilde{\mathbf{y}}_{\mathrm{d}} \in \mathbb{R}^{r_{\mathrm{w}}}$. *Then, starting with bounded initial state* $\mathbf{x}(0)$, *all the state trajectories of AFN system (11.1) remain bounded and the tracking-error vector* $\tilde{\mathbf{e}}$ *converges to zero exponentially, by using the following conventional tracking controller (designed by the IOL method with* $L_h L_g^{r_{\mathrm{w}}-1} y$ *required to be nonzero):*

$$u = \frac{1}{L_h L_g^{r_{\mathrm{w}}-1} y} \left(-L_g^{r_{\mathrm{w}}} y + y_{\mathrm{d}}^{(r_{\mathrm{w}})} - \tilde{\mathbf{k}}^{\mathrm{T}} \tilde{\mathbf{e}} \right), \tag{11.2}$$

where $\tilde{\mathbf{k}} = [k_1,\ k_2,\ \cdots,\ k_{r_{\mathrm{w}}}]^{\mathrm{T}} \in \mathbb{R}^{r_{\mathrm{w}}}$ *satisfies the condition that all roots of the characteristic polynomial* $\tilde{P}(s) = s^{r_{\mathrm{w}}} + k_{r_{\mathrm{w}}} s^{r_{\mathrm{w}}-1} + \cdots + k_2 s + k_1$ *are in the open left-half complex plane.*

Proof See Appendix 1 for details. □

However, the limitation of the conventional IOL controller (11.2) is evident. If the system fails to have a well-defined relative degree over the region of interest, this IOL controller is not valid. To be more specific, if the divisor $L_h L_g^{r_{\mathrm{w}}-1} y$ in IOL controller (11.2) passes zero, this tracking controller cannot be implemented since the magnitude of the control input becomes infinite. When encountering the DBZ points, the IOL controller fails to solve the tracking-control problem of AFN system (11.1). As the DBZ problem may be frequently encountered in practice, this chapter [4] aims at providing a structurally simple, practically effective and theoretically complete solution method for solving this kind of DBZ problem by incorporating the GD into IOL.

11.3 GD-Aided IOL Controller Design and Analyses

The design procedure of GD-aided IOL controller is presented in this section. In addition, the detailed theoretical analyses on tracking-error bound and convergence performance for the tracking control of AFN system equipped with the GD-aided IOL controller are provided.

11.3.1 Loose Condition on Relative Degree

To lay a basis for further discussion, we firstly present the loose condition on relative degree. Specifically, AFN system (11.1) is said to have the loose condition on relative degree r_l in the region of interest denoted by $\mathbb{L}$ if the following two properties hold true:

- $\forall \mathbf{x} \in \mathbb{L}$, $L_h L_g^i y(\mathbf{x}) = 0$, for $0 \leq i < r_l - 1$;
- $\exists \mathbf{x} \in \mathbb{L}$, $L_h L_g^{r_l - 1} y(\mathbf{x}) \neq 0$.

Note that the well-defined relative degree r_w is assumed to be n, which is the same as the order of AFN system (11.1). Besides, it is worth pointing out that we mainly focus on AFN system (11.1) with loose condition on relative degree $r_l = n$ in the region of interest denoted by $\mathbb{L}$. Although the value of r_l is equal to the value of r_w in the general case, the difference between them is apparent and significant. In Definition 11.1, the time-varying value of $L_h L_g^{r_w - 1} y(\mathbf{x})$ is assumed to be always nonzero for the region of interest denoted by $\mathbb{W}$ (i.e., $\forall \mathbf{x} \in \mathbb{W}$, $L_h L_g^{r_w - 1} y(\mathbf{x}) \neq 0$), which is difficult or even sometimes impossible to be implemented in practice. By contrast, in the loose condition on relative degree r_l, the existence of DBZ point of $L_h L_g^{r_l - 1} y(\mathbf{x}) = 0$ is allowed. In mathematics, $\exists \mathbf{x} \in \mathbb{L}$, $L_h L_g^{r_l - 1} y(\mathbf{x}) \neq 0$; i.e., it is sufficient to have one point of $L_h L_g^{r_l - 1} y(\mathbf{x}) \neq 0$ for region $\mathbb{L}$, which improves the feasibility of controller design. Specifically, let us discuss the regions of interest. On the one hand, the region of interest denoted by $\mathbb{W}$ is a solid region without any DBZ point in it, which is a perfect, dense but quite small region. In addition, region $\mathbb{W}$ may actually be time-varying and can sometimes be a subset of either $\mathbb{R}^+$ or $\mathbb{R}^-$ without zero point for $L_h L_g^{r_l - 1} y(\mathbf{x})$. If some trajectories of region $\mathbb{W}$ encounter DBZ points, some parts of the region with such DBZ points have to be set aside from $\mathbb{W}$, which makes the region smaller. On the other hand, the region of interest denoted by $\mathbb{L}$ can be a sparse region, in which some or even countless DBZ points exist. Region $\mathbb{L}$ may also be time-varying and can be a subset of the whole $\mathbb{R}$ with zero points for $L_h L_g^{r_l - 1} y(\mathbf{x})$. In a strict sense, $\mathbb{L}$ can be a subset of $\mathbb{R}$ excluding the zero line that is $L_h L_g^{r_l - 1} y(\mathbf{x}) = 0$ for all time. This condition means that $\mathbb{L}$ can be much larger than $\mathbb{W}$ in practice. That is the reason why we use different symbols r_w and r_l to denote the relative degree in different conditions. Besides, the proposition of the loose condition on relative degree is the starting point for solving the DBZ problem of AFN system and is also the key point of this chapter [4].

11.3.2 Design of GD-Aided IOL Controller

Note that the potential possibility of DBZ mostly lies in the division operation of IOL controller (11.2). To conquer these DBZ points, an effective controller should be designed to avoid the division operation. Thus, in the process of designing a division-free controller, the basic idea is to transform the division operation into a time-varying minimization problem solving. Specifically, based on IOL controller (11.2), the GD [21, 22] can be used to get rid of the division operation and then conquer the DBZ problem of IOL. The design procedure of such a DBZ-conquering controller is presented as follows.

Step 1 Consider AFN system (11.1), which satisfies the loose condition on relative degree r_1 in the region of interest denoted by $\mathbb{L}$. Let $\hat{\mathbf{y}} = [y,\ \dot{y},\ \cdots,\ y^{(r_1-1)}]^{\mathrm{T}} \in \mathbb{R}^{r_1}$ and $\hat{\mathbf{y}}_{\mathrm{d}} = [y_{\mathrm{d}},\ \dot{y}_{\mathrm{d}},\ \cdots,\ y_{\mathrm{d}}^{(r_1-1)}]^{\mathrm{T}} \in \mathbb{R}^{r_1}$. Besides, the tracking-error vector is defined as $\hat{\mathbf{e}} = [e,\ \dot{e},\ \cdots,\ e^{(r_1-1)}]^{\mathrm{T}} = \hat{\mathbf{y}} - \hat{\mathbf{y}}_{\mathrm{d}} \in \mathbb{R}^{r_1}$. In addition, $\hat{\mathbf{k}} = [k_1,\ k_2,\ \cdots,\ k_{r_1}]^{\mathrm{T}} \in \mathbb{R}^{r_1}$ satisfies the condition that all roots of the characteristic polynomial $\hat{P}(s) = s^{r_1} + k_{r_1}s^{r_1-1} + \cdots + k_2 s + k_1$ are in the open left-half complex plane. Then, according to (11.2), a time-varying EF is defined as

$$\epsilon = \frac{\phi^2}{2} = \frac{(\alpha_1 + \alpha_2 u)^2}{2}, \tag{11.3}$$

where $\alpha_1 = L_g^{r_1} y - y_{\mathrm{d}}^{(r_1)} + \hat{\mathbf{k}}^{\mathrm{T}}\hat{\mathbf{e}}$ and $\alpha_2 = L_h L_g^{r_1-1} y$. That is, the direct computational manner of controller (11.2) is transformed to the manner of minimizing the time-varying EF (11.3).

Step 2 Based on the GD method [21, 22], the following GD design formula is employed to minimize EF (11.3):

$$\dot{u} = -\gamma \frac{\partial \epsilon}{\partial u}, \tag{11.4}$$

where the design parameter $\gamma \in \mathbb{R}^+$ is used to scale the convergence rate of the GD solution. The value of design parameter γ needs to be set sufficiently large or selected appropriately for simulative purposes.

Step 3 From (11.3) and (11.4), the DBZ-conquering controller for AFN system (11.1) is designed as

$$\dot{u} = -\gamma\alpha_2(\alpha_1 + \alpha_2 u) = -\gamma L_h L_g^{r_1-1} y \left(L_g^{r_1} y - y_{\mathrm{d}}^{(r_1)} + \hat{\mathbf{k}}^{\mathrm{T}}\hat{\mathbf{e}} + u L_h L_g^{r_1-1} y \right). \tag{11.5}$$

This tracking controller successfully gets rid of the division operation, which implies that it is not restricted by the condition of IOL that the system must have a well-defined relative degree. For convenience, GD-aided controller (11.5) based on IOL is termed GD-aided IOL controller.

11.3.3 Convergence Performance Analyses

In this subsection, the convergence performance analyses on GD-aided IOL controller (11.5) for the tracking control of AFN system (11.1) are presented.

Theorem 11.1 *Consider AFN system (11.1) equipped with GD-aided IOL controller (11.5) for smooth and bounded desired trajectory y_d, which satisfies the loose condition on relative degree r_1 in the region of interest denoted by $\mathbb{L}$ and the UBIBS stability property. Starting with bounded initial state $\mathbf{x}(0) \in \mathbb{R}^n$ and control input $u(0) \in \mathbb{R}$, the following results are achieved on a large scale for the tracking control of the system.*

- *For the case of $\alpha_2 \neq 0$ (i.e., the non-DBZ case), the tracking error of the system converges toward or stays within the error bound $\chi\varrho/(k_1\gamma\varsigma)$, provided that i) $\sqrt{\varsigma} \leq |\alpha_2| \leq \chi$, $\exists 0 < \sqrt{\varsigma} \leq \chi < +\infty$, and ii) $|\dot{u}^*| = |\mathrm{d}u^*/\mathrm{d}t| \leq \varrho$, $\exists 0 \leq \varrho < +\infty$.*
- *For the case of $\alpha_2 = 0$ (i.e., the DBZ case), the tracking error of the system is bounded.*

Besides, all the state trajectories of the system remain bounded.

Proof For the case of $\alpha_2 \neq 0$ (i.e., the non-DBZ case), let us define a solution error of GD-aided IOL controller (11.5) for solving the time-varying minimization problem of ϵ as $\psi = u - u^*$ with $u^* = -\alpha_1/\alpha_2$ being the desired time-varying solution. According to [21, 22], we can prove that the solution error of controller (11.5) for solving the time-varying minimization problem of ϵ is upper bounded by $\varrho/(\gamma\varsigma)$, i.e.,

$$\limsup_{t\to+\infty} |\psi| = \limsup_{t\to+\infty} |u - u^*| \leq \frac{\varrho}{\gamma\varsigma}, \tag{11.6}$$

where $t \geq t_\mathrm{e}$ with t_e being large enough. For completeness, the proof of (11.6) is given in Appendix 2. Besides, from (11.6), we know that GD-aided IOL controller (11.5) generates the control input u, which converges to the desired control input u^* with a bounded maximal steady-state deviation, i.e., $\varrho/(\gamma\varsigma)$. Moreover, given that AFN system (11.1) starts with bounded initial state $\mathbf{x}(0)$, $u^*(0)$ is bounded and further $u(0)$ is bounded. In addition, since AFN system (11.1) satisfies the UBIBS stability property, the bounded input $u(0)$ produces the bounded state at the next time instant. Then, we can recursively derive that u^* and u are bounded.

According to [7], we have $y^{(r_1)} = L_g^{r_1} y + (L_h L_g^{r_1-1} y)u = L_g^{r_1} y + \alpha_2 u$. Since $\alpha_1 = L_g^{r_1} y - y_\mathrm{d}^{(r_1)} + \hat{\mathbf{k}}^\mathrm{T}\hat{\mathbf{e}}$ and $e^{(r_1)} = y^{(r_1)} - y_\mathrm{d}^{(r_1)}$, we have

$$e^{(r_1)} + \hat{\mathbf{k}}^\mathrm{T}\hat{\mathbf{e}} = \alpha_1 + \alpha_2 u = \alpha_2(-u^* + u) = \alpha_2\psi.$$

Thus, according to (11.6), when $t \geq t_{\mathrm{e}}$, we have

$$-\frac{\chi\varrho}{\gamma\varsigma} \leq -|\alpha_2||\psi| \leq e^{(r_1)} + \hat{\mathbf{k}}^{\mathrm{T}}\hat{\mathbf{e}} \leq |\alpha_2||\psi| \leq \frac{\chi\varrho}{\gamma\varsigma}. \quad (11.7)$$

Let $\hat{\mathbf{s}} = [s_1,\ s_2,\ \cdots,\ s_l]$ denote all roots of $\hat{P}(s) = s^{r_1} + k_{r_1}s^{r_1-1} + \cdots + k_2 s + k_1$. Besides, the multiplicities of $s_1, s_2, \cdots, s_l$ are $m_1, m_2, \cdots, m_l$, respectively, with multiplicity vector $\hat{\mathbf{m}} = [m_1,\ m_2,\ \cdots,\ m_l]$ defined correspondingly. Note that $l \leq r_1$, $\sum_{i=1}^{l} m_i = r_1$, and $\hat{P}(s) = \prod_{i=1}^{l}(s - s_i)^{m_i}$. According to (11.7), since the induced error dynamics is in the canonical form [7], we have

$$|e| \leq \left|\sum_{i=1}^{l}\exp(\mathrm{real}(s_i)t)p_i\right| + \frac{\chi\varrho}{k_1\gamma\varsigma},\ \forall t \geq t_{\mathrm{e}}, \quad (11.8)$$

where $p_i = \sum_{j=1}^{m_i} c_j t^{j-1}$ with $\{c_j\}$ being constants. Note that $\hat{\mathbf{k}}$ is selected such that all roots of $\hat{P}(s)$ are in the open left-half complex plane; i.e., $\mathrm{real}(s_i) < 0$ with $i = 1, 2, \cdots, l$. Then, based on [26], it can be obtained that

$$\left|\sum_{i=1}^{l}\exp(\mathrm{real}(s_i)t)p_i\right| \leq \bar{\alpha}\exp(-\bar{\beta}t),\ \exists\bar{\alpha} > 0,\ \bar{\beta} > 0. \quad (11.9)$$

Thus, combining (11.8) and (11.9), we have

$$\begin{aligned}\limsup_{t\to+\infty}|e| &\leq \limsup_{t\to+\infty}\left|\sum_{i=1}^{l}\exp(\mathrm{real}(s_i)t)p_i\right| + \frac{\chi\varrho}{k_1\gamma\varsigma} \\ &\leq \limsup_{t\to+\infty}\left(\bar{\alpha}\exp(-\bar{\beta}t)\right) + \frac{\chi\varrho}{k_1\gamma\varsigma} = \frac{\chi\varrho}{k_1\gamma\varsigma}.\end{aligned} \quad (11.10)$$

From (11.10), we can conclude that the tracking error of AFN system (11.1) converges toward or stays within the error bound $\chi\varrho/(k_1\gamma\varsigma)$.

For the case of $\alpha_2 = 0$ (i.e., the DBZ case), we can readily know that $\lim_{\alpha_2\to 0}\dot{u} = 0$ in view of $\dot{u} = -\gamma\alpha_2(\alpha_1 + \alpha_2 u)$. Therefore, the control input at the DBZ time instant t_{s} is the same as that at the previous time instant t_{s^-}, i.e., $u(t_{\mathrm{s}}) = u(t_{\mathrm{s}^-})$. Similarly, at t_{s^+} (which is the time instant after the DBZ point), we have $u(t_{\mathrm{s}}) = u(t_{\mathrm{s}^+})$. Note that, at the time instant t_{s^-}, the value of control input $u(t_{\mathrm{s}^-})$ is bounded and the tracking error converges toward or stays within an error bound. Then, we have the result that $u(t_{\mathrm{s}^-}) = u(t_{\mathrm{s}}) = u(t_{\mathrm{s}^+})$ is bounded. For UBIBS system (11.1), the bounded control input produces the bounded state and bounded output. Since the desired trajectory y_{d} is bounded, the tracking error is thus bounded at the time instants t_{s^-}, t_{s} and t_{s^+}.

According to the above analyses on the non-DBZ and DBZ cases, we know that the control input remains bounded in both cases. Since AFN system (11.1) satisfies the UBIBS stability property, the bounded input leads to the bounded state. Therefore, all the state trajectories of AFN system (11.1) remain bounded. The proof is thus completed. □

Based on Theorem 11.1, the following results on the exponential convergence of AFN system (11.1) equipped with GD-aided IOL controller (11.5) to a relatively loose tracking-error bound are obtained.

Corollary 11.1 *Consider AFN system (11.1) equipped with GD-aided IOL controller (11.5) for smooth and bounded desired trajectory* y_{d}*, which satisfies the loose condition on relative degree* r_{l} *in the region of interest denoted by* $\mathbb{L}$ *and the UBIBS stability property. Starting with bounded initial state* $\mathbf{x}(0) \in \mathbb{R}^n$ *and control input* $u(0) \in \mathbb{R}$*, the following results are achieved on a large scale for the tracking control of the system.*

- *For the case of* $\alpha_2 \neq 0$ *(i.e., the non-DBZ case), the tracking error of the system exponentially converges toward or stays within the error bound* $\chi\varrho/(\omega k_1 \gamma \varsigma)$ *with loosening parameter* $\omega \in (0, 1)$*, provided that (1)* $\sqrt{\varsigma} \leq |\alpha_2| \leq \chi$*,* $\exists 0 < \sqrt{\varsigma} \leq \chi < +\infty$*, and (2)* $|\dot{u}^*| \leq \varrho$*,* $\exists 0 \leq \varrho < +\infty$*.*
- *For the case of* $\alpha_2 = 0$ *(i.e., the DBZ case), the tracking error of the system is bounded.*

Besides, all the state trajectories of the system remain bounded.

Proof It can be generalized from the proof of Theorem 11.1. □

Remark 11.1 The trajectories of AFN system (11.1) are guaranteed to leave the DBZ set, in which $\alpha_2 = L_h L_g^{r-1} y = 0$ when they fall into the DBZ set. Now, let us consider that the trajectories of the AFN system fall into the DBZ set of $\alpha_2(\mathbf{x}, y, u, t) = 0$ at a time instant t_{s}. Besides, let $\mathbf{x}_{\mathrm{s}}$, y_{s} and u_{s} denote the values of $\mathbf{x}$, y and u at t_{s}, respectively. That is, $\alpha_2(\mathbf{x}_{\mathrm{s}}, y_{\mathrm{s}}, u_{\mathrm{s}}, t_{\mathrm{s}}) = 0$. It follows from the Taylor series expansion of $\alpha_2(\mathbf{x}, y, u, t)$ with respect to the DBZ point $(\mathbf{x}_{\mathrm{s}}, y_{\mathrm{s}}, u_{\mathrm{s}}, t_{\mathrm{s}})$ that

$$
\begin{aligned}
&\alpha_2(\mathbf{x}, y, u, t) = \alpha_2(\mathbf{x}_{\mathrm{s}}, y_{\mathrm{s}}, u_{\mathrm{s}}, t_{\mathrm{s}})\\
&+\left((\mathbf{x}-\mathbf{x}_{\mathrm{s}})\frac{\partial}{\partial \mathbf{x}} + (y-y_{\mathrm{s}})\frac{\partial}{\partial y} + (u-u_{\mathrm{s}})\frac{\partial}{\partial u} + (t-t_{\mathrm{s}})\frac{\partial}{\partial t}\right)\alpha_2(\mathbf{x}, y, u, t)|_{(\mathbf{x}_{\mathrm{s}}, y_{\mathrm{s}}, u_{\mathrm{s}}, t_{\mathrm{s}})}\\
&+\frac{1}{2!}\left((\mathbf{x}-\mathbf{x}_{\mathrm{s}})\frac{\partial}{\partial \mathbf{x}} + (y-y_{\mathrm{s}})\frac{\partial}{\partial y} + (u-u_{\mathrm{s}})\frac{\partial}{\partial u} + (t-t_{\mathrm{s}})\frac{\partial}{\partial t}\right)^2\alpha_2(\mathbf{x}, y, u, t)|_{(\mathbf{x}_{\mathrm{s}}, y_{\mathrm{s}}, u_{\mathrm{s}}, t_{\mathrm{s}})}\\
&+\frac{1}{3!}\left((\mathbf{x}-\mathbf{x}_{\mathrm{s}})\frac{\partial}{\partial \mathbf{x}} + (y-y_{\mathrm{s}})\frac{\partial}{\partial y} + (u-u_{\mathrm{s}})\frac{\partial}{\partial u} + (t-t_{\mathrm{s}})\frac{\partial}{\partial t}\right)^3\alpha_2(\mathbf{x}, y, u, t)|_{(\mathbf{x}_{\mathrm{s}}, y_{\mathrm{s}}, u_{\mathrm{s}}, t_{\mathrm{s}})}\\
&+\cdots
\end{aligned}
$$

Note that the time-varying problem is different from the time-invariant one because the former changes with time; e.g., the objective function relates to time t that is a unidirectional uniform (or say, even) stream parameter [27]. For the AFN system equipped with GD-aided IOL controller (11.5) to track a desired trajectory $y_d(t)$ that is smoothly time-varying, we know that the objective function $\alpha_2(t)$ must contain the explicit expression of time t because of the introduction of $y_d(t)$ to u, $\mathbf{x}$ and y, i.e., $\alpha_2(t) = \alpha_2(\mathbf{x}, y, u, t)$. Thus, the objective function with the explicit expression of time t keeps changing (even in the DBZ set). Otherwise, $y_d(t)$ has to be a time-invariant function but not a time-varying one with respect to time t, which contradicts the tracking-control problem itself. According to the Taylor series expansion of $\alpha_2(\mathbf{x}, y, u, t)$ with respect to the DBZ point $(\mathbf{x}_s, u_s, y_s, t_s)$, we know that $\alpha_2(\mathbf{x}, u, y, t)$ has the terms of the partial derivatives with respect to t, i.e., $(\partial\alpha_2(\mathbf{x}_s, y_s, u_s, t_s)/\partial t)(t-t_s)$, $(\partial^2\alpha_2(\mathbf{x}_s, y_s, u_s, t_s)/\partial t^2)((t-t_s)^2/2!)$, $\cdots$, with each generally time-varying (i.e., containing the explicit expression of t). Thus, there must exist at least one of these partial-derivative terms being nonzero (otherwise, the facts that $y_d(t)$ is time-varying and that $\alpha_2(\mathbf{x}, y, u, t)$ explicitly relates to time t cannot hold true). Hence, the trajectories of AFN system (11.1) are guaranteed to leave the DBZ set.

11.4 Simulation, Verification and Comparison

In this section, the computer simulations are conducted for the tracking control of AFN system. The corresponding simulation results are provided to illustrate that GD-aided IOL controller (11.5) can accomplish the tracking-control task and conquer the DBZ problem successfully with satisfactory tracking accuracy.

According to [28], the modified Chen chaotic system can be described as

$$\begin{cases} \dot{x}_1 = f_1(x_1, x_2) = a(x_2 - x_1), \\ \dot{x}_2 = f_2(x_1, x_2, x_3) = dx_1 - x_1x_3 + cx_2, \\ \dot{x}_3 = f_3(x_1, x_2, x_3) = x_1x_2 - bx_3. \end{cases} \tag{11.11}$$

Besides, when $a = 35$, $b = 3$, $c = 12$ and $d = 7$, system (11.11) shows chaotic dynamic behaviors.

The tracking-control problem of the above modified Chen chaotic system with a single control input is investigated here. Specifically, by adding a control input to the second equation of (11.11), this example can be further formulated as AFN system (11.1) with

$$\mathbf{g}(\mathbf{x}) = \begin{bmatrix} a(x_2 - x_1) \\ dx_1 - x_1x_3 + cx_2 \\ x_1x_2 - bx_3 \end{bmatrix}, \ \mathbf{h}(\mathbf{x}) = \begin{bmatrix} 0 \\ 1 \\ 0 \end{bmatrix}, \ y(\mathbf{x}) = x_3, \tag{11.12}$$

where the relative degree $r_1 = 3$. Note that IOL controller (11.2) cannot solve this tracking-control problem because it may encounter the DBZ case of $L_hL_g^2y = x_1 = 0$. To achieve the DBZ-conquering tracking control, GD-aided IOL controller (11.5)

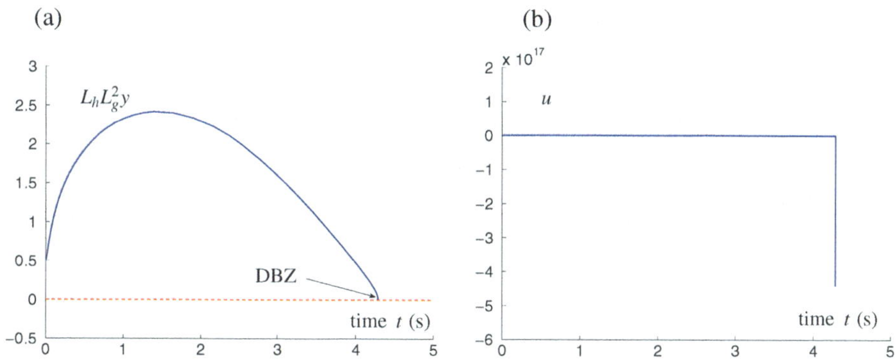

Fig. 11.1 Crash of AFN system (11.12) equipped with conventional IOL controller (11.2) for desired trajectory $y_d = \sin(t)+1.05$ encountering DBZ point. (**a**) Trajectory of $L_h L_g^2 y$. (**b**) Control input

is designed with $\alpha_1 = 0.5x_1\cos(2x_1x_2 + 2x_3^2) + 4x_1\cos(x_1x_2 + x_3^2) + x_1^3 - x_3^3 + x_2^2 + (k_1 + 4.5)x_1 + k_2x_2 + k_3x_3 - \dddot{y}_d - k_3\ddot{y}_d - k_2\dot{y}_d - k_1y_d$ and $\alpha_2 = L_hL_g^2y = x_1$.

The simulation results of AFN system (11.12) equipped with the conventional IOL controller (11.2) for tracking $y_d = \sin(t) + 1.05$ are illustrated in Fig. 11.1. Without loss of generality, we set $\mathbf{x}(0) = [0.5,\ 0.5,\ 0.5]^T$, $u(0) = 0$, $\hat{\mathbf{k}} = [5,\ 5,\ 5]^T$ and $\gamma = 10^5$. As shown in Fig. 11.1a, b, AFN system (11.12) equipped with IOL controller (11.2) crashes when approaching the DBZ plane $L_hL_g^2y = 0$. By contrast, as shown in Fig. 11.2a, AFN system (11.12) equipped with GD-aided IOL controller (11.5) encounters many DBZ points during the tracking-control process but still runs well. Besides, Fig. 11.2b illustrates that the values of control input u with GD-aided IOL controller (11.5) are bounded at the DBZ points and are acceptable. In addition, Fig. 11.2c, d show the output trajectory and the absolute tracking error with GD-aided IOL controller (11.5) for tracking $y_d = \sin(t) + 1.05$, respectively. Such two subfigures illustrate that the output trajectory of AFN system (11.12) equipped with GD-aided IOL controller (11.5) tracks well the desired trajectory, and that the absolute tracking error is quite small. The above comparative simulation results substantiate the DBZ-conquering superiority of GD-aided IOL controller (11.5), as compared with the conventional IOL controller (11.2). Furthermore, in order to investigate and compare the tracking performance of GD-aided IOL controller (11.5) using different values of parameter γ, we further perform simulation tests. As shown in Fig. 11.3, when the value of γ increases from 10^2 to 10^5, the maximal absolute value of the steady-state tracking error decreases from about 4×10^{-2} to 2×10^{-3}, respectively. That is, the tracking performance of AFN system (11.12) equipped with GD-aided IOL controller (11.5) can be improved by properly increasing the value of parameter γ.

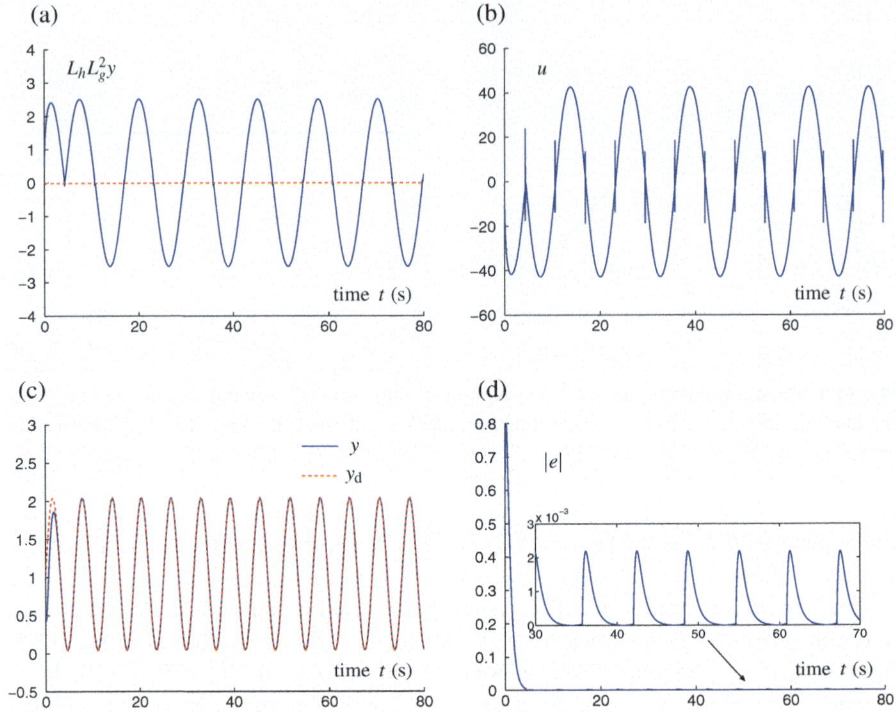

Fig. 11.2 Tracking performance of AFN system (11.12) equipped with GD-aided IOL controller (11.5) for desired trajectory $y_d = \sin(t) + 1.05$ encountering DBZ points. (**a**) Trajectory of $L_hL_g^2y$. (**b**) Control input. (**c**) Output trajectory and desired trajectory. (**d**) Absolute tracking error

11.5 Chapter Summary

By incorporating the GD into IOL, this chapter has presented the GD-aided IOL controller to conquer the DBZ problem of IOL for the tracking control of AFN system. Then, the detailed theoretical results and analyses on the tracking-error bound and convergence performance of the GD-aided IOL controller have been provided to guarantee the efficacy for the DBZ-conquering tracking control. As substantiated by the comparative simulation results, the GD-aided IOL controller has successfully conquered the DBZ problem encountered during the tracking-control process, with the satisfactory tracking accuracy achieved.

At last, it is worth pointing out that the presented GD-aided IOL controller is also within the framework of the ZG method. On the one hand, if the IOL method and the ZD method are regarded as two independent methods that are not related to each other, the GD-aided IOL controller can be deemed as a z0g1 controller, since it is obtained by applying the GD method once and without using the ZD method. On the other hand, if we look for the link between the IOL method and the ZD method, the latter can be viewed as a special case of the former, and then the presented GD-aided

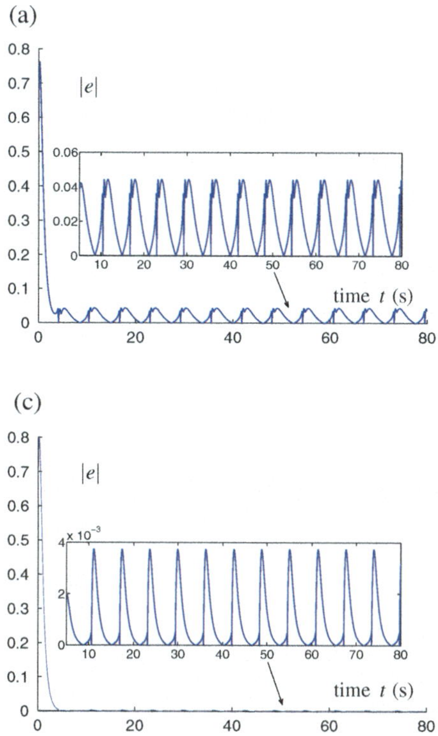

Fig. 11.3 Absolute tracking errors of AFN system (11.12) equipped with GD-aided IOL controller (11.5) using different values of parameter γ for desired trajectory $y_d = \sin(t) + 1.05$ encountering DBZ points. (**a**) With $\gamma = 10^2$. (**b**) With $\gamma = 10^3$. (**c**) With $\gamma = 10^4$. (**d**) With $\gamma = 10^5$

IOL controller can be deemed as a zr_lg1 controller. Particularly, when the relative degree r_l is equal to n, the presented GD-aided IOL controller can be further deemed as a zng1 controller.

Appendix 1: Proof of Lemma 11.1

Note that $u(0)$ is bounded and the AFN system starts with bounded initial state $\mathbf{x}(0)$, and that $\mathbf{g}(\mathbf{x})$, $\mathbf{h}(\mathbf{x})$ and $y(\mathbf{x})$ are smooth and bounded for bounded $\mathbf{x}$. Since AFN system (11.1) satisfies the UBIBS stability property, the bounded input $u(0)$ produces the bounded state at the next time instant. Thus, we can recursively derive that the control input u is bounded and all the state trajectories of the AFN system remain bounded during the whole tracking-control process, provided that the controller's divisor term $L_h L_g^{r_w - 1} y$ is always nonzero.

Besides, with IOL controller (11.2) applied, we have

$$\left(L_h L_g^{r_w-1} y\right) u + L_g^{r_w} y - y_d^{(r_w)} + \tilde{\mathbf{k}}^T \tilde{\mathbf{e}} = 0.$$

According to [7], we have $y^{(r_w)} = (L_h L_g^{r_w-1} y)u + L_g^{r_w} y$. Then, $e^{(r_w)} + \tilde{\mathbf{k}}^T \tilde{\mathbf{e}} = 0$. Since $\tilde{\mathbf{k}} = [k_1, \ k_2, \ \cdots, \ k_{r_w}]^T \in \mathbb{R}^{r_w}$ satisfies the condition that all roots of the characteristic polynomial $\tilde{P}(s) = s^{r_w} + k_{r_w} s^{r_w-1} + \cdots + k_2 s + k_1$ are in the open left-half complex plane, the tracking-error vector $\tilde{\mathbf{e}}$ exponentially converges to zero. The proof is thus completed. □

Appendix 2: Proof of Inequality (11.6)

Since $\psi = u - u^*$ with $u^* = -\alpha_1/\alpha_2$, we have $u = \psi + u^*$ and its time derivative as

$$\dot{u} = \dot{\psi} + \dot{u}^*. \tag{11.13}$$

Substituting (11.5) into (11.13) yields $-\gamma\alpha_2(\alpha_1 + \alpha_2 u) = \dot{\psi} + \dot{u}^*$ and further

$$\dot{\psi} = -\gamma\alpha_2^2\psi - \dot{u}^*.$$

For further analysis, let us define a Lyapunov function candidate as $L = \psi^2/2$. Evidently, L is positive definite in view of $L = \psi^2/2 > 0$ for $\psi \neq 0$ and $L = 0$ only for $\psi = 0$. Then, taking its time derivative yields

$$\dot{L} = \psi\dot{\psi} = \psi\left(-\gamma\alpha_2^2\psi - \dot{u}^*\right) = -\gamma\alpha_2^2\psi^2 - \psi\dot{u}^*. \tag{11.14}$$

There are two terms in the right side of (11.14), i.e, $-\gamma\alpha_2^2\psi^2$ and $-\psi\dot{u}^*$. Let us handle these two terms individually. For the first term $-\gamma\alpha_2^2\psi^2$, we have

$$-\gamma\alpha_2^2\psi^2 \leq -\gamma\varsigma\psi^2, \tag{11.15}$$

where $\varsigma > 0$ is defined previously by $\alpha_2^2 \geq \varsigma$. For the second term $-\psi\dot{u}^*$, we obtain the following result based on Cauchy inequality [21, 22]:

$$-\psi\dot{u}^* \leq |\psi||\dot{u}^*| \leq \varrho|\psi|, \tag{11.16}$$

where ϱ is defined previously by $|\dot{u}^*| \leq \varrho$. Then, substituting (11.15) and (11.16) into (11.14) yields

$$\dot{L} = -\gamma\alpha_2^2\psi^2 - \psi\dot{u}^* \leq -\gamma\varsigma\psi^2 + \varrho|\psi| = -|\psi|(\gamma\varsigma|\psi| - \varrho).$$

Similar to the proof process of (10.7) in Theorem 10.2, we can conclude that the solution error of controller (11.5) for solving the time-varying minimization problem of ϵ is upper bounded by $\varrho/(\gamma\varsigma)$, i.e.,

$$\limsup_{t\to+\infty}|\psi| = \limsup_{t\to+\infty}|u - u^*| \leq \frac{\varrho}{\gamma\varsigma},$$

where $t \geq t_\mathrm{e}$ with t_e being large enough. The proof is thus completed. □

References

1. Duan C, Wu F (2012) Output-feedback control for switched linear systems subject to actuator saturation. Int J Control 85(10):1532–1545
2. Gasparetto A, Zanotto V (2008) A technique for time-jerk optimal planning of robot trajectories. Rob Comput Integr Manuf 24(3):415–426
3. Madhavan R (1999) Tracking control of a pendulum using a numerically efficient state estimation approach for polynomic non-linear systems. Int J Control 72(17):1565–1591
4. Zhang Y, Chen D, Jin L, Zhang Y, Yin Y (2016) GD-aided IOL (input-output linearisation) controller for handling affine-form nonlinear system with loose condition on relative degree. Int J Control 89(4):757–769
5. Hua MD, Hamel T, Morin P, Samson C (2013) Introduction to feedback control of underactuated VTOL vehicles: a review of basic control design ideas and principles. IEEE Control Syst Mag 33(1):61–75
6. Dorleansa P, Massieua JF, Ahmed-Alia T (2011) High-gain observer design with sampled measurements: application to inverted pendulum. Int J Control 84(4):801–807
7. Slotine JE, Li W (1991) Applied nonlinear control. Prentice Hall, New Jersey
8. Hwang C, Kuo C (2001) A stable adaptive fuzzy sliding-mode control for affine nonlinear systems with application to four-bar linkage systems. IEEE Trans Fuzzy Syst 9(2):238–252
9. Sun Y (2007) Necessary and sufficient condition for global controllability of planar affine nonlinear systems. IEEE Trans Autom Control 52(8):1454–1460
10. Yang Q, Jagannathan S (2012) Reinforcement learning controller design for affine nonlinear discrete-time systems using online approximators. IEEE Trans Syst Man Cybern B Cybern 42(2):1257–1269
11. Kim E, Lee CH, Cho YW (2005) Analysis and design of an affine fuzzy system via bilinear matrix inequality. IEEE Trans Fuzzy Syst 13(1):115–123
12. Sahnoun M, Andrieu V, Nadri M (2012) Nonlinear and locally optimal controllers design for input affine locally controllable systems. Int J Control 85(2):159–170
13. Zhao Q, Xu H, Jagannathan S (2015) Neural network-based finite-horizon optimal control of uncertain affine nonlinear discrete-time systems. IEEE Trans Neural Netw Learn Syst 26(3):486–499
14. Zhang Y, Yu X, Yin Y, Peng C, Fan Z (2014) Singularity-conquering ZG controllers of z2g1 type for tracking control of the IPC system. Int J Control 87(9):1729–1746
15. Zhang Y, Yi C (2011) Zhang neural networks and neural-dynamic method. Nova Science Publishers, New York
16. Guo D, Zhang Y (2014) Acceleration-level inequality-based MAN scheme for obstacle avoidance of redundant robot manipulators. IEEE Trans Ind Electron 61(12):6903–6914
17. Hauser J, Sastry S, Kokotovic P (1992) Nonlinear control via approximate input-output linearization: the ball and beam example. IEEE Trans Autom Control 37(3):392–398
18. Kulkarni A, Purwar S (2009) Wavelet based adaptive backstepping controller for a class of nonregular systems with input constraints. Expert Syst Appl 36(3):6686–6696

19. Tomlin CJ, Sastry SS (1997) Switching through singularities. In: Proceedings of the 36th IEEE conference on decision and control, pp 1–6
20. Zhang Y, Yu X, Yin Y, Xiao L, Fan Z (2013) Using GD to conquer the singularity problem of conventional controller for output tracking of nonlinear system of a class. Phys Lett A 377(25):1611–1614
21. Zhang Y, Chen K, Tan H (2009) Performance analysis of gradient neural network exploited for online time-varying matrix inversion. IEEE Trans Autom Control 54(8):1940–1945
22. Zhang Y, Yang Y, Ruan G (2011) Performance analysis of gradient neural network exploited for online time-varying quadratic minimization and equality-constrained quadratic programming. Neurocomputing 74(10):1710–1719
23. Kaplan R (2000) The nothing that is: a natural history of zero. Oxford University Press, New York
24. Bacciotti A, Mazzi L (2000) A necessary and sufficient condition for bounded-input bounded-state stability of nonlinear systems. SIAM J Control Optim 39(2):478–491.
25. Isidori A (2013) The zero dynamics of a nonlinear system: from the origin to the latest progresses of a long successful story. Eur J Control 19(5):369–378
26. Zhang Z, Zhang Y (2013) Design and experimentation of acceleration-level drift-free scheme aided by two recurrent neural networks. IET Control Theory Appl 7(1):25–42
27. Liao B, Zhang Y (2014) Different complex ZFs leading to different complex ZNN models for time-varying complex generalized inverse matrices. IEEE Trans Neural Netw Learn Syst 25(9):1621–1631
28. Li Y, Tang W, Chen G (2005) Generating hyperchaos via state feedback control. Int J Bifurcat Chaos 15(10):3367–3375.

Chapter 12
ZG Trajectory Generation of Van der Pol Oscillator

Abstract In this chapter, a classic nonlinear system of Van der Pol oscillator in the affine-control form is investigated. The ZG method is utilized to design a ZG controller for the trajectory generation of the aforementioned nonlinear oscillator. Simulation results further illustrate the feasibility and efficacy of the ZG controller with the DBZ problem conquered. In addition, the effects of ZD and GD design parameters on the performance of ZG controller are investigated.

12.1 Introduction

Trajectory generation is one of the classic and important issues related to almost all kinds of oscillators and systems in control field, which has also been investigated in various scientific and engineering applications, e.g., robot control [1], flight control [2], and motor control [3]. Generally speaking, the trajectory generation is to design a controller in terms of control input u for the oscillator such that its output y generates the desired trajectory y_{d}.

Nonlinear problems are of interest in engineering because most systems are inherently nonlinear. The general form of an affine-control nonlinear system can be formulated as the following vector-form differential equation:

$$\dot{\mathbf{x}} = \mathbf{f}(\mathbf{x}, u) = \mathbf{g}(\mathbf{x}) + \mathbf{h}(\mathbf{x})u,$$

where $\mathbf{x} \in \mathbb{R}^n$ is the state vector, $u \in \mathbb{R}$ is the control input of the nonlinear system. Besides, $\mathbf{g}(\mathbf{x})$ and $\mathbf{h}(\mathbf{x})$ are known continuous functions of $\mathbf{x}$.

The Van der Pol oscillator is a classic nonlinear system with self-oscillatory affine-control form, which is often viewed as an important mathematical model that can be used in complicated and modified systems as well as in the control field [4]. The Van der Pol oscillator can be further expressed as the following second-order nonlinear differential equations:

$$\begin{cases} \dot{x}_1 = f_1(x_2) = x_2, \\ \dot{x}_2 = f_2(x_1, x_2, u) = -x_1 - \sigma(1 - x_1^2)x_2 + x_1 u, \end{cases} \tag{12.1}$$

Y. Zhang et al., *Zhang-Gradient Control*,
https://doi.org/10.1007/978-981-15-8257-8_12

where x_1 and x_2 are the state variables, and the parameter σ describes the strength of the damping effect, which is set as $\sigma = 0.5$ in this chapter [5].

The conventional method of tracking control for nonlinear systems is IOL [6, 7]. However, the IOL method cannot handle the DBZ problem. Motivated by this, it is important and meaningful to develop an effective method for solving such a difficult problem. Recently, both the ZD and GD methods have been widely investigated for online time-varying problem solving [8–12]. By combining the ZD and GD methods together, this chapter [5] utilizes the ZG method to design a ZG controller for the trajectory generation of Van der Pol oscillator in the affine-control form. Note that, by applying only the ZD method, the resultant ZD controller for Van der Pol oscillator (12.1) encounters the DBZ problem, possibly resulting in the system crash. For such a troublesome problem, the presented ZG controller can conquer it successfully (in addition to the effective trajectory generation).

12.2 Design of ZD Controller

In this section, a ZD controller is designed and investigated for the trajectory generation of Van der Pol oscillator in the affine-control form (12.1). Without loss of generality, the output of Van der Pol oscillator (12.1) is set as $y = x_1$ throughout this chapter [5].

Firstly, we construct the first ZF [8–12] as below:

$$z_1 = y - y_{\mathrm{d}} \in \mathbb{R},$$

where y_{d} is the desired trajectory for y. Note that the tracking error is also formulated as $e = y - y_{\mathrm{d}}$. In order to force the tracking error e to converge to zero, the following ZD design formula [8–12] is used:

$$\dot{z}_1 = -\lambda z_1,$$

where the design parameter $\lambda \in \mathbb{R}^+$ is used to scale the convergence rate of the ZD solution. Thus, we have

$$\dot{x}_1 - \dot{y}_{\mathrm{d}} = -\lambda(x_1 - y_{\mathrm{d}}).$$

From (12.1), we further have

$$x_2 - \dot{y}_{\mathrm{d}} = -\lambda(x_1 - y_{\mathrm{d}}).$$

Secondly, constructing the second ZF as

$$z_2 = x_2 - \dot{y}_{\mathrm{d}} + \lambda(x_1 - y_{\mathrm{d}}),$$

and applying the ZD design formula again as $\dot{z}_2 = -\lambda z_2$, we have

$$\dot{x}_2 - \ddot{y}_d + \lambda(\dot{x}_1 - \dot{y}_d) = -\lambda(x_2 - \dot{y}_d + \lambda(x_1 - y_d)). \tag{12.2}$$

Finally, by substituting (12.1) into (12.2), the ZD controller in the form of u is obtained as below:

$$u = \frac{x_1 + 0.5(1 - x_1^2)x_2 + \ddot{y}_d - 2\lambda(x_2 - \dot{y}_d) - \lambda^2(x_1 - y_d)}{x_1}. \tag{12.3}$$

Since ZD controller (12.3) is obtained by applying the ZD method twice and without applying the GD method, it is also termed z2g0 controller.

From (12.3), we know that the divisor of z2g0 controller is x_1. Evidently, if x_1 is close to zero, the control input u tends to infinity. Note that, in practical applications, it is impossible or difficult to implement z2g0 controller (12.3). Thus, z2g0 controller (12.3) for the trajectory generation of Van der Pol oscillator (12.1) cannot get through the DBZ point. In other words, z2g0 controller (12.3) can only run well when x_1 is far away from zero. In view of the limitation of ZD controller, it is necessary to develop a DBZ-conquering controller for the trajectory generation of Van der Pol oscillator in the affine-control form (12.1).

12.3 Design of ZG Controller

In order to handle the DBZ problem in z2g0 controller (12.3), the GD method [8, 9] is thus introduced and exploited to develop a DBZ-conquering controller. Specifically, by applying the ZG method, a ZG controller is designed for the trajectory generation of Van der Pol oscillator (12.1) with the DBZ problem conquered.

From (12.3), we further define

$$\phi = x_1 u - x_1 - 0.5(1 - x_1^2)x_2 - \ddot{y}_d + 2\lambda(x_2 - \dot{y}_d) + \lambda^2(x_1 - y_d),$$

which should theoretically be zero. According to the GD method [8, 9], an EF can be defined as $\epsilon = \phi^2/2$. Then, we apply the following GD design formula:

$$\dot{u} = -\gamma\frac{\partial \epsilon}{\partial u},$$

where the design parameter $\gamma \in \mathbb{R}^+$ is used to scale the convergence rate of the GD solution. Thus, the ZG controller in the form of $\dot{u}$ is obtained as

$$\dot{u} = -\gamma x_1 \phi. \tag{12.4}$$

Since ZG controller (12.4) is obtained by applying the ZD method twice and applying the GD method once, it is also termed z2g1 controller. According to (12.4), we know that the z2g1 controller has no division operation. Therefore, z2g1 controller (12.4) can effectively conquer the DBZ problem encountered in z2g0 controller (12.3) (which can also be seen in the ensuing section).

12.4 Simulation, Verification and Comparison

In this section, we conduct the computer simulations and compare the performance of z2g0 controller (12.3) and z2g1 controller (12.4). Besides, we investigate the effects of both GD design parameter γ and ZD design parameter λ on z2g1 controller (12.4) for the trajectory generation of Van der Pol oscillator (12.1).

12.4.1 Comparison Between ZD and ZG Controllers

In this subsection, the computer simulations of generating desired trajectories $y_\mathrm{d} = \sin(t)$ and $y_\mathrm{d} = \cos(2t)\exp(0.1t)$ are performed by means of z2g0 controller (12.3) and z2g1 controller (12.4). The initial states are set as $x_1(0) = 0.1$ and $x_2(0) = 0.5$ for such two controllers, as well as the initial control input $u(0) = 0$ for z2g1 controller (12.4). Besides, we set $\lambda = 5$ for such two controllers and $\gamma = 10^5$ for z2g1 controller (12.4).

As shown in Figs. 12.1 and 12.2, for z2g0 controller (12.3), as time t evolves to around 1.873 s, the magnitude of control input u becomes extremely large (specifically, see Fig. 12.2a), which finally leads to the crash of the oscillator in the form of simulation crash. In contrast, the output trajectory with z2g1 controller (12.4) tracks the desired trajectory $y_\mathrm{d} = \sin(t)$ uninterruptedly, accurately and successfully. As seen from Fig. 12.1f, the absolute tracking error of Van der Pol oscillator (12.1) equipped with z2g1 controller (12.4) is small enough. Besides, Fig. 12.2b shows that the control input u with z2g1 controller (12.4) is bounded and small. The above simulation results indicate that z2g1 controller (12.4) can effectively conquer the DBZ problem.

To further illustrate the superiority of z2g1 controller (12.4) in handling the DBZ problem, another example with relatively complex desired trajectory $y_\mathrm{d} = \cos(2t)\exp(0.1t)$ is simulated, and the corresponding simulation results are shown in Figs. 12.3 and 12.4. As seen from Figs. 12.3a, c, e, and 12.4a, the trajectory-generation task of Van der Pol oscillator (12.1) equipped with z2g0 controller (12.3) stops when time t approaches around 1.126 s. In contrast, z2g1 controller (12.4) can get through the DBZ points of $x_1 = 0$. That is, the z2g1 controller can conquer the DBZ problem successfully. In summary, the above comparative simulation results substantiate well that z2g1 controller (12.4) can accomplish the trajectory-generation task of Van der Pol oscillator (12.1) with the DBZ problem conquered successfully.

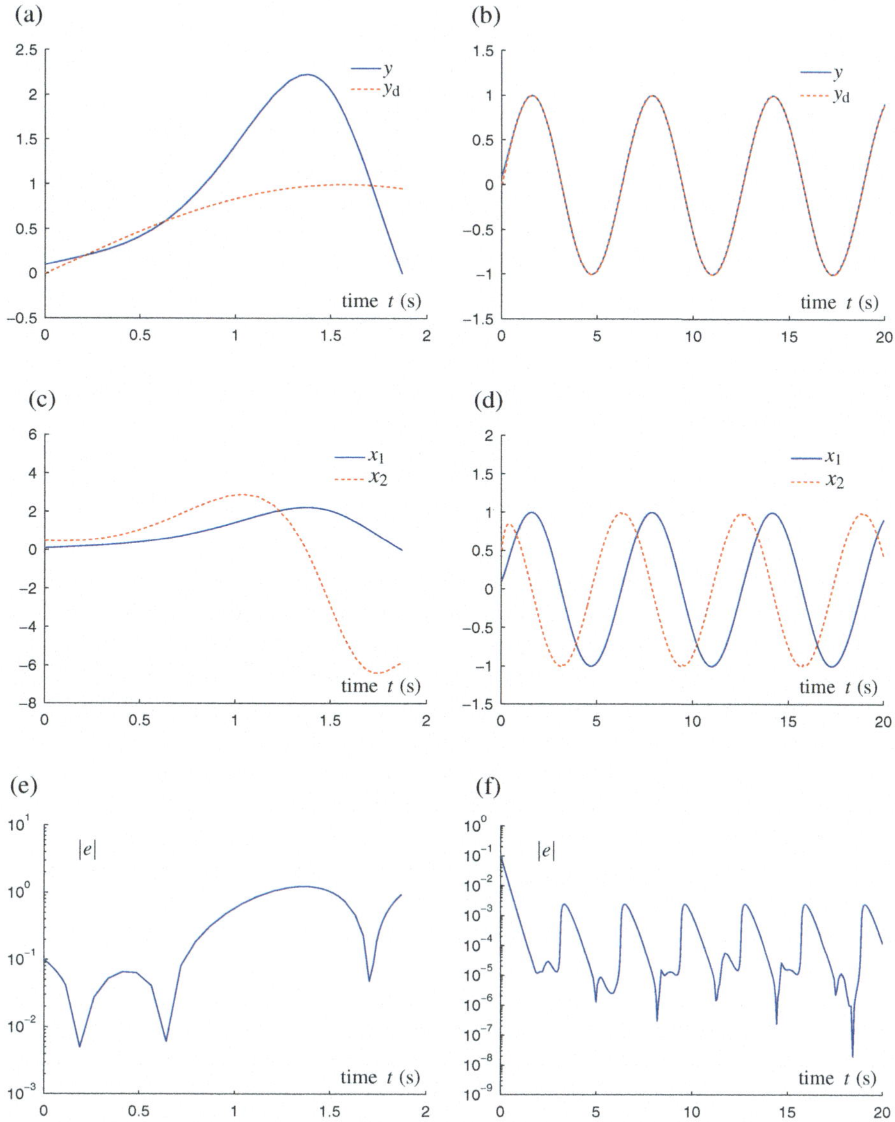

Fig. 12.1 Trajectory-generation performance of Van der Pol oscillator (12.1) equipped with z2g0 controller (12.3) and z2g1 controller (12.4), respectively, for desired trajectory $y_d = \sin(t)$. (**a**) Output trajectory with z2g0 controller (12.3) and desired trajectory. (**b**) Output trajectory with z2g1 controller (12.4) and desired trajectory. (**c**) System states with z2g0 controller (12.3). (**d**) System states with z2g1 controller (12.4). (**e**) Absolute tracking error with z2g0 controller (12.3). (**f**) Absolute tracking error with z2g1 controller (12.4)

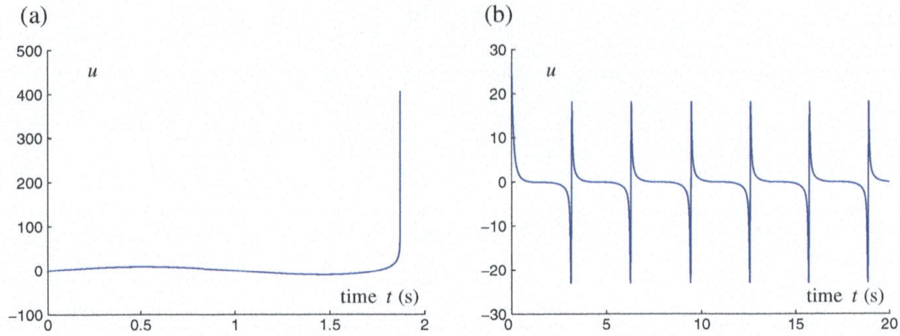

Fig. 12.2 Control inputs of Van der Pol oscillator (12.1) equipped with z2g0 controller (12.3) and z2g1 controller (12.4), respectively, for desired trajectory $y_d = \sin(t)$. (**a**) Control input with z2g0 controller (12.3). (**b**) Control input with z2g1 controller (12.4)

12.4.2 Effect of ZD Design Parameter on ZG Controller

As presented above, z2g1 controller (12.4) has much better performance than z2g0 controller (12.3) for the DBZ-containing trajectory generation of Van der Pol oscillator (12.1). In this subsection, we investigate the effect of ZD design parameter λ on z2g1 controller (12.4) for the trajectory generation of Van der Pol oscillator (12.1). The computer simulations of generating desired trajectory $y_d = \sin(t)$ are performed by means of z2g1 controller (12.4). The initial states are set as $x_1(0) = 0.1$ and $x_2(0) = 0.5$, and the initial control input is $u(0) = 0$ for z2g1 controller (12.4). The simulation results are shown in Figs. 12.5 and 12.6, where the GD design parameter is fixed as $\gamma = 10^5$, and the ZD design parameter λ is selected as $\lambda = 5$, 10, 20 or 30. Specifically, Fig. 12.5 shows the absolute tracking errors of Van der Pol oscillator (12.1) equipped with z2g1 controller (12.4) using different values of ZD design parameter λ. As seen from Fig. 12.5, with the increase of ZD design parameter λ, the maximal steady-state value of $|e|$ becomes smaller and smaller. Evidently, we can improve the output of z2g1 controller (12.4) by increasing properly the value of ZD design parameter λ. Besides, Fig. 12.6 shows the control inputs of Van der Pol oscillator (12.1) equipped with z2g1 controller (12.4) using different values of ZD design parameter λ. As shown in Fig. 12.6, when the value of ZD design parameter λ turns to be larger, the magnitude of the peak value of control input u becomes larger, and the peak value becomes more sharp so as to achieve the results of more accurate trajectory generation.

Therefore, the tradeoff of selecting an appropriate value of ZD design parameter λ for the z2g1 controller is between convenient control input generation and desired output trajectory generation.

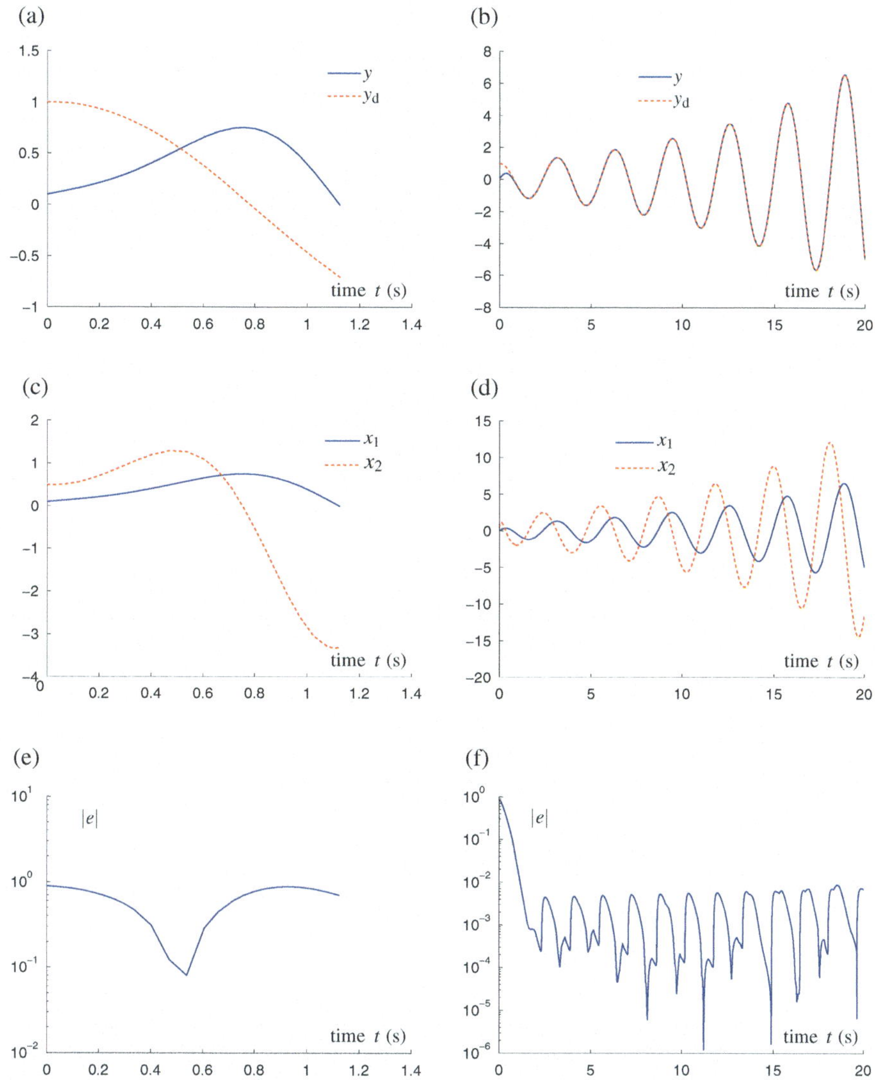

Fig. 12.3 Trajectory-generation performance of Van der Pol oscillator (12.1) equipped with z2g0 controller (12.3) and z2g1 controller (12.4), respectively, for desired trajectory $y_{\rm d} = \cos(2t)\exp(0.1t)$. (**a**) Output trajectory with z2g0 controller (12.3) and desired trajectory. (**b**) Output trajectory with z2g1 controller (12.4) and desired trajectory. (**c**) System states with z2g0 controller (12.3). (**d**) System states with z2g1 controller (12.4). (**e**) Absolute tracking error with z2g0 controller (12.3). (**f**) Absolute tracking error with z2g1 controller (12.4)

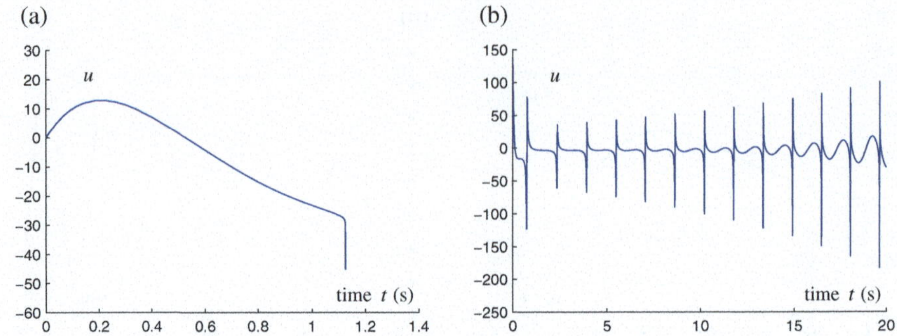

Fig. 12.4 Control inputs of Van der Pol oscillator (12.1) equipped with z2g0 controller (12.3) and z2g1 controller (12.4), respectively, for desired trajectory $y_d = \cos(2t)\exp(0.1t)$. (**a**) Control input with z2g0 controller (12.3). (**b**) Control input with z2g1 controller (12.4)

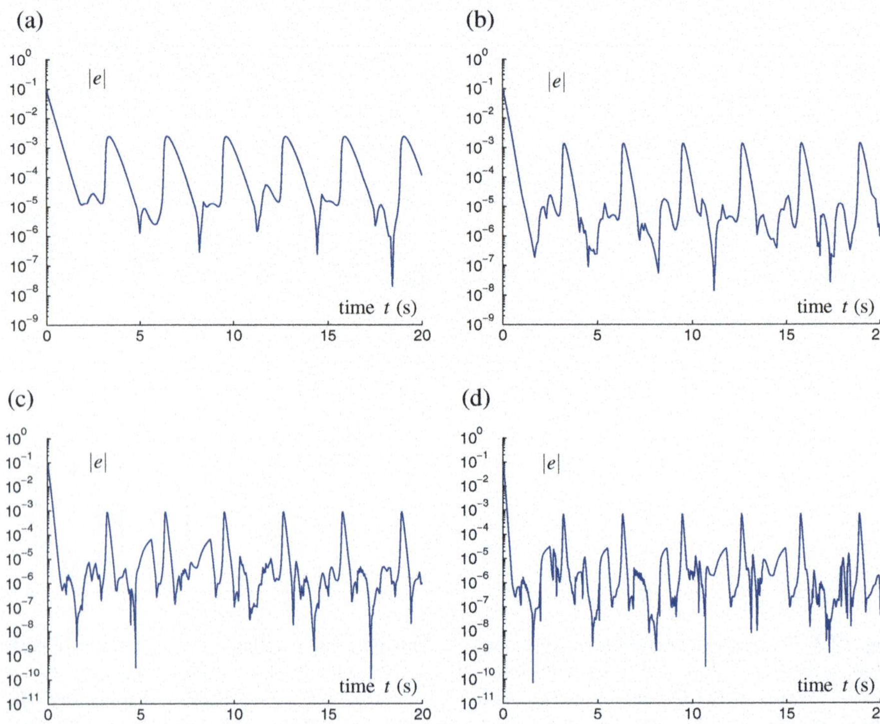

Fig. 12.5 Absolute tracking errors of Van der Pol oscillator (12.1) equipped with z2g1 controller (12.4) using different values of ZD design parameter λ for desired trajectory $y_d = \sin(t)$. (**a**) Absolute tracking error with $\lambda = 5$. (**b**) Absolute tracking error with $\lambda = 10$. (**c**) Absolute tracking error with $\lambda = 20$. (**d**) Absolute tracking error with $\lambda = 30$

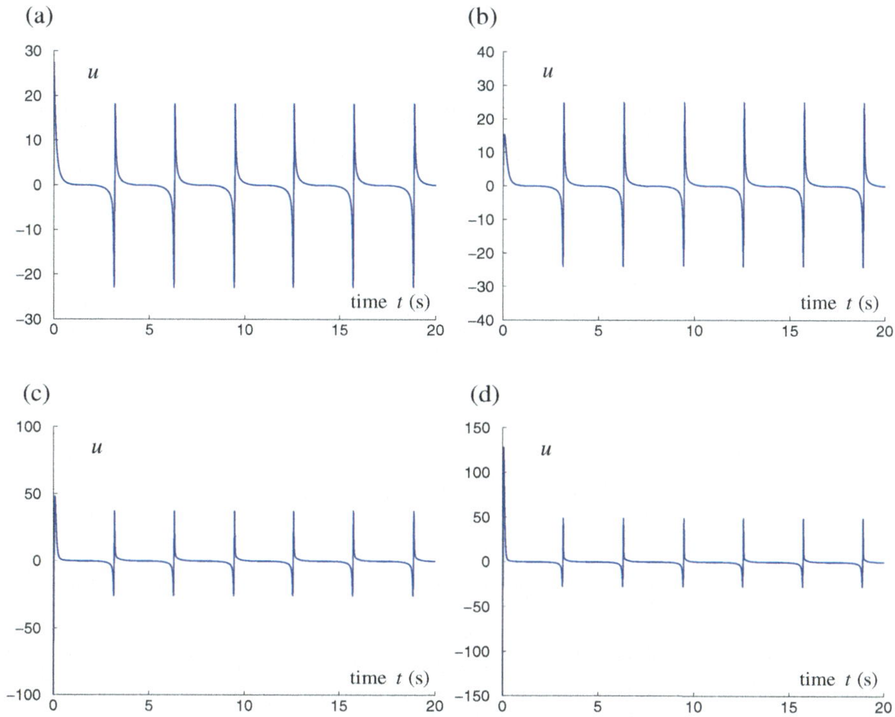

Fig. 12.6 Control inputs of Van der Pol oscillator (12.1) equipped with z2g1 controller (12.4) using different values of ZD design parameter λ for desired trajectory $y_d = \sin(t)$. (**a**) Control input with $\lambda = 5$. (**b**) Control input with $\lambda = 10$. (**c**) Control input with $\lambda = 20$. (**d**) Control input with $\lambda = 30$

12.4.3 Effect of GD Design Parameter on ZG Controller

In this subsection, we investigate the effect of GD design parameter γ on z2g1 controller (12.4) for the trajectory generation of Van der Pol oscillator (12.1).

The computer simulations of generating desired trajectory $y_d = \cos(2t)\exp(0.1t)$ are performed by means of z2g1 controller (12.4). Again, the initial states are set as $x_1(0) = 0.1$ and $x_2(0) = 0.5$, and the initial control input is $u(0) = 0$ for z2g1 controller (12.4). The simulation results are displayed in Figs. 12.7 and 12.8, where the ZD design parameter is fixed as $\lambda = 5$, and the GD design parameter is selected as $\gamma = 10, 10^3, 10^5$ or 10^7. Specifically, Fig. 12.7 shows the absolute tracking errors of Van der Pol oscillator (12.1) equipped with z2g1 controller (12.4) using different values of GD design parameter γ. As seen from Fig. 12.7, with the increase of the value of GD design parameter γ, the maximal steady-state value of $|e|$ becomes smaller and smaller. Evidently, we can improve the output of z2g1 controller (12.4) by increasing properly the value of ZG design parameter γ.

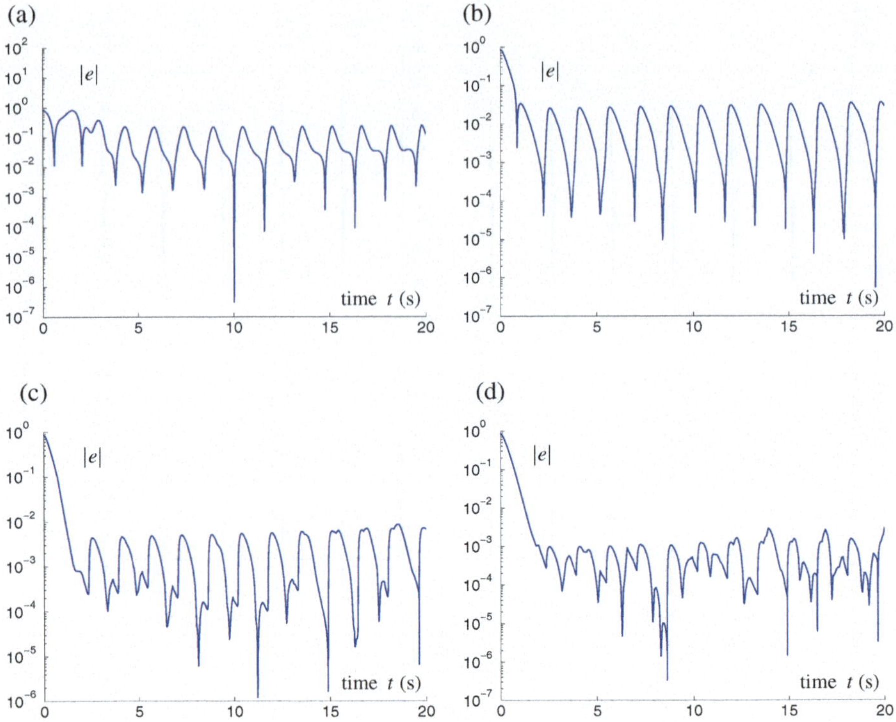

Fig. 12.7 Absolute tracking errors of Van der Pol oscillator (12.1) equipped with z2g1 controller (12.4) using different values of GD design parameter γ for desired trajectory $y_\mathrm{d} = \cos(2t)\exp(0.1t)$. (**a**) Absolute tracking error with $\gamma = 10$. (**b**) Absolute tracking error with $\gamma = 10^3$. (**c**) Absolute tracking error with $\gamma = 10^5$. (**d**) Absolute tracking error with $\gamma = 10^7$

Moreover, as shown in Fig. 12.8, when the value of GD design parameter γ turns to be larger, the magnitude of control input u becomes larger and the peak value becomes more sharp (specifically, Fig. 12.8d) so as to achieve the results of more accurate trajectory generation. It is known that the signal with large amplitude and sharp peak is hard to produce by common instruments and devices (i.e., signal generator). Thus, we are prone to select some appropriately small values of γ so as to obtain the signal of control input u more conveniently.

Considering the effect of GD design parameter γ on the tracking error $|e|$ and the control input u, we need to select a compromise value of GD design parameter γ in view of convenient control input signal and desired output.

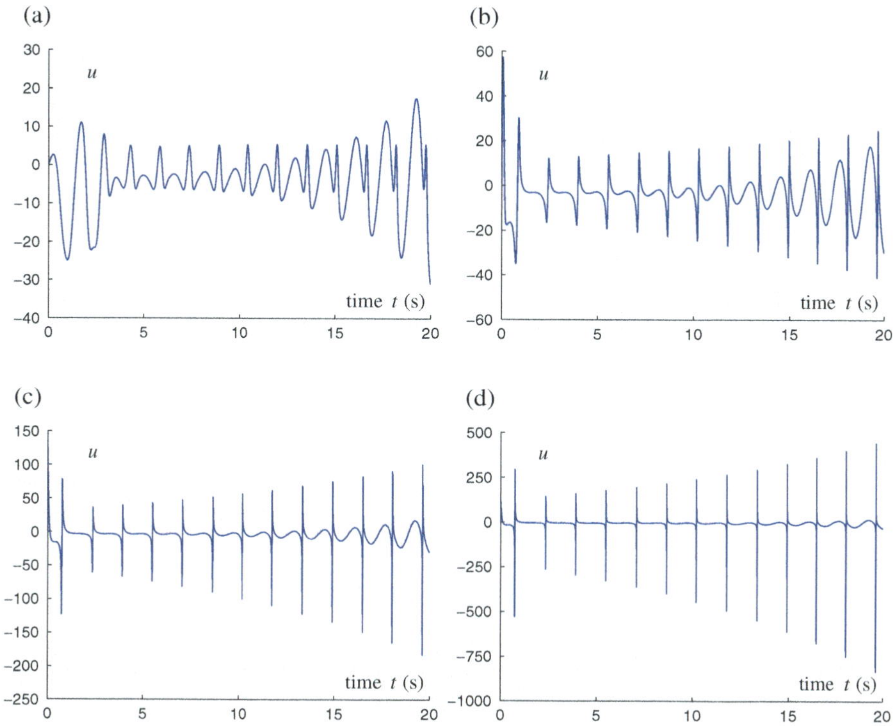

Fig. 12.8 Control inputs of Van der Pol oscillator (12.1) equipped with z2g1 controller (12.4) using different values of GD design parameter γ for desired trajectory $y_{\rm d} = \cos(2t)\exp(0.1t)$. (**a**) Control input with $\gamma = 10$. (**b**) Control input with $\gamma = 10^3$. (**c**) Control input with $\gamma = 10^5$. (**d**) Control input with $\gamma = 10^7$

12.5 Chapter Summary

In this chapter, the ZG method has been exploited to design z2g0 controller (12.3) and z2g1 controller (12.4) for the trajectory generation of Van der Pol oscillator (12.1). The comparative simulation results have shown the efficacy and superiority of z2g1 controller (12.4) in conquering the DBZ problem for the trajectory generation. Moreover, the simulation results have indicated that the ZD design parameter λ and the GD design parameter γ both have effects on the performance of z2g1 controller (12.4) for the trajectory generation of Van der Pol oscillator (12.1). Therefore, the values of design parameters λ and γ in the z2g1 controller should be appropriately selected in practical design and application.

References

1. Cai B, Zhang Y (2009) Equivalence of velocity-level and acceleration-level redundancy-resolution of manipulators. Phys Lett A 373(37):3450–3453
2. Ye D, Yang G (2006) Adaptive fault-tolerant tracking control against actuator faults with application to flight control. IEEE Trans Control Syst Technol 14(6):1088–1096
3. Braembussche PVD, Swevers V, Brussel HV, Vanherckl P (1996) Accurate tracking control of linear synchronous motor machine tool axes. Mechantronics 6(5):507–521
4. Sassano M, Astolfi A (2010) Dynamic solution of the HJB equation and the optimal control of nonlinear systems. In: Proceedings of the 49th IEEE conference on decision and control, pp 3271–3276
5. Hu C, Wang Y, Kang X, Guo D, Zhang Y (2014) ZG trajectory generation of Van der Pol oscillator in affine-control form with division-by-zero problem handled. In: Proceedings of the 10th international conference on natural computation, pp 1082–1087
6. Slotine JJE, Li W (1991) Applied nonlinear control. Prentice Hall, New Jersey
7. Guardabassi GO, Savaresi SM (2001) Approximate linearization via feedback – an overview. Automatica 37(1):1–15
8. Zhang Y, Yi C, Guo D, Zheng J (2011) Comparison on Zhang neural dynamics and gradient-based neural dynamics for online solution of nonlinear time-varying equation. Neural Comput Appl 20(1):1–7
9. Zhang Y, Yu X, Yin Y, Xiao L, Fan Z (2013) Using GD to conquer the singularity problem of conventional controller for output tracking of nonlinear system of a class. Phys Lett A 377(25):1611–1614
10. Cai B, Zhang Y (2012) Different-level redundancy resolution and it equivalent relationship analysis for robot manipulators using gradient-descent and Zhang et al.'s neural-dynamic methods. IEEE Trans Ind Electron 59(8):3146–3155
11. Zhang Y, Wang Y, Yin Y, Jin L, Chen D (2013) ZG controllers for output tracking of nonlinear mass-spring-damper mechanical system with division-by-zero problem solved. In: Proceedings of IEEE international conference on robotics and biomimetics, pp 1845–1850
12. Zhang Y, Yi C (2011) Zhang neural networks and neural-dynamic method. Nova Science Publishers, New York

Chapter 13
ZD, ZG and IOL Controllers for AFN System

Abstract In this chapter, by following the ZG method, a ZD controller and a ZG controller are presented for the tracking control of AFN system, which may encounter the DBZ problem. For comparison, the conventional IOL controller is also presented. The ZD, ZG and IOL controllers are compared in different relative-degree cases, i.e., the standard relative-degree case, the pseudo-DBZ (PDBZ) relative-degree case, and the true-DBZ (TDBZ) relative-degree case. Note that the ZG controller is valid in the three relative-degree cases, while the ZD and IOL controllers are valid only in the standard relative-degree case and the PDBZ relative-degree case. In addition, theoretical results and analyses on the ZD and ZG controllers are provided. Corresponding computer simulations are further performed to illustrate the tracking performance of the ZD, ZG and IOL controllers, as well as to show the superiority of the ZG controller in conquering the TDBZ problem for the tracking control of AFN system.

13.1 Introduction

In light of the complexity and uncertainty of nonlinear systems, much effort has been devoted to solving the nonlinear system problems [1–3]. As an important part of nonlinear system problems, the tracking control of nonlinear systems has attracted much attention in recent decades [3–5]. For example, in [4], Chang et al. addressed the problem of designing robust tracking control for a large class of uncertain nonlinear chaotic systems. In [5], Niu et al. solved the problem of robust stabilization and tracking control for a class of switched nonlinear systems via the approach of multiple Lyapunov functions.

For decades, the tracking-control problem has been investigated since it is frequently encountered in practical applications [1–3]. For solving this problem, a conventional method is IOL [1]. However, the limitation of the IOL controller is evident. To be more specific, if the AFN system encounters the DBZ problem, the IOL controller cannot be implemented. The DBZ problem is a challenging subject in the area of tracking control [2, 3, 6, 7].

Y. Zhang et al., *Zhang-Gradient Control*,
https://doi.org/10.1007/978-981-15-8257-8_13

In this chapter [8], we further investigate the tracking-control problem of AFN system depicted in (11.1). Specifically, by applying the ZG method, both a ZD controller and a ZG controller are designed for the tracking control of AFN system (11.1) with the DBZ problem conquered.

13.2 Design of Controllers

In this section, the detailed design procedures of ZD and ZG controllers are provided for AFN system (11.1).

13.2.1 Design of ZD Controller

To monitor and control the tracking process of the system output y for the desired trajectory y_d, a series of ZFs are constructed until the control input u appears explicitly. The number of ZFs constructed for obtaining the explicit expression of u is intuitively termed the u-appearing index. For presentation convenience, let z_j denote the jth ZF of y for y_d, where j is a positive integer. The specific design procedures of ZD Controller are shown as follows.

The first ZF of y is constructed as

$$z_1 = y - y_\mathrm{d} \in \mathbb{R}. \tag{13.1}$$

Then, the following ZD design formula [9, 10] is adopted:

$$\dot{z}_1 = -\lambda z_1, \tag{13.2}$$

where the design parameter $\lambda \in \mathbb{R}^+$ is used to scale the convergence rate of the ZD solution. If (13.2) does not contain u explicitly, more ZFs need to be constructed by using ZD design formulas. The second ZF is constructed as $z_2 = \dot{z}_1 + \lambda z_1$, and the corresponding ZD design formula $\dot{z}_2 = -\lambda z_2$ is adopted. By following the similar constructing procedure, the jth ZF is recursively constructed as

$$z_j = \dot{z}_{j-1} + \lambda z_{j-1}, \quad \text{with} \quad j = 2, 3, \cdots.$$

Through the derivation, the ith ZF of y for y_d is expressed as

$$z_i = \sum_{l=0}^{i-1} \mathrm{C}_{i-1}^{l} \lambda^{i-1-l} \left(y^{(l)} - y_\mathrm{d}^{(l)} \right), \quad \text{with} \ i = 1, 2, \cdots, \tag{13.3}$$

where $\mathrm{C}_{i-1}^{l} = (i-1)!/(l!(i-1-l)!)$ denotes a binomial coefficient; $y^{(l)}$ and $y_\mathrm{d}^{(l)}$ denote the lth-order time derivatives of y and y_d, respectively. Note that the derivation procedure of (13.3) is given in Appendix 1.

We suppose that the control input u appears explicitly in the νth ZF, where ν is called the u-appearing index, which means that the following two properties hold true in the region of interest denoted by $\mathbb{I} \subseteq \mathbb{R}^n$ (or say, the possible region that state vector $\mathbf{x}$ can reach):

- $\forall \mathbf{x} \in \mathbb{I}$, $\partial z_i/\partial u = 0$, for $2 \leq i \leq \nu - 1$;
- $\exists \mathbf{x} \in \mathbb{I}$, $\partial z_\nu/\partial u \neq 0$.

Note that we start i with 2 for the following presentation convenience, as, with i being 1, $\partial z_1/\partial u = \partial y/\partial u = \partial y(\mathbf{x})/\partial u = 0$ always holds true for AFN system (11.1). In view of (13.3), we have

$$\frac{\partial z_i}{\partial u} = \sum_{l=0}^{i-1} \mathrm{C}_{i-1}^{l} \lambda^{i-1-l} \left(\frac{\partial y^{(l)}}{\partial u}\right).$$

Then, the above two properties become

- $\forall \mathbf{x} \in \mathbb{I}$, $\partial z_i/\partial u = \partial y^{(i-1)}/\partial u = 0$, for $2 \leq i \leq \nu - 1$;
- $\exists \mathbf{x} \in \mathbb{I}$, $\partial z_\nu/\partial u = \partial y^{(\nu-1)}/\partial u \neq 0$.

According to AFN system (11.1), we know

$$y^{(1)} = \frac{\partial y(\mathbf{x})}{\partial \mathbf{x}}\mathbf{g}(\mathbf{x}) + \frac{\partial y(\mathbf{x})}{\partial \mathbf{x}}\mathbf{h}(\mathbf{x})u = L_g y(\mathbf{x}) + L_h y(\mathbf{x})u,$$

where $L_g y(\mathbf{x})$ denotes the first Lie derivative of $y(\mathbf{x})$ with respect to $\mathbf{g}(\mathbf{x})$ [1]. Thus, if $\partial y^{(1)}/\partial u = 0$, we have $L_h y(\mathbf{x}) = 0$ and $y^{(1)} = L_g y(\mathbf{x})$. Then, we have

$$y^{(2)} = \frac{\partial L_g y(\mathbf{x})}{\partial \mathbf{x}}\mathbf{g}(\mathbf{x}) + \frac{\partial L_g y(\mathbf{x})}{\partial \mathbf{x}}\mathbf{h}(\mathbf{x})u = L_g^2 y(\mathbf{x}) + L_h L_g y(\mathbf{x})u,$$

where $L_g^2 y(\mathbf{x})$ denotes the second Lie derivative of $y(\mathbf{x})$ with respect to $\mathbf{g}(\mathbf{x})$. If $\partial y^{(2)}/\partial u = 0$, we have $L_h L_g y(\mathbf{x}) = 0$ and $y^{(2)} = L_g^2 y(\mathbf{x})$. From the above analysis, we can obtain $L_h L_g^{i-2} y(\mathbf{x}) = 0$ and $y^{(i)} = L_g^i y(\mathbf{x}) + L_h L_g^{i-1} y(\mathbf{x})u$, if $\partial y^{(0)}/\partial u = \partial y^{(1)}/\partial u = \cdots = \partial y^{(i-1)}/\partial u = 0$. Note that $L_g^0 y(\mathbf{x}) = y(\mathbf{x})$. Then, the above two properties become

- $\forall \mathbf{x} \in \mathbb{I}$, $\partial z_i/\partial u = \partial y^{(i-1)}/\partial u = L_h L_g^{i-2} y = 0$, for $2 \leq i \leq \nu - 1$;
- $\exists \mathbf{x} \in \mathbb{I}$, $\partial z_\nu/\partial u = \partial y^{(\nu-1)}/\partial u = L_h L_g^{\nu-2} y \neq 0$.

Furthermore, in view of (13.3), $z_\nu = \sum_{l=0}^{\nu-1} \mathrm{C}_{\nu-1}^{l} \lambda^{\nu-1-l}\left(y^{(l)} - y_{\mathrm{d}}^{(l)}\right)$, in which $y^{(\nu-1)}$ is the only element that contains u explicitly. Then, $z_\nu = \sum_{l=0}^{\nu-2} \mathrm{C}_{\nu-1}^{l} \lambda^{\nu-1-l}\left(y^{(l)} - y_{\mathrm{d}}^{(l)}\right) + L_g^{\nu-1} y + (L_h L_g^{\nu-2} y)u - y_{\mathrm{d}}^{(\nu-1)}$. To make z_ν converge to zero, the ZD controller can be designed as

$$u = -\frac{1}{L_h L_g^{\nu-2} y}\left(L_g^{\nu-1} y - y_{\mathrm{d}}^{(\nu-1)} + \sum_{l=0}^{\nu-2} \mathrm{C}_{\nu-1}^{l} \lambda^{\nu-1-l}\left(y^{(l)} - y_{\mathrm{d}}^{(l)}\right)\right). \tag{13.4}$$

Note that the above two properties are necessary conditions instead of sufficient conditions for the validity of ZD controller (13.4), since we force z_ν to zero and then the term $L_h L_g^{\nu-2} y$ becomes the divisor. Thus, the term $\mathbf{h}(\mathbf{x})u = (\mathbf{h}(\mathbf{x})\cdots)/L_h L_g^{\nu-2} y$ may encounter the DBZ problem. By excluding the DBZ case, ZD controller (13.4) is valid if either of the following two cases holds true:

- standard relative-degree case: $\forall \mathbf{x} \in \mathbb{S}$, $L_h L_g^{i-2} y = 0$, for $2 \le i \le \nu - 1$, and $\forall \mathbf{x} \in \mathbb{S}$, $L_h L_g^{\nu-2} y \neq 0$;
- pseudo-DBZ (PDBZ) relative-degree case: $\forall \mathbf{x} \in \mathbb{P}$, $L_h L_g^{i-2} y = 0$, for $2 \le i \le \nu - 1$, and $\exists \mathbf{x} \in \mathbb{P}$, $L_h L_g^{\nu-2} y = 0$ and $\exists \mathbf{x} \in \mathbb{P}$, $L_h L_g^{\nu-2} y \neq 0$, where there exists $\mathbf{h}(\mathbf{x}) = \Psi(\mathbf{x}) L_h L_g^{\nu-2} y$ with $\Psi(\mathbf{x}) \in \mathbb{R}^n$ being bounded.

Remark 13.1 In terms of the PDBZ relative-degree case, it appears that $u \to +\infty$ when $L_h L_g^{\nu-2} y \to 0$, and thus ZD controller (13.4) crashes. However, we know $\dot{\mathbf{x}} = \mathbf{g}(\mathbf{x}) + \mathbf{h}(\mathbf{x})u$ from AFN system (11.1). If $\mathbf{h}(\mathbf{x}) = \Psi(\mathbf{x}) L_h L_g^{\nu-2} y$, $\mathbf{h}(\mathbf{x})u$ is bounded, although $u \to +\infty$. Moreover, in practice, the computation is conducted discretely by an ODE solver in control process or simulation process, such as MATLAB ode15s, which is employed in the simulation part of this chapter [8]. Specifically, if $L_h L_g^{\nu-2} y|_{t=t_s} = 0$, $\mathbf{h}(\mathbf{x})u$ does not tend to infinity as $t \to t_s$, and thus the error generated by ode15s solver does not increase. Then, the step size Δt does not decrease. Finally, ZD controller (13.4) gets through the time instant t_s. Note that, if Δt continually decreases with $t \to t_s$ for desired calculative precision, ZD controller (13.4) crashes when Δt cannot be smaller. Besides, related computer simulations are conducted in the ensuing Sect. 13.5.2.

For comparison purposes, the conventional IOL controller [1] is also presented. Let $\tilde{\mathbf{y}} = [y, \dot{y}, \cdots, y^{(\nu-2)}]^{\mathrm{T}} \in \mathbb{R}^{\nu-1}$ and $\tilde{\mathbf{y}}_{\mathrm{d}} = [y_{\mathrm{d}}, \dot{y}_{\mathrm{d}}, \cdots, y_{\mathrm{d}}^{(\nu-2)}]^{\mathrm{T}} \in \mathbb{R}^{\nu-1}$. Besides, the tracking-error vector is defined as $\tilde{\mathbf{e}} = [e, \dot{e}, \cdots, e^{(\nu-2)}]^{\mathrm{T}} = \tilde{\mathbf{y}} - \tilde{\mathbf{y}}_{\mathrm{d}} \in \mathbb{R}^{\nu-1}$. Then, the IOL controller is presented as

$$u = \frac{1}{L_h L_g^{\nu-2} y}\left(-L_g^{\nu-1} y + y_{\mathrm{d}}^{(\nu-1)} - \tilde{\mathbf{k}}^{\mathrm{T}} \tilde{\mathbf{e}}\right), \tag{13.5}$$

where $\tilde{\mathbf{k}} = [k_1, k_2, \cdots, k_{\nu-1}]^{\mathrm{T}}$ is a parameter vector. It is worth mentioning here that IOL controller (13.5) exactly corresponds to the one given in Lemma 11.1 of Chap. 11, i.e., (11.2). To be more specific, the relation between the value of the u-appearing index ν and the value of the well-defined relative degree r_{w} is $\nu = r_{\mathrm{w}} + 1$.

Moreover, by comparing ZD controller (13.4) with IOL controller (13.5), the connection and difference can be obtained as follows.

- The ZD controller can be viewed as a more concise form of the IOL controller, because the latter has the same expression as the former if the parameter vector $\tilde{\mathbf{k}}$ is selected as $[\mathrm{C}_{\nu-1}^{0}\lambda^{\nu-1}, \mathrm{C}_{\nu-1}^{1}\lambda^{\nu-2}, \cdots, \mathrm{C}_{\nu-1}^{\nu-2}\lambda]^{\mathrm{T}}$.
- The ZD controller has only one design parameter, i.e., λ, which can be simply set as a positive number, while the parameter vector $\tilde{\mathbf{k}}$ in the IOL controller is required to be skillfully selected such that all roots of the characteristic

polynomial $\tilde{P}(s) = s^{\nu-1} + k_{\nu-1}s^{\nu-2} + \cdots + k_2 s + k_1$ are in the open left-half complex plane.

In terms of the IOL controller, the characteristic polynomial may have different roots. Although all roots are in the open left-half complex plane, there is a root, which relatively approaches zero compared with other roots. Thus, the root relatively approaching zero affects the convergence performance. In contrast, all roots of the characteristic polynomial of the ZD controller are the same (i.e., $-\lambda$), which could be the best simultaneously. Thus, from the perspective of the roots, the ZD controller may be a specific IOL controller with the best parameter vector chosen. In view of the connection between the ZD and IOL controllers, the IOL controller is also valid in the standard relative-degree case and the PDBZ relative-degree case.

13.2.2 Design of ZG Controller

Let us now consider the true-DBZ (TDBZ) relative-degree case:

$\forall \mathbf{x} \in \mathbb{T}$, $L_h L_g^{i-2} y = 0$, for $2 \leq i \leq \nu - 1$, and $\exists \mathbf{x} \in \mathbb{T}$, $L_h L_g^{\nu-2} y = 0$ and $\exists \mathbf{x} \in \mathbb{T}$, $L_h L_g^{\nu-2} y \neq 0$, where there does not exist bounded $\Psi(\mathbf{x})$ such that $\mathbf{h}(\mathbf{x}) = \Psi(\mathbf{x}) L_h L_g^{\nu-2} y$.

Note that the ZD and IOL controllers are not valid in the above TDBZ relative-degree case. Based on the GD method and ZD controller (13.4), a ZG controller is further designed to conquer the DBZ problem as follows.

Specifically, a square-based EF is firstly defined as

$$\epsilon = \frac{\phi^2}{2} \tag{13.6}$$

with $\phi = z_\nu = \sum_{l=0}^{\nu-2} \mathrm{C}_{\nu-1}^{l} \lambda^{\nu-1-l} \left(y^{(l)} - y_{\mathrm{d}}^{(l)} \right) + L_g^{\nu-1} y + (L_h L_g^{\nu-2} y) u - y_{\mathrm{d}}^{(\nu-1)}$. Then, we adopt the following GD design formula [11]:

$$\dot{u} = -\gamma \frac{\partial \epsilon}{\partial u},$$

where the design parameter $\gamma \in \mathbb{R}^+$ is used to scale the convergence rate of the GD solution. Therefore, a division-free controller (i.e., the ZG controller) for the tracking control of AFN system (11.1) is designed as

$$\dot{u} = -\gamma (L_h L_g^{\nu-2} y)\phi = -\gamma L_h L_g^{\nu-2} y \Bigg(L_g^{\nu-1} y + (L_h L_g^{\nu-2} y) u - y_{\mathrm{d}}^{(\nu-1)} + \sum_{l=0}^{\nu-2} \mathrm{C}_{\nu-1}^{l} \lambda^{\nu-1-l} \left(y^{(l)} - y_{\mathrm{d}}^{(l)} \right) \Bigg). \tag{13.7}$$

Note that the validity of ZG controller (13.7) is based on the assumption of the existence of the u-appearing index ν. Besides, ZG controller (13.7) is valid not only in the standard relative-degree case and the PDBZ relative-degree case, but also in the TDBZ relative-degree case.

13.3 Convergence Performance Analysis on ZD Controller

The convergence performance analysis on ZD controller (13.4) for the tracking control of AFN system (11.1) is given by the following theorem.

Theorem 13.1 *Suppose that AFN system (11.1) has a u-appearing index* ν *and satisfies the UBIBS stability property. Starting with bounded initial state* $\mathbf{x}(0) \in \mathbb{R}^n$*, all the state trajectories of the system equipped with ZD controller (13.4) remain bounded and the tracking error exponentially converges to zero on a large scale, provided that there exists no TDBZ case during the tracking-control process.*

Proof Since the AFN system starts with a bounded initial state $\mathbf{x}(0)$, $u(0)$ is bounded in consideration of the fact that $\mathbf{g}(\mathbf{x})$, $\mathbf{h}(\mathbf{x})$ and $y(\mathbf{x})$ are continuous and smooth, which remain bounded for bounded $\mathbf{x}$. In addition, since AFN system (11.1) satisfies the UBIBS stability property, the bounded input $u(0)$ produces the bounded state $\mathbf{x}(t)$ at the next time instant. Therefore, it can be recursively derived that the control input u is bounded and all the state trajectories of the system remain bounded during the whole tracking-control process.

According to the expression of ZFs, i.e., (13.3), we readily obtain

$$z_\nu = \sum_{l=0}^{\nu-1} \mathrm{C}_{\nu-1}^{l} \lambda^{\nu-1-l}\left(y^{(l)} - y_{\mathrm{d}}^{(l)}\right) = \sum_{l=0}^{\nu-1} \mathrm{C}_{\nu-1}^{l} \lambda^{\nu-1-l} e^{(l)}. \tag{13.8}$$

By applying ZD controller (13.4), the νth ZF z_ν is forced to zero. That is,

$$\sum_{l=0}^{\nu-1} \mathrm{C}_{\nu-1}^{l} \lambda^{\nu-1-l} e^{(l)} = 0.$$

Solving the above differential equation, we have

$$e = \exp(-\lambda t) \sum_{\kappa=0}^{\nu-2} c_\kappa t^\kappa, \tag{13.9}$$

where c_κ with $\kappa = 0, 1, \cdots, \nu-2$ are constants on the basis of $e(0), \dot{e}(0), \cdots,$ $e^{(\nu-2)}(0)$. According to [12], we know

$$\left| \exp(-\lambda t) \sum_{\kappa=0}^{\nu-2} c_\kappa t^\kappa \right| \leq \bar{\alpha} \exp(-\bar{\beta} t),\ \exists \bar{\alpha} > 0,\ \bar{\beta} > 0.$$

The proof is thus completed. □

13.4 Convergence Performance Analyses on ZG Controller

The convergence performance analyses on ZG controller (13.7) for the tracking control of AFN system (11.1) are presented in this section.

13.4.1 Tight Error Bound

As previous sections show, the basic idea of ZG controller (13.7) is to transform the direct computational manner of ZD controller (13.4) to the manner of minimizing time-varying EF (13.6). Then, ZG controller (13.7) can be viewed as a solver for solving the time-varying minimization problem. To lay a basis for further analysis, ZG controller (13.7) is rewritten as

$$\dot{u} = -\gamma\alpha_1(\alpha_2 + \alpha_1 u), \tag{13.10}$$

with $\alpha_1 = L_h L_g^{\nu-2} y$ and $\alpha_2 = L_g^{\nu-1} y - y_{\mathrm{d}}^{(\nu-1)} + \sum_{l=0}^{\nu-2} \mathrm{C}_{\nu-1}^{l} \lambda^{\nu-1-l}\left(y^{(l)} - y_{\mathrm{d}}^{(l)}\right)$. Besides, let $u^* = -\alpha_2/\alpha_1$ denote the desired time-varying solution (or say, the desired control input) for the time-varying minimization problem. According to [11], we have the following lemma on ZG controller (13.7) for solving the time-varying minimization problem.

Lemma 13.1 *Consider the time-varying minimization problem of EF (13.6). By using ZG controller (13.7), the steady-state solution error is bounded with an enough large time instant t_{e}:*

$$|u - u^*| \leq \frac{\varrho}{\gamma\varsigma}, \quad \forall t \geq t_{\mathrm{e}},$$

provided that $\sqrt{\varsigma} \leq |\alpha_1|, \exists 0 < \sqrt{\varsigma} < +\infty$, and $|\dot{u}^*| = |\mathrm{d}u^*/\mathrm{d}t| \leq \varrho, \exists 0 \leq \varrho < +\infty$.

Proof See Appendix 2 for details. □

Theorem 13.2 *Consider AFN system (11.1) equipped with ZG controller (13.7) for smooth and bounded desired trajectory y_{d}, which has a u-appearing index ν and satisfies the UBIBS stability property. Starting with bounded initial state $\mathbf{x}(0) \in \mathbb{R}^n$ and control input $u(0) \in \mathbb{R}$, the following results are achieved on a large scale for the tracking control of the system.*

- *For the standard relative-degree case or the PDBZ relative-degree case, the tracking error of the system converges toward or stays within the error bound $\chi\varrho/(\lambda^{\nu-1}\gamma\varsigma)$, provided that (1) $\sqrt{\varsigma} \leq |\alpha_1| \leq \chi, \exists 0 < \sqrt{\varsigma} \leq \chi < +\infty$, and (2) $|\dot{u}^*| \leq \varrho, \exists 0 \leq \varrho < +\infty$.*
- *For the TDBZ relative-degree case, the tracking error of the system is bounded.*

Besides, all the state trajectories of the system remain bounded.

Proof For the standard relative-degree case or the PDBZ relative-degree case, from Lemma 13.1, we know that the solution error $\psi = u - u^*$ enters the steady state when $t \geq t_e$, i.e.,

$$|\psi| = |u - u^*| \leq \frac{\varrho}{\gamma\varsigma}.$$

In addition, since $u(0)$ is bounded and AFN system (11.1) satisfies the UBIBS stability property, at the next time instant, $\mathbf{x}(0_+)$ is bounded and further $u^*(0_+)$ is bounded. The control input u converges toward u^* with a maximal deviation tolerance of $\pm\varrho/(\gamma\varsigma)$, and then $u(0_+)$ is also bounded. Therefore, it can be recursively derived that $\mathbf{x}$ and u remain bounded in the standard relative-degree case or the PDBZ relative-degree case.

From the design procedures of ZG controller (13.7) and Eq. (13.8), we have

$$z_\nu = \sum_{l=0}^{\nu-1} \mathrm{C}_{\nu-1}^{l} \lambda^{\nu-1-l} e^{(l)} = \alpha_2 + \alpha_1 u = \alpha_1(-u^* + u) = \alpha_1\psi.$$

According to the result of Lemma 13.1, when $t \geq t_e$, we have

$$-\frac{\chi\varrho}{\gamma\varsigma} \leq -|\alpha_1||\psi| \leq \sum_{l=0}^{\nu-1} \mathrm{C}_{\nu-1}^{l} \lambda^{\nu-1-l} e^{(l)} = \alpha_1\psi \leq |\alpha_1||\psi| \leq \frac{\chi\varrho}{\gamma\varsigma}. \tag{13.11}$$

Then, based on Gronwall inequality [13], the two-sided inequality

$$-\exp(-\lambda t)\sum_{\kappa=0}^{\nu-2} \hat{c}_\kappa t^\kappa - \frac{\chi\varrho}{\lambda^{\nu-1}\gamma\varsigma} \leq e \leq \exp(-\lambda t)\sum_{\kappa=0}^{\nu-2} \tilde{c}_\kappa t^\kappa + \frac{\chi\varrho}{\lambda^{\nu-1}\gamma\varsigma} \tag{13.12}$$

holds true when $t \geq t_e$, where $\hat{c}_\kappa$ and $\tilde{c}_\kappa$ (with $\kappa = 0, 1, \cdots, \nu-2$) are constants on the basis of $e(t_e), \dot{e}(t_e), \cdots, e^{(\nu-2)}(t_e)$ and $\chi\varrho/(\lambda^{\nu-1}\gamma\varsigma)$. Note that the proof of (13.12) is given in Appendix 3. Furthermore, we have

$$|e| \leq \exp(-\lambda t)\sum_{\kappa=0}^{\nu-2} |c_\kappa| t^\kappa + \frac{\chi\varrho}{\lambda^{\nu-1}\gamma\varsigma},$$

where $t \geq t_e$, and $|c_\kappa| = \max\{|\hat{c}_\kappa|, |\tilde{c}_\kappa|\}$ with $\kappa = 0, 1, \cdots, \nu-2$. In light of the analysis on Eq. (13.9), we have

$$\limsup_{t\to+\infty} |e| \leq \frac{\chi\varrho}{\lambda^{\nu-1}\gamma\varsigma}.$$

Therefore, the tracking error of the system converges toward or stays with the error bound $\chi\varrho/(\lambda^{\nu-1}\gamma\varsigma)$ in the standard relative-degree case or the PDBZ relative-degree case.

For the TDBZ relative-degree case, it can be readily derived that $\lim_{t\to t_s}\dot{u} = \lim_{\alpha_1\to 0}\dot{u} = 0$ in view of $\dot{u} = -\gamma\alpha_1(\alpha_2+\alpha_1 u)$. Therefore, the control input at the DBZ time instant t_s is the same as that at the previous time instant t_{s^-}, which implies that $u(t_s) = u(t_{s^-})$. Similarly, at t_{s^+} (which is the time instant after t_s), $u(t_s) = u(t_{s^+})$. Note that, at the time instant t_{s^-}, the control input $u(t_{s^-})$ is bounded and the tracking error converges toward or stays within an error bound. Then, we have the result that $u(t_{s^-}) = u(t_s) = u(t_{s^+})$ is bounded. For AFN system (11.1), the bounded control input produces the bounded state and the bounded output. Since the desired trajectory y_d is bounded, the tracking error is bounded at the time instants t_{s^-}, t_s and t_{s^+}.

By the above analyses, the control input remains bounded in the three relative-degree cases. Since AFN system (11.1) satisfies the UBIBS stability property, the bounded input leads to the bounded state. Thus, all the state trajectories of the system remain bounded. The proof is thus completed. □

13.4.2 Exponential Convergence Rate

In Theorem 13.2, the tracking error is just asymptotic convergence, which may be less desirable in practical applications. In this subsection, we proves that the exponential convergence of ZG controller (13.7) to a relatively loose tracking-error bound. To lay a basis for further analysis, we firstly present the following lemma based on [11].

Lemma 13.2 *Consider the time-varying minimization problem of EF (13.6). By using ZG controller (13.7), the solution error* $|\psi| = |u - u^*|$ *exponentially converges toward or stays within the error bound of* $\varrho/(\omega\gamma\varsigma)$ *with loosening parameter* $\omega \in (0,1)$, *where the exponential convergence rate is* $(1-\omega)\gamma\varsigma$, *and the convergence time* t_c *is* $\ln(\omega\gamma\varsigma|\psi(0)|/\varrho)/((1-\omega)\gamma\varsigma)$ *for* $|\psi(0)| \geq \varrho/(\omega\gamma\varsigma)$ *or zero for* $|\psi(0)| \leq \varrho/(\omega\gamma\varsigma)$.

Proof See Appendix 4 for details. □

Theorem 13.3 *Consider AFN system (11.1) equipped with ZG controller (13.7) for smooth and bounded desired trajectory* y_d, *which has a u-appearing index* ν *and satisfies the UBIBS stability property. Starting with bounded initial state* $\mathbf{x}(0) \in \mathbb{R}^n$ *and control input* $u(0) \in \mathbb{R}$, *the following results are achieved on a large scale for the tracking control of the system.*

- *For the standard relative-degree case or the PDBZ relative-degree case, the tracking error of the system exponentially converges toward or stays within the error bound* $\chi\varrho/(\omega\lambda^{\nu-1}\gamma\varsigma)$ *with* $\omega \in (0,1)$, *provided that 1)* $\sqrt{\varsigma} \leq |\alpha_1| \leq \chi$, $\exists 0 < \sqrt{\varsigma} \leq \chi < +\infty$, *and 2)* $|\dot{u}^*| \leq \varrho$, $\exists 0 \leq \varrho < +\infty$.

- *For the TDBZ relative-degree case, the tracking error of the system is bounded. Besides, all the state trajectories of the system remain bounded.*

Proof For the standard relative-degree case or the PDBZ relative-degree case, from Lemma 13.2, we know that the solution error $|\psi| = |u - u^*|$ exponentially converges toward or stays within the error bound $\varrho/(\omega\gamma\varsigma)$ with $\omega \in (0, 1)$, i.e.,

$$|\psi| = |u - u^*| \leq \frac{\varrho}{\omega\gamma\varsigma}, \ \forall t \geq t_c.$$

In addition, similar to the analysis in Theorem 13.2, the control input u in this case remains bounded. Then,

$$|e| \leq \exp(-\lambda t) \sum_{\kappa=0}^{\nu-2} |c_\kappa| t^\kappa + \frac{\chi\varrho}{\omega\lambda^{\nu-1}\gamma\varsigma} \tag{13.13}$$

holds true for $t \geq t_c$, where c_κ with $\kappa = 0, 1, \cdots, \nu - 2$ are constants on the basis of $e(t_c), \dot{e}(t_c), \cdots, e^{(\nu-2)}(t_c)$ and $\chi\varrho/(\omega\lambda^{\nu-1}\gamma\varsigma)$. The term $\exp(-\lambda t)\sum_{\kappa=0}^{\nu-2} |c_\kappa| t^\kappa$ of (13.13) exponentially converges toward zero in light of the analysis on (13.9), and then the right side of (13.13) exponentially converges toward $\chi\varrho/(\omega\lambda^{\nu-1}\gamma\varsigma)$. Therefore, the tracking error of the system exponentially converges toward or stays within the error bound $\chi\varrho/(\omega\lambda^{\nu-1}\gamma\varsigma)$ in the standard relative-degree case or the PDBZ relative-degree case.

For the TDBZ relative-degree case, similar to the analysis in Theorem 13.2, the control input and the tracking error of the system in this case are bounded.

By the above analyses, the control input remains bounded in the three relative-degree cases. Since AFN system (11.1) satisfies the UBIBS stability property, the bounded input leads to the bounded state. Thus, all the state trajectories of the system remain bounded. The proof is thus completed. □

13.5 Simulation, Verification and Comparison

In this section, three illustrative examples of AFN system (11.1) equipped with IOL, ZD and ZG controllers, respectively, in different relative-degree cases (i.e., the standard relative-degree case, the PDBZ relative-degree case, and the TDBZ relative-degree case) for the tracking control are presented and discussed.

13.5.1 Standard Relative-Degree Case

Let us consider a practical system of ship steering. The linear Nomoto model [14] is presented as below:

$$\tau\dot{\upsilon} + \upsilon = \varpi\theta,$$

where υ is the yaw rate, θ is the rudder angle, τ is the time-scale constant, and ϖ is the rudder gain. For the availability of the steering equation for rapid and large rudder angles, a nonlinear characteristic for υ is added and the following AFN system is obtained [15]:

$$\tau\dot{\upsilon} + \upsilon + \eta\upsilon^3 = \varpi\theta,$$

where η is the Norrbin coefficient. For the heading angle δ, we have $\dot{\delta} = \upsilon$, where δ is the output of ship course control system. Then, the above system can be described by AFN system (11.1) with

$$\mathbf{g}(\mathbf{x}) = \begin{bmatrix} x_2 \\ \sigma_1 x_2 + \sigma_2 x_2^3 \end{bmatrix}, \ \mathbf{h}(\mathbf{x}) = \begin{bmatrix} 0 \\ \sigma_0 \end{bmatrix}, \ \text{and } y(\mathbf{x}) = x_1, \tag{13.14}$$

where $x_1 = \delta$, $x_2 = \upsilon$, $\sigma_0 = \varpi/\tau$, $\sigma_1 = -1/\tau$ and $\sigma_2 = -\eta/\tau$. Besides, the control input $u = \theta$.

To make the output y track the desired trajectory $y_\mathrm{d} = \sin(t)$, we need to construct the specific ZD and ZG controllers. Firstly, we know $L_h y(\mathrm{x}) = 0$ and $L_g y(\mathrm{x}) = x_2$ by computation. Then, $L_h L_g y(\mathrm{x})$ needs to be calculated and we obtain $L_h L_g y(\mathrm{x}) = \sigma_0$, which satisfies the standard relative-degree case. Thus, the u-appearing index $\nu = 3$, and the specific ZD and ZG controllers are obtained respectively as

$$u = -\frac{1}{\sigma_0}\Big(\sigma_1 x_2 + \sigma_2 x_2^3 + \cos(t) + \lambda^2(x_1 - \sin(t)) + 2\lambda(x_2 - \cos(t))\Big)$$

and

$$\dot{u} = -\gamma\sigma_0\big(u\sigma_0 + \big(\sigma_1 x_2 + \sigma_2 x_2^3 + \cos(t) + \lambda^2(x_1 - \sin(t)) + 2\lambda(x_2 - \cos(t))\big)\big).$$

For such two controllers, we select parameters $\lambda = 2$ and $\gamma = 10^5$, as well as $\sigma_0 = \sigma_1 = \sigma_2 = 100$.

The simulation results are shown in Fig. 13.1. For comparison, the results generated by the conventional IOL controller are also presented. Evidently, we can observe that the IOL, ZD and ZG controllers perform well with the absolute tracking errors $|e| = |y - y_\mathrm{d}|$ all approaching zero. In addition, it can be seen that the absolute tracking errors with the ZD and ZG controllers exponentially converge to zero.

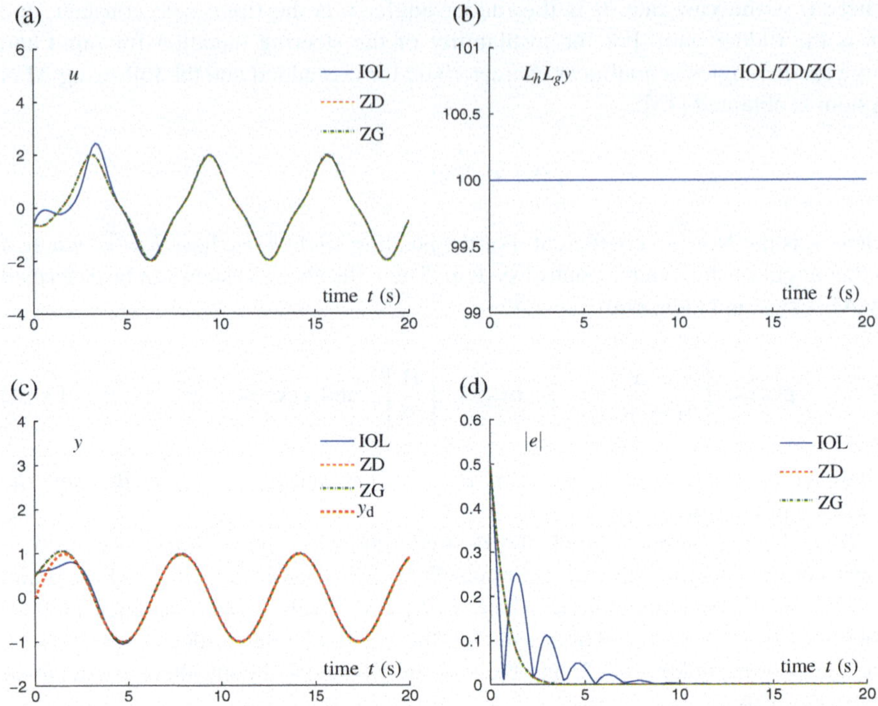

Fig. 13.1 Tracking performance of AFN system (13.14) equipped with IOL, ZD and ZG controllers, respectively, for desired trajectory $y_{\mathrm{d}} = \sin(t)$. (**a**) Control inputs. (**b**) Trajectories of $L_h L_g y$. (**c**) Output trajectories and desired trajectory. (**d**) Absolute tracking errors

13.5.2 PDBZ Relative-Degree Case

Let us consider a relatively complex system from [16] described by AFN system (11.1) with

$$\mathbf{g}(\mathbf{x}) = \begin{bmatrix} -x_1 + x_2 + 2x_2^3 \\ -0.5(x_1 + x_2) + 0.5x_2(1 + 2x_2^2)\sin^2 x_1 \end{bmatrix},$$
$$\mathbf{h}(\mathbf{x}) = \begin{bmatrix} 0 \\ \sin x_1 \end{bmatrix}, \text{ and } y(\mathbf{x}) = x_1. \tag{13.15}$$

Then, we construct the specific ZD and ZG controllers such that the output y tracks the desired trajectory $y_{\mathrm{d}} = \sin(t)$. Firstly, we calculate that $L_h y(\mathbf{x}) = 0$ and $L_g y(\mathbf{x}) = -x_1 + x_2 + 2x_2^3$. Then, we calculate that $L_h L_g y(\mathbf{x}) = (1 + 6x_2^2)\sin x_1$, which satisfies the PDBZ relative-degree case. Specifically, $\forall \mathbf{x} \in \mathbb{P}$, $L_h y(\mathbf{x}) = 0$, and $\exists \mathbf{x}|_{x_1=0} \in \mathbb{P}$, $L_h L_g y(\mathbf{x}) = (1 + 6x_2^2)\sin x_1 = 0$ and $\exists \mathbf{x}|_{x_1=\pi/2} \in \mathbb{P}$,

$L_hL_gy(\mathbf{x}) = (1+6x_2^2) \neq 0$ with

$$\mathbf{h}(\mathbf{x}) = \begin{bmatrix} 0 \\ \sin x_1 \end{bmatrix} = \begin{bmatrix} 0 \\ 1/(1+6x_2^2) \end{bmatrix} L_hL_gy(\mathbf{x}),$$

where $1/(1+6x_2^2)$ is evidently bounded. Thus, the u-appearing index $\nu = 3$, and the specific ZD and ZG controllers are obtained respectively as

$$u = -\frac{1}{(1+6x_2^2)\sin x_1}\Big(x_1 - x_2 + (1+6x_2^2)(-0.5(x_1+x_2) + 0.5x_2(1+2x_2^2)\sin^2 x_1) \\ - 2x_2^3 + \sin(t) + \lambda^2(x_1 - \sin(t)) + 2\lambda(-x_1 + x_2 + 2x_2^3 - \cos(t))\Big)$$

and

$$\begin{aligned}\dot{u} =& -\gamma\big((1+6x_2^2)\sin x_1\big)\big(((1+6x_2^2)\sin x_1)u + x_1 - x_2 - 2x_2^3 \\ &+ (1+6x_2^2)(-0.5(x_1+x_2) + 0.5x_2(1+2x_2^2)\sin^2 x_1) + \sin(t) \\ &+ \lambda^2(x_1 - \sin(t)) + 2\lambda(-x_1 + x_2 + 2x_2^3 - \cos(t))\big).\end{aligned}$$

The simulation results are shown in Fig. 13.2. As shown in this figure, although the trajectories of L_hL_gy pass zero several times, all of the three controllers perform well for fulfilling the tracking-control task. Besides, from Fig. 13.2a, we can observe that the control inputs with IOL and ZD controllers change sharply, while the control input with the ZG controller changes smoothly. This subfigure shows the better performance of the ZG controller in the PDBZ relative-degree case.

13.5.3 TDBZ Relative-Degree Case

Let us consider the tracking control of Lu chaotic system [17], of which the chaotic characteristic is presented in Fig. 13.3a. Such a Lu chaotic system can be described by AFN system (11.1) with

$$\mathbf{g}(\mathbf{x}) = \begin{bmatrix} a(x_2 - x_1) \\ -x_1x_3 + cx_2 \\ x_1x_2 - bx_3 \end{bmatrix}, \ \mathbf{h}(\mathbf{x}) = \begin{bmatrix} 0 \\ 1 \\ 0 \end{bmatrix}, \text{ and } y(\mathbf{x}) = x_3, \tag{13.16}$$

where $a = 36$, $b = 3$ and $c = 20$. We construct the specific ZD and ZG controllers such that the output y tracks the desired trajectory $y_\mathrm{d} = \sin(t) + 1.05$. Firstly, we calculate that $L_hy(\mathbf{x}) = 0$ and $L_gy(\mathbf{x}) = x_1x_2 - bx_3$. Then, we calculate that $L_hL_gy(\mathbf{x}) = x_1$, which satisfies the TDBZ relative-degree case. Specifically, $\forall \mathbf{x} \in \mathbb{T}$, $L_hy(\mathbf{x}) = 0$, and $\exists \mathbf{x}|_{x_1=0} \in \mathbb{T}$, $L_hL_gy(\mathbf{x}) = x_1 = 0$ and $\exists \mathbf{x}|_{x_1\neq 0} \in \mathbb{T}$, $L_hL_gy(\mathbf{x}) = x_1 \neq 0$, where there does not exist a bounded $\Psi(\mathbf{x})$ to satisfy

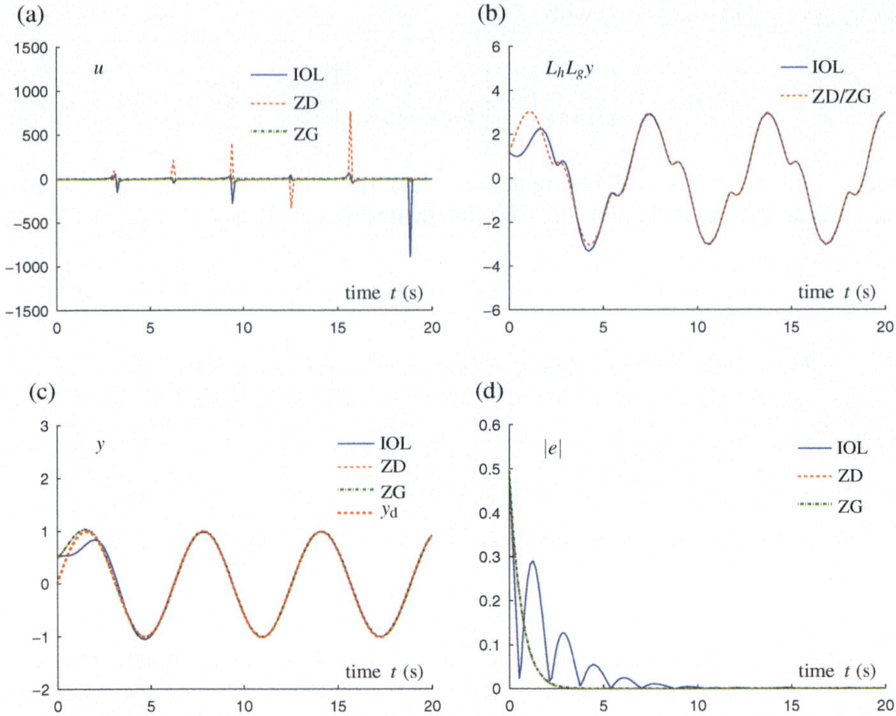

Fig. 13.2 Tracking performance of AFN system (13.15) equipped with IOL, ZD and ZG controllers, respectively, for desired trajectory $y_d = \sin(t)$. (**a**) Control inputs. (**b**) Trajectories of L_hL_gy. (**c**) Output trajectories and desired trajectory. (**d**) Absolute tracking errors

$\mathbf{h}(\mathbf{x}) = \Psi(\mathbf{x})L_hL_gy(\mathbf{x})$. Thus, the u-appearing index $\nu = 3$, and the specific ZD and ZG controllers are obtained respectively as

$$\begin{aligned}u =& -\frac{1}{x_1}\Big(ax_2(x_2 - x_1) + x_1(-x_1x_3 + cx_2) - b(x_1x_2 - bx_3) + \sin(t)\\&+ \lambda^2\big(x_3 - \sin(t) - 1.05\big) + 2\lambda\big(x_1x_2 - bx_3 - \cos(t)\big)\Big)\end{aligned}$$

and

$$\begin{aligned}\dot{u} =& -\gamma x_1\big(ax_2(x_2 - x_1) + x_1(-x_1x_3 + cx_2) - b(x_1x_2 - bx_3) + x_1u\\&+ \sin(t) + \lambda^2\big(x_3 - \sin(t) - 1.05\big) + 2\lambda\big(x_1x_2 - bx_3 - \cos(t)\big)\big).\end{aligned}$$

The simulation results are shown in Fig. 13.3. Specifically, from Fig. 13.3b, we can observe that the trajectories of $L_hL_gy(\mathbf{x})$ with the IOL and ZD controllers cannot pass zero, and that the IOL and ZD controllers crash at around 3.766 and 4.306 s, respectively. In contrast, the ZG controller gets through the DBZ points

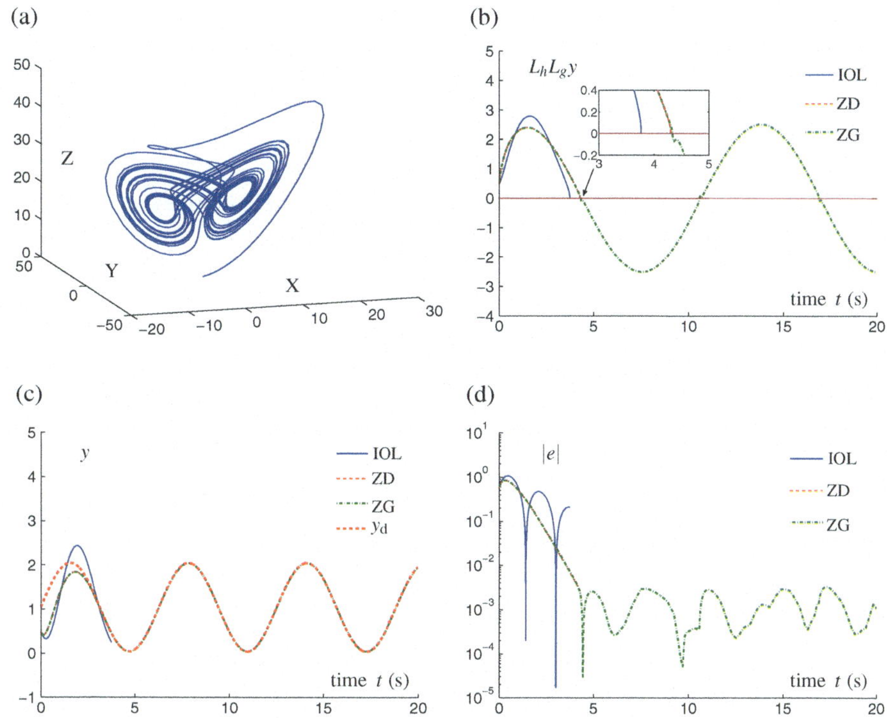

Fig. 13.3 Chaotic characteristic and tracking performance of Lu chaotic system (13.16) equipped with IOL, ZD and ZG controllers, respectively, for desired trajectory $y_\mathrm{d} = \sin(t) + 1.05$. (**a**) Chaotic characteristic. (**b**) Trajectories of L_hL_gy. (**c**) Output trajectories and desired trajectory. (**d**) Absolute tracking errors

of $L_hL_gy(\mathbf{x}) = 0$ successfully during the tracking-control process. In addition, as seen from Fig. 13.3c, d, the output y with the ZG controller tracks well the desired trajectory y_d and the absolute tracking error $|e|$ with the ZG controller is quite small. Therefore, the above simulation results illustrate well the superiority of the ZG controller in the TDBZ relative-degree case, as compared with the IOL and ZD controllers.

13.6 Chapter Summary

In this chapter, based on the ZG method, both the ZD controller and the ZG controller have been designed for the tracking control of AFN system, which may encounter the DBZ problem. For comparison, the conventional IOL controller has also been presented. Besides, the ZD, ZG and IOL controllers have been compared in different relative-degree cases (i.e., the standard relative-degree case, the PDBZ relative-degree case, and the TDBZ relative-degree case). Moreover, the theoretical results and analyses on the ZD and ZG controllers have been provided, which have shown the superiority of the ZG controller in conquering the TDBZ problem. The corresponding computer simulations with three illustrative examples have also been conducted to illustrate the tracking performance of the ZD, ZG and IOL controllers in the three different relative-degree cases.

At last, it is worth pointing out that, from the perspective of relative degree for the AFN system, this chapter can be regarded as a further in-depth research of Chap. 11. To be more specific, in addition to the well-defined relative-degree case in Chap. 11 corresponding to the standard relative-degree case in this chapter, the loose condition on relative degree in Chap. 11 can be further subdivided into the PDBZ relative-degree case and the TDBZ relative-degree case in this chapter. Furthermore, in the ensuing Chap. 14, the PDBZ and TDBZ problems will be further discussed, distinguished and solved systematically, methodologically and comparatively. In the ensuing Chaps. 16 and 17, the PDBZ phenomena will also be further investigated and shown for the stabilization of time-varying linear (TVL) system as well as the output tracking of TVL and time-varying nonlinear (TVN) systems.

Appendix 1: Derivation of Equation (13.3)

Suppose that the mth ZF is constructed as

$$z_m = \sum_{l=0}^{m-1} \mathrm{C}_{m-1}^{l} \lambda^{m-1-l}\left(y^{(l)} - y_{\mathrm{d}}^{(l)}\right), \tag{13.17}$$

where m is a positive integer.

To obtain the $(m+1)$th ZF, the ZD design formula $\dot{z}_m = -\lambda z_m$ is adopted. Then, based on (13.17), the $(m+1)$th ZF $z_{m+1} = \dot{z}_m + \lambda z_m$ can be expanded as

$$\begin{aligned} z_{m+1} &= \sum_{l=0}^{m-1} \mathrm{C}_{m-1}^{l} \lambda^{m-1-l}\left(y^{(l+1)} - y_{\mathrm{d}}^{(l+1)}\right) + \lambda \sum_{l=0}^{m-1} \mathrm{C}_{m-1}^{l} \lambda^{m-1-l}\left(y^{(l)} - y_{\mathrm{d}}^{(l)}\right) \\ &= \mathrm{C}_{m-1}^{0} \lambda^{m}(y - y_{\mathrm{d}}) + \left(\mathrm{C}_{m-1}^{1} + 1\right)\lambda^{m-1}(\dot{y} - \dot{y}_{\mathrm{d}}) + \cdots \\ &\quad + \left(\mathrm{C}_{m-1}^{k} + \mathrm{C}_{m-1}^{k-1}\right)\lambda^{m-k}\left(y^{(k)} - y_{\mathrm{d}}^{(k)}\right) + \cdots + \left(y^{(m)} - y_{\mathrm{d}}^{(m)}\right) \\ &= \mathrm{C}_{m}^{0} \lambda^{m}(y - y_{\mathrm{d}}) + \mathrm{C}_{m}^{1} \lambda^{m-1}(\dot{y} - \dot{y}_{\mathrm{d}}) + \cdots + \mathrm{C}_{m}^{m} \lambda^{0}\left(y^{(m)} - y_{\mathrm{d}}^{(m)}\right) \\ &= \sum_{l=0}^{m} \mathrm{C}_{m}^{l} \lambda^{m-l}\left(y^{(l)} - y_{\mathrm{d}}^{(l)}\right) \\ &= \sum_{l=0}^{(m+1)-1} \mathrm{C}_{(m+1)-1}^{l} \lambda^{(m+1)-1-l}\left(y^{(l)} - y_{\mathrm{d}}^{(l)}\right). \end{aligned} \tag{13.18}$$

Based on Eqs. (13.1), (13.17) and (13.18), as well as the mathematical induction [18], the ith ZF (with $i = 1, 2, \cdots$) can be constructed as

$$z_i = \sum_{l=0}^{i-1} \mathrm{C}_{i-1}^{l} \lambda^{i-1-l}\left(y^{(l)} - y_{\mathrm{d}}^{(l)}\right), \quad \text{with } i = 1, 2, \cdots,$$

The derivation of Eq. (13.3) is thus completed. □

Appendix 2: Proof of Lemma 13.1

Let $\psi = u - u^*$ denote the solution error of ZG controller (13.7) for solving the time-varying minimization problem of EF (13.6) with $u^* = -\alpha_2/\alpha_1$ being the desired time-varying solution. Then, $u = \psi + u^*$ and its time derivative is

$$\dot{u} = \dot{\psi} + \dot{u}^*. \tag{13.19}$$

Substituting (13.10) into (13.19) yields $-\gamma\alpha_1(\alpha_2 + \alpha_1 u) = \dot{\psi} + \dot{u}^*$, which can be further written as

$$\dot{\psi} = -\gamma\alpha_1^2\psi - \dot{u}^*.$$

Let us define a Lyapunov function candidate as $L = \psi^2/2$. Evidently, L is positive definite in view of $L = \psi^2/2 > 0$ for $\psi \neq 0$ and $L = 0$ only for $\psi = 0$. Then, taking its time derivative yields

$$\dot{L} = \psi\dot{\psi} = \psi(-\gamma\alpha_1^2\psi - \dot{u}^*) = -\gamma\alpha_1^2\psi^2 - \psi\dot{u}^*. \tag{13.20}$$

There are two terms in the time derivative $\dot{L}$, i.e, $-\gamma\alpha_1^2\psi^2$ and $-\psi\dot{u}^*$. Let us handle these two terms individually. For the first term $-\gamma\alpha_1^2\psi^2$, we have

$$-\gamma\alpha_1^2\psi^2 \leq -\gamma\varsigma\psi^2, \tag{13.21}$$

where $\varsigma > 0$ is defined previously by $\alpha_1^2 \geq \varsigma$. Besides, for the second term $-\psi\dot{u}^*$, we have the following result based on Cauchy inequality [19]:

$$-\psi\dot{u}^* \leq |\psi||\dot{u}^*| \leq \varrho|\psi|, \tag{13.22}$$

where ϱ is defined previously by $|\dot{u}^*| \leq \varrho$. Then, substituting (13.21) and (13.22) into (13.20) yields

$$\dot{L} = -\gamma\alpha_1^2\psi^2 - \psi\dot{u}^* \leq -\gamma\varsigma\psi^2 + \varrho|\psi| = -|\psi|(\gamma\varsigma|\psi| - \varrho). \tag{13.23}$$

During the time evolution of solution error ψ, (13.23) falls into one of the three situations: (1) $\gamma\varsigma|\psi| - \varrho > 0$; (2) $\gamma\varsigma|\psi| - \varrho = 0$; (3) $\gamma\varsigma|\psi| - \varrho < 0$. These three situations are analyzed in detail as follows.

- In the first situation (i.e., $|\psi| > \varrho/(\gamma\varsigma)$), $\dot{L} < 0$, which implies that ψ approaches zero (i.e., u approaches u^*) as time evolves.
- In the second situation (i.e., $|\psi| = \varrho/(\gamma\varsigma)$), $\dot{L} \leq 0$, which implies that ψ approaches zero (i.e., u approaches u^*) or ψ stays on the ball surface with $|\psi| = \varrho/(\gamma\varsigma)$ (i.e., $|u - u^*| = \varrho/(\gamma\varsigma)$), in view of $\dot{L} \leq 0$ containing sub-situations $\dot{L} < 0$ and $\dot{L} = 0$, respectively. That is, ψ will not go outside the ball of $\varrho/(\gamma\varsigma)$ in this situation.
- In the third situation (i.e., $|\psi| < \varrho/(\gamma\varsigma)$), $\dot{L}$ is less than a positive scalar (containing sub-situations $\dot{L} \leq 0$ and $\dot{L} > 0$), and thus ψ may not decrease again. Let us analyze the worst case, i.e., sub-situation $\dot{L} > 0$. In this case, with $\gamma\varsigma|\psi| - \varrho < 0$, L and $|\psi|$ would increase, which implies that $\gamma\varsigma|\psi| - \varrho$ increases as well. Evidently, there exists a certain time instant such that $\gamma\varsigma|\psi| - \varrho = 0$, which returns to the second situation, i.e., $\dot{L} \leq 0$.

By summarizing the above three situations, the solution error for solving the time-varying minimization problem of EF (13.6) via ZG controller (13.7) or (13.10) is upper bounded by $\varrho/(\gamma\varsigma)$ when the solving process enters steady state. That is, in mathematics, the steady-state solution error

$$|\psi| = |u - u^*| \leq \frac{\varrho}{\gamma\varsigma}, \quad \forall t \geq t_{\mathrm{e}},$$

where t_{e} is large enough. The proof is thus completed. □

Appendix 3: Proof of Inequality (13.12)

From (13.11), we have

$$\sum_{l=0}^{\nu-1} \mathrm{C}_{\nu-1}^{l} \lambda^{\nu-1-l} e^{(l)} = \dot{z}_{\nu-1} + \lambda z_{\nu-1} \leq \frac{\chi\varrho}{\gamma\varsigma},$$

where $t \geq t_{\mathrm{e}}$. Making use of Gronwall inequality [15], we have

$$\begin{aligned} z_{\nu-1} &\leq z_{\nu-1}(t_{\mathrm{e}}) \exp(-\lambda t) + \int_{t_{\mathrm{e}}}^{t} \exp(-\lambda(t-\iota)) \frac{\chi\varrho}{\gamma\varsigma} \mathrm{d}\iota \\ &\leq \bar{c}_0 \exp(-\lambda t) + \frac{\chi\varrho}{\lambda\gamma\varsigma}, \end{aligned} \tag{13.24}$$

where $t \geq t_{\mathrm{e}}$ and $\bar{c}_0$ is a constant regarding $z_{\nu-1}(t_{\mathrm{e}})$ and $\chi\varrho/(\lambda\gamma\varsigma)$. As $z_{\nu-1} = \dot{z}_{\nu-2} + \lambda z_{\nu-2}$, we have

$$\dot{z}_{\nu-2} + \lambda z_{\nu-2} \leq \Omega, \tag{13.25}$$

where $\Omega = \bar{c}_0 \exp(-\lambda t) + \chi\varrho/(\lambda\gamma\varsigma)$. Then, making use of Gronwall inequality again, we have

$$\begin{aligned} z_{\nu-2} &\leq z_{\nu-2}(t_{\mathrm{e}}) \exp(-\lambda t) + \int_{t_{\mathrm{e}}}^{t} \exp(-\lambda(t-\iota))\Omega(\iota)\mathrm{d}\iota \\ &\leq \acute{c}_0 \exp(-\lambda t) + \acute{c}_1 t \exp(-\lambda t) + \frac{\chi\varrho}{\lambda^2\gamma\varsigma}, \end{aligned} \tag{13.26}$$

where $t \geq t_{\mathrm{e}}$, and $\acute{c}_0$ and $\acute{c}_1$ are constants in terms of $z_{\nu-2}(t_{\mathrm{e}})$, $\dot{z}_{\nu-2}(t_{\mathrm{e}})$ and $\chi\varrho/(\lambda^2\gamma\varsigma)$. Note that the derivations of (13.25) and (13.26) are available only when $\nu \geq 3$, which is presented for better illustration. Similar to the derivation procedures of (13.24) and (13.26), we finally have

$$z_1 = e \leq \exp(-\lambda t) \sum_{\kappa=0}^{\nu-2} \tilde{c}_\kappa t^\kappa + \frac{\chi\varrho}{\lambda^{\nu-1}\gamma\varsigma}, \tag{13.27}$$

where $t \geq t_{\mathrm{e}}$, and $\tilde{c}_\kappa$ with $\kappa = 0, 1, \cdots, \nu-2$ are constants regarding $e(t_{\mathrm{e}}), \dot{e}(t_{\mathrm{e}}), \cdots, e^{(\nu-2)}(t_{\mathrm{e}})$ and $\chi\varrho/(\lambda^{\nu-1}\gamma\varsigma)$.

In addition, from (13.11), we have

$$-\sum_{l=0}^{\nu-1} \mathrm{C}_{\nu-1}^{l} \lambda^{\nu-1-l} e^{(l)} = -\dot{z}_{\nu-1} - \lambda z_{\nu-1} \leq \frac{\chi\varrho}{\gamma\varsigma}.$$

Similar to the derivation procedure of (13.27), we have

$$e \geq -\exp(-\lambda t) \sum_{\kappa=0}^{\nu-2} \hat{c}_\kappa t^\kappa - \frac{\chi\varrho}{\lambda^{\nu-1}\gamma\varsigma},$$

where $t \geq t_{\mathrm{e}}$, and $\hat{c}_\kappa$ with $\kappa = 0, 1, \cdots, \nu-2$ are constants regarding $e(t_{\mathrm{e}}), \dot{e}(t_{\mathrm{e}}), \cdots, e^{(\nu-2)}(t_{\mathrm{e}})$ and $\chi\varrho/(\lambda^{\nu-1}\gamma\varsigma)$. The proof of (13.12) is thus completed. □

Appendix 4: Proof of Lemma 13.2

From (13.23), we have

$$\dot{L} \leq -\gamma\varsigma\psi^2 + \varrho|\psi| = -(1-\omega)\gamma\varsigma\psi^2 + (-\omega\gamma\varsigma\psi^2 + \varrho|\psi|), \tag{13.28}$$

where loosening parameter $\omega \in (0, 1)$. On the right side of (13.28), the first term $-(1-\omega)\gamma\varsigma\psi^2 \leq 0$, and the second term $-\omega\gamma\varsigma\psi^2+\varrho|\psi|$ falls into one of the three situations as time evolves: (1) $-\omega\gamma\varsigma\psi^2+\varrho|\psi| > 0$; (2) $-\omega\gamma\varsigma\psi^2+\varrho|\psi| = 0$; (3) $-\omega\gamma\varsigma\psi^2+\varrho|\psi| < 0$.

- In the first situation (i.e., $|\psi| < \varrho/(\omega\gamma\varsigma)$), $\dot{L} \leq 0$ or $\dot{L} > 0$. If $\dot{L} \leq 0$, $|\psi|$ decreases or still stays. If $\dot{L} > 0$, both L and $|\psi|$ increase; this leads to the increase of $|\psi|(\omega\gamma\varsigma|\psi| - \varrho)$ and the decrease of $-\omega\gamma\varsigma\psi^2 + \varrho|\psi|$, and then there exists a certain time instant such that the second term $-\omega\gamma\varsigma\psi^2+\varrho|\psi| = 0$, which enters the second situation.
- In the second situation (i.e., $|\psi| = \varrho/(\omega\gamma\varsigma)$), $\dot{L} \leq -(1-\omega)\gamma\varsigma\psi^2 < 0$. This means that $|\psi|$ decreases, which leads to the increase of $-\omega\gamma\varsigma\psi^2+\varrho|\psi|$. Then, the second term returns to the first situation.
- In the third situation (i.e., $|\psi| > \varrho/(\omega\gamma\varsigma)$), $\dot{L} \leq -(1-\omega)\gamma\varsigma\psi^2+(-\omega\gamma\varsigma\psi^2+\varrho|\psi|) < -(1-\omega)\gamma\varsigma\psi^2 = -2(1-\omega)\gamma\varsigma L$. Then, $L \leq L(t_0)\exp(-2(1-\omega)\gamma\varsigma t)$, where t_0 is the time instant when the second term enters this situation. In view of the above analyses, the second term will not escape from the first and second situations. Thus, $t_0 = 0$ is the only possible case for the second term to enter this situation. Therefore, we have

$$|\psi| \leq |\psi(0)|\exp(-(1-\omega)\gamma\varsigma t), \forall t \in [0, t_\mathrm{c}],$$

  where the exponential convergence rate is $(1-\omega)\gamma\varsigma$, and the convergence time $t_\mathrm{c} = \ln(\omega\gamma\varsigma|\psi(0)|/\varrho)/((1-\omega)\gamma\varsigma)$ by solving $|\psi(0)|\exp(-(1-\omega)\gamma\varsigma t_\mathrm{c}) = \varrho/(\omega\gamma\varsigma)$. Note that the second term returns to the second situation at time t_c.

  According to the above analyses, we have the following facts.

- If $|\psi(0)| \geq \varrho/(\omega\gamma\varsigma)$, then

$$|\psi| \begin{cases} \leq |\psi(0)|\exp(-(1-\omega)\gamma\varsigma t), \forall t \in [0, t_\mathrm{c}], \\ \leq \varrho/(\omega\gamma\varsigma), \forall t \in [t_\mathrm{c}, +\infty). \end{cases}$$

- If $|\psi(0)| \leq \varrho/(\omega\gamma\varsigma)$, then $|\psi| \leq \varrho/(\omega\gamma\varsigma), \forall t \in [0, +\infty)$.

Therefore, the solution error $|\psi| = |u - u^*|$ exponentially converges toward or stays within the error bound $\varrho/(\omega\gamma\varsigma)$ with $\omega \in (0, 1)$. The exponential convergence rate is $(1-\omega)\gamma\varsigma$; the convergence time is $\ln(\omega\gamma\varsigma|\psi(0)|/\varrho)/((1-\omega)\gamma\varsigma)$ for $|\psi(0)| \geq \varrho/(\omega\gamma\varsigma)$ or zero for $|\psi(0)| \leq \varrho/(\omega\gamma\varsigma)$. The proof is thus completed. □

References

1. Isidori A (1989) Nonlinear control systems: an introduction, 2nd edn. Springer, New York
2. Hauser J, Sastry S, Kokotovic P (1992) Nonlinear control via approximate input-output linearization: the ball and beam example. IEEE Trans Autom Control 37(3):392–398

3. Guardabassi GO, Savaresi SM (2001) Approximate linearization via feedback – an overview. Automatica 37(1):1–15
4. Chang Y (2013) A robust tracking control for a class of uncertain chaotic systems. Asian J Control 15(6):1752–1763
5. Niu B, Zhao J (2013) Robust stabilization and tracking control for a class of switched nonlinear systems. Asian J Control 15(5):246–252
6. Tomlin CJ, Sastry SS (1997) Switching through singularities. Syst Control Lett 35(3):145–154
7. Zhang Y, Yu X, Yin Y, Xiao L, Fan Z (2013) Using GD to conquer the singularity problem of conventional controller for output tracking of nonlinear system of a class. Phys Lett A 377(25–27):1611–1614
8. Li J, Mao M, Zhang Y, Chen D, Yin Y (2017) ZD, ZG and IOL controllers and comparisons for nonlinear system output tracking with DBZ problem conquered in different relative-degree cases. Asian J Control 19(4):1482–1495
9. Zhang Y, Yi C (2011) Zhang neural networks and neural-dynamic method. Nova Science Publishers, New York
10. Zhang Y, Cai B, Yin J, Zhang L (2012) Two/infinity norm criteria resolution of manipulator redundancy at joint-acceleration level using primal-dual neural network. Asian J Control 14(4):1036–1046
11. Zhang Y, Chen K, Tan H (2009) Performance analysis of gradient neural network exploited for online time-varying matrix inversion. IEEE Trans Autom Control 54(8):1940–1945
12. Zhang Z, Zhang Y (2013) Design and experimentation of acceleration-level drift-free scheme aided by two recurrent neural networks. IET Control Theory Appl 7(1):25–42
13. Chu S, Metcalf F (1967) On Gronwall's inequality. Proc Amer Math Soc 18(3):439–440
14. Jia X, Yang Y (1998) Ship motion mathematic model. Dalian Maritime University Press, Dalian
15. Du J, Guo C (2005) Nonlinear adaptive design for course-tracking control of ship without a priori knowledge of control gain. Control Theory Appl 22(2):316–320
16. Lewis F, Vrabie D, Vamvoudakis K (2012) Reinforcement learning and feedback control: using natural decision methods to design optimal adaptive controllers. IEEE Control Syst Mag 32(6):76–105
17. Lu J, Lu J (2003) Controlling uncertain Lu system using linear feedback. Chaos Solitons Fractals 17(1):127–133
18. Hu C (2005) Mathematic induction and Peano's axioms. Math Theory Appl 25(4):152–154
19. Abramowitz M, Stegun IA (1972) Handbook of mathematical functions with formulas, graphs, and mathematical tables. Dover Publications, New York

Chapter 14
PDBZ and TDBZ Problem Solving and Comparing

14.1 Introduction

The controller design of nonlinear systems has become a heated investigative topic in recent decades [1–4]. One of the difficult problems occurring in the controller-design process for nonlinear systems is the DBZ problem, which is quite destructive for computations and practical applications [5, 6]. It may lead to controller saturation and even damage the actuators. In terms of control systems, the controller with possible DBZ problem is not desirable [5–9].

There are many approaches that can be adopted to design controllers for nonlinear systems. Among them, the feedback linearization method is known [1, 6, 10]. With the aid of feedback, that method transforms a nonlinear system into a linear system in the operating region and then the design of controllers can be fulfilled under the framework of the systematic linear control theory. However, it should not be neglected that the controllers designed by that method or extended methods (e.g., adaptive feedback linearization method) are limited because they may lead to the DBZ problem [5, 6]. In recent years, the backstepping method has also drawn considerable attention in fulfilling the task of designing controllers for nonlinear systems [11–14]. For example, in [12], the backstepping method is successfully adopted to solve the robust output feedback control problem for a class of time delayed nonlinear systems in the strict-feedback form. Due to the fact that the controller-design process based on the backstepping method is a series of recursive selections of Lyapunov functions, the stability of the control system can be readily guaranteed. However, the controller design process is quite difficult to follow when it comes to complex nonlinear systems. Besides, it is worth pointing out that the controllers designed by the backstepping method may also be limited by the DBZ problem [11, 13].

Y. Zhang et al., *Zhang-Gradient Control*,
https://doi.org/10.1007/978-981-15-8257-8_14

To deal with the DBZ problem, various methods have been presented and investigated [9, 15–18]. For example, in [17], it is pointed out that, by superimposing a dead-zone on division controllers, the DBZ problem can be avoided. However, it may lead to the controller discontinuity, which is not acceptable in many practical applications. For this reason, these researchers further presented two continuous division controllers to approximate the discontinuous ones. In [18], the possible DBZ problem in the presented controller is avoided by adopting the integral-type Lyapunov function, which makes it complicated to apply the controller in practice. In [15, 16], the switching control method is presented as an alternative to avoid the DBZ problem. Based on the switching control method, the controlled system needs to switch from one controller to another controller when the DBZ problem is detected, so the approach costs much for physical implementations. Besides, some of the other methods for dealing with the DBZ problem are shown and discussed in [19] with various weaknesses. Therefore, a simple and effective method is demanded. Moreover, although the DBZ problem has drawn much attention from practitioners and researchers, to the best of the authors' knowledge, there is still no literature classifying the DBZ problem. It is believed that the classification of the problem may provide a new perspective in dealing with it.

In recent years, the ZD method and the ZG method have been developed and presented successively for the design of controllers. The former is generalized from the Zhang neural network, which have been extensively studied and successfully adopted to solve various time-varying problems [20–23]. The ZD method can be easily used to design controllers for both linear and nonlinear systems in a quite simple manner [24, 25]. It was later found that the controllers designed only by the ZD method may be limited by the DBZ problem, and thus the ZG method [7, 8] was presented, which is the combination of the ZD method and the GD method. The ZG method can elegantly conquer the DBZ problem with high tracking accuracy, and this method is quite simple to be understood and utilized in practice. It is worth pointing out that the ZG method, unlike the conventional methods, does not try to avoid the DBZ points but can successfully get through them by transforming division into multiplication and minimization. Owing to the above facts, this chapter [26] focuses on the classification, presentation and solution of the DBZ problem under the framework of the ZG method.

14.2 DBZ Analysis and Classification

For better understanding the DBZ problem, let us consider the following forced AFN system:

$$\begin{cases} \dot{x}_i = f_i(x_{i+1}) = x_{i+1}, \\ \dot{x}_n = f_n(\mathbf{x}, u) = g(\mathbf{x}) + h(\mathbf{x})u, \\ y = y(\mathbf{x}), \end{cases} \tag{14.1}$$

where $i = 1, 2, \cdots, n-1$; $\mathbf{x} = [x_1, x_2, \cdots, x_n]^{\mathrm{T}} \in \mathbb{R}^n$; $x_i \in \mathbb{R}$ is the ith system state; $u \in \mathbb{R}$ is the control input; $y \in \mathbb{R}$ denotes the system output; $g(\mathbf{x}) : \mathbb{R}^n \to \mathbb{R}$, $h(\mathbf{x}) : \mathbb{R}^n \to \mathbb{R}$, and $y(\mathbf{x}) : \mathbb{R}^n \to \mathbb{R}$ are smooth functions.

If the system output is defined as $y = x_1$, with the aid of the feedback linearization method, the commonly used control structure is $u = (-g(\mathbf{x}) + \mu)/h(\mathbf{x})$, with μ being a new control variable [6]. Evidently, the controller is limited by the DBZ problem. Specifically, if $h(\mathbf{x}) \to 0$, the magnitude of u becomes infinite, which may lead to the crash of the forced AFN system. By substituting the controller expression into the system equation with $y = x_1$, the forced AFN system (14.1) can be transformed into the following form:

$$\begin{cases} \dot{x}_i = f_i(x_{i+1}) = x_{i+1}, \\ \dot{x}_n = f_n(\mu) = \mu, \\ y = x_1, \end{cases}$$

where the possible DBZ operation is not included. That is, in this case, the DBZ problem in the controller does not lead to infinity for any state of the system. Therefore, we have the following proposition.

Proposition 14.1 *If there is no DBZ operation in the state description of the system, which is transformed from the forced AFN system equipped with the controller containing a DBZ problem, then the DBZ problem is termed the PDBZ problem.*

If a controller designed for the forced AFN system (14.1) with $h(\mathbf{x}) = 1$ for all $\mathbf{x} \in \mathbb{R}^n$ is expressed in the form of $u = 1/q(\mathbf{x}) - g(\mathbf{x})$ with $q(\mathbf{x})$ being a smooth function, where the DBZ problem arises for $q(\mathbf{x}) = 0$, then the forced AFN system (14.1) can be transformed into the following form:

$$\begin{cases} \dot{x}_i = f_i(x_{i+1}) = x_{i+1}, \\ \dot{x}_n = f_n(\mathbf{x}) = 1/q(\mathbf{x}), \\ y = y(\mathbf{x}), \end{cases}$$

where the possible DBZ operation is still involved.

Proposition 14.2 *If there exists DBZ operation in the state description of the system, which is transformed from the forced AFN system equipped with the controller containing a DBZ problem, then the DBZ problem is termed the TDBZ problem.*

It is worth pointing out that the difference between the PDBZ problem and the TDBZ problem can also be readily distinguished in MATLAB-based computer simulations of control systems by means of the ODE tool, such as ode45 and ode15s. Specifically, the simulations of control systems equipped with the controllers containing the PDBZ problem run well without interruption. By contrast, the simulations of control systems equipped with the controllers containing the TDBZ problem stop at a DBZ point.

14.3 PDBZ Example

In this section, the tracking-control problem of a specific nonlinear system is solved and discussed as an example to explain the concept of PDBZ problem. To solve the tracking-control problem, two controllers are designed by the ZD method and the ZG method, respectively.

14.3.1 Problem Description

Without loss of generality, the tracking control of the following system is considered and investigated:

$$\begin{cases} \dot{x}_1 = f_1(x_2) = x_2, \\ \dot{x}_2 = f_2(x_1, x_2, u) = 10x_2 + 3x_2^2 + u \sin x_1, \\ y = x_1, \end{cases} \tag{14.2}$$

where $x_1 \in \mathbb{R}$ and $x_2 \in \mathbb{R}$ are the system states; $u \in \mathbb{R}$ is the control input; $y \in \mathbb{R}$ is the output. The tracking-control problem of the system is to design a suitable control input u such that the output y can track the desired trajectory $y_\mathrm{d} \in \mathbb{R}$. In other words, the absolute value of the tracking error, i.e., $|e| = |y - y_\mathrm{d}|$, should be acceptably small or even zero.

14.3.2 Design of ZD and ZG Controllers

By defining ZF $z_1 = y - y_\mathrm{d}$ and adopting the ZD design formula $\dot{z}_1 = -\lambda z_1$ with (14.2) taken into account, the following equation can be obtained:

$$x_2 - \dot{y}_\mathrm{d} + \lambda(x_1 - y_\mathrm{d}) = 0,$$

where control input u is not explicitly included. According to the ZD method, ZFs need to be defined successively until the expression explicitly including u is obtained. Therefore, another ZF is defined as

$$z_2 = x_2 - \dot{y}_\mathrm{d} + \lambda(x_1 - y_\mathrm{d}).$$

By utilizing the ZD design formula $\dot{z}_2 = -\lambda z_2$, the expression explicitly including u is derived as

$$10x_2 + 3x_2^2 + u \sin x_1 - \ddot{y}_d + 2\lambda(x_2 - \dot{y}_d) + \lambda^2(x_1 - y_d) = 0.$$

As a result, the ZD controller is obtained as

$$u = \frac{\ddot{y}_d - 2\lambda(x_2 - \dot{y}_d) - \lambda^2(x_1 - y_d) - 10x_2 - 3x_2^2}{\sin x_1}. \quad (14.3)$$

This controller is termed z2g0 controller in view of the fact that, during the controller design process, the ZD design formula is adopted for two times while the GD design formula is not used. Evidently, from a traditional viewpoint, there exists the DBZ problem in z2g0 controller (14.3) when $\sin x_1 = 0$, which may lead to the crash of the control system. Thus, this controller is less favorable. However, substituting (14.3) into (14.2) yields the following system:

$$\begin{cases} \dot{x}_1 = x_2, \\ \dot{x}_2 = \ddot{y}_d - 2\lambda(x_2 - \dot{y}_d) - \lambda^2(x_1 - y_d), \\ y = x_1, \end{cases}$$

where the DBZ operation is not included. According to Proposition 14.1, this DBZ problem is a PDBZ problem.

According to the ZG method, the ZG controller can be further obtained. Specifically, an EF is defined as $\epsilon_p = \phi_p^2/2$ with $\phi_p = \ddot{y}_d - 2\lambda(x_2 - \dot{y}_d) - \lambda^2(x_1 - y_d) - 10x_2 - 3x_2^2 - u \sin x_1$. By utilizing the GD design formula [24], the ZG controller is obtained as

$$\dot{u} = -\gamma \frac{\partial \epsilon_p}{\partial u} = \gamma \sin x_1 \phi_p, \quad (14.4)$$

where no DBZ problem exists. To obtain the ZG controller, the ZD and GD methods are adopted for two times and one time, respectively. Therefore, controller (14.4) is termed z2g1 controller.

Remark 14.1 For z2g1 controller (14.4), at the DBZ points $x_1 = k\pi$ with $k = 0, 1, 2, \cdots$, we have $\dot{u} = 0$. In other words, the value of the control input at the corresponding time instant is the same as that at the previous time instant. Unless the system is trapped at the origin (i.e., $x_1 = x_2 = 0$), the system keeps changing and the state x_1 will leave the DBZ points at the next time instant. Then, $\dot{u}$ will not always be zero, which guarantees the control performance. That is, the tracking error may increase when the system approaches the DBZ points, and it decreases at the next time instant when the system leaves the DBZ points.

14.3.3 Simulation, Verification and Comparison

To verify the efficacy of the presented controllers (i.e., z2g0 controller (14.3) and z2g1 controller (14.4)) and to further show the characteristics of PDBZ problem, the simulation results of system (14.2) equipped with the two controllers are provided and discussed. In the simulations, the parameters and initial values for z2g0 controller (14.9) and z2g1 controller (14.10) are set as $\lambda = 10$, $\gamma = 10^5$, $x_1(0) = 1$, $x_2(0) = 0$ and $u(0) = 0$ illustratively. Without loss of generality, the desired trajectory $y_{\mathrm{d}} = \sin(0.5t)$ is considered. In what follows, $|e| = |y - y_{\mathrm{d}}|$ represents the absolute tracking error.

The tracking performance of the system equipped with z2g0 controller (14.3) and z2g1 controller (14.4), respectively, is comparatively presented in Fig. 14.1. Specifically, Fig. 14.1a shows that the trajectories of the system states are smooth and bounded for the z2g0 controller, while those for the z2g1 controller oscillate in the neighborhood of the DBZ points (i.e., $\sin x_1 = 0$), which can be seen from Fig. 14.1b. Besides, Fig. 14.1c, d present that the output trajectory of the system quickly tracks the desired trajectory with the aid of both controllers. As seen from Fig. 14.1e, the maximal steady-state tracking error with z2g0 controller (14.3) is of order 10^{-4}, which is smaller than that with z2g1 controller (14.4) (being of order 10^{-2}) shown in Fig. 14.1f. In addition, in accordance with Remark 14.1, except for the time instants when the z2g1 controller is close to the DBZ points (i.e., around $t = 2k\pi$ s with k being 1, 2 and 3), the steady-state tracking errors with the z2g1 controller are of order 10^{-4}.

In view of the system states and the absolute tracking error, the tracking control task of the system is completed in a quite satisfactory manner by means of the z2g0 controller with the PDBZ problem. For further comparison, the corresponding trajectories of the control inputs are shown in Fig. 14.2, which indicate that, in terms of the PDBZ problem solving, the magnitude of the control input for the z2g1 controller is much smaller than that for the z2g0 controller. That is, the z2g1 controller can be used to reduce the magnitude of the control input for the control system that may be limited by the PDBZ problem.

In summary, the specific example of PDBZ problem is shown and solved to some extent by means of the ZG method. Although the PDBZ problem leads to large magnitude of the control input, the system does not crash during the simulation. The simulation results clearly illustrate the characteristics of the PDBZ problem and the efficacy of the ZG method.

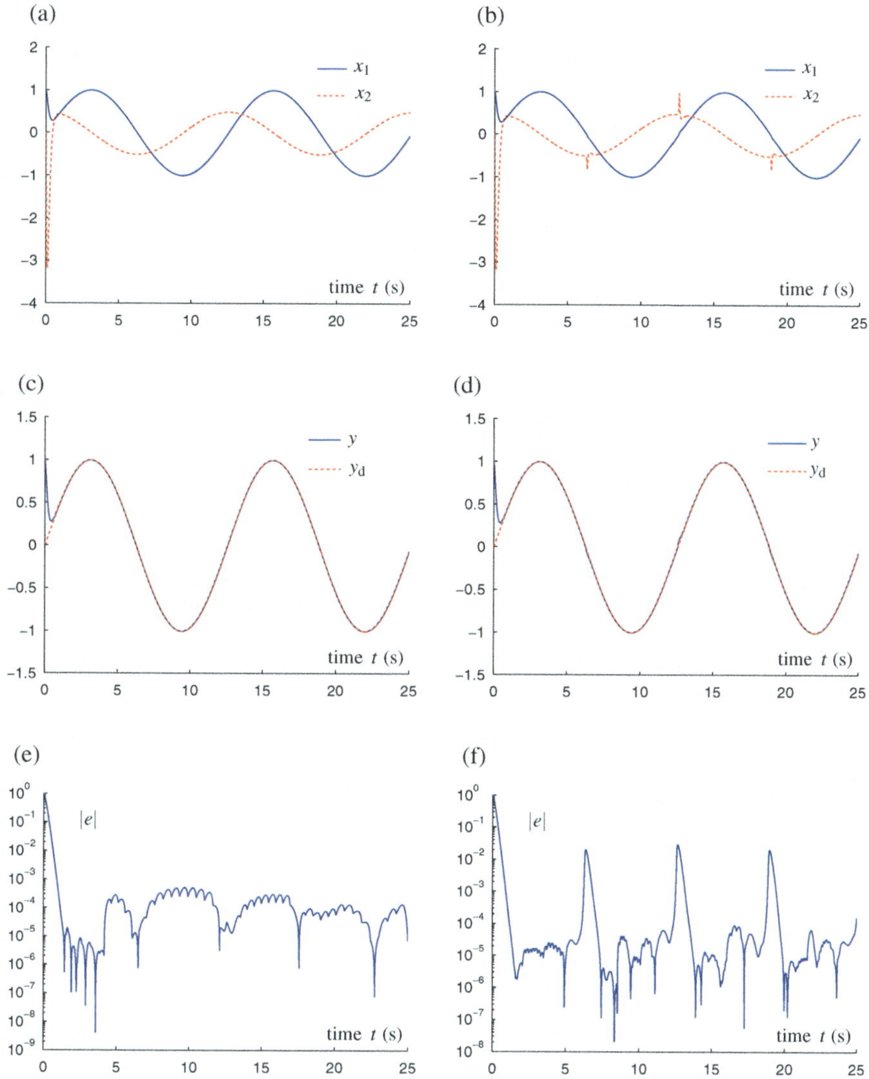

Fig. 14.1 Tracking performance of system (14.2) equipped with z2g0 controller (14.3) and z2g1 controller (14.4), respectively, for desired trajectory $y_d = \sin(0.5t)$. (**a**) System states with z2g0 controller (14.3). (**b**) System states with z2g1 controller (14.4). (**c**) Output trajectory with z2g0 controller (14.3) and desired trajectory. (**d**) Output trajectory with z2g1 controller (14.4) and desired trajectory. (**e**) Absolute tracking error with z2g0 controller (14.3). (**f**) Absolute tracking error with z2g1 controller (14.4)

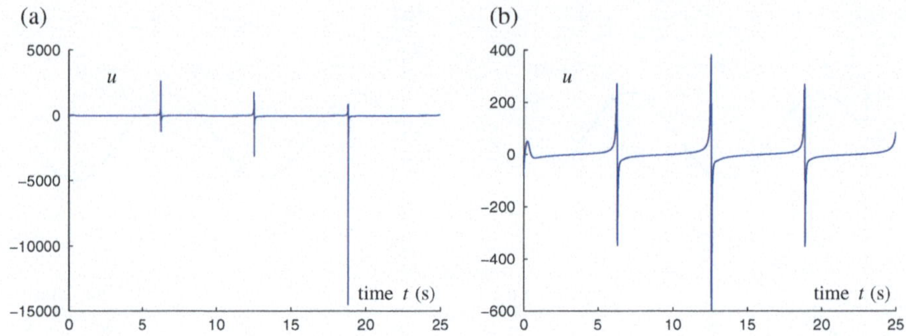

Fig. 14.2 Control inputs of system (14.2) equipped with z2g0 controller (14.3) and z2g1 controller (14.4), respectively, for desired trajectory $y_d = \sin(0.5t)$. (**a**) Control input with z2g0 controller (14.3). (**b**) Control input with z2g1 controller (14.4)

14.4 TDBZ Example

In this section, the tracking-control problem of a specific nonlinear system is solved and discussed as an example to explain the concept of the TDBZ problem. To solve the tracking-control problem, another two controllers are designed by the ZD method and the ZG method, respectively.

14.4.1 Problem Description

The tracking control of the following system is considered:

$$\begin{cases} \dot{x}_1 = f_1(x_2) = x_2, \\ \dot{x}_2 = f_2(x_1, x_2, u) = x_1 x_2 + 3\sin x_1 - x_2 + u, \\ y = x_1^2 + x_1, \end{cases} \tag{14.5}$$

where the symbols are defined as aforementioned. The tracking control problem of the system is to design a suitable control input u such that the output trajectory y can track the smooth desired trajectory $y_d \in \mathbb{R}$. The tracking performance is evaluated by the absolute tracking error, i.e., $|e| = |y - y_d|$.

14.4.2 Design of ZD and ZG Controllers

To solve the tracking-control problem of system (14.5), controllers can be designed according to the ZD and ZG methods. To design the ZD controller, a ZF is defined as $z_1 = y - y_d$, i.e., $z_1 = x_1^2 + x_1 - y_d$. By using the ZD design formula $\dot{z}_1 = -\lambda z_1$, the following equation is obtained:

$$2x_1\dot{x}_1 + \dot{x}_1 - \dot{y}_d + \lambda(x_1^2 + x_1 - y_d) = 0. \tag{14.6}$$

Based on (14.5), (14.6) can be rewritten as

$$2x_1x_2 + x_2 - \dot{y}_d + \lambda(x_1^2 + x_1 - y_d) = 0, \tag{14.7}$$

where the control input u is not explicitly included. Thus, we define another ZF based on (14.7):

$$z_2 = 2x_1x_2 + x_2 - \dot{y}_d + \lambda(x_1^2 + x_1 - y_d).$$

Then, ZD design formula $\dot{z}_2 = -\lambda z_2$ is adopted again with $\dot{x}_1 = x_2$ taken into account, which yields

$$2x_2^2 + 2x_1\dot{x}_2 + \dot{x}_2 - \ddot{y}_d + \lambda(2x_1x_2 + x_2 - \dot{y}_d) + \lambda z_2 = 0. \tag{14.8}$$

In view of (14.5), by defining $\alpha_1 = x_1x_2 + 3\sin x_1 - x_2$, (14.8) can be rewritten as

$$2x_2^2 + (2x_1 + 1)(\alpha_1 + u) - \ddot{y}_d + \lambda(2x_1x_2 + x_2 - \dot{y}_d) + \lambda z_2 = 0.$$

Therefore, the ZD controller for the tracking control of system (14.5) is obtained as follows:

$$u = \frac{\alpha_2}{\alpha_3} - \alpha_1, \tag{14.9}$$

where $\alpha_2 = \ddot{y}_d - 4\lambda x_1x_2 - 2\lambda x_2 + 2\lambda\dot{y}_d - \lambda^2(x_1^2 + x_1 - y_d) - 2x_2^2$ and $\alpha_3 = 2x_1 + 1$. During the derivation procedure, the ZD design formula is adopted for two times and the GD design formula is not adopted. Thus, controller (14.9) is termed z2g0 controller.

Evidently, similar to z2g0 controller (14.3), z2g0 controller (14.9) is also limited by the DBZ problem. If α_3 is near zero, then the magnitude of u would be extremely large. However, the substitution of (14.9) into (14.5) yields the following system:

$$\begin{cases} \dot{x}_1 = x_2, \\ \dot{x}_2 = \dfrac{\alpha_2}{\alpha_3}, \\ y = x_1^2 + x_1, \end{cases}$$

where the DBZ problem still exists. This situation, which belongs to the TDBZ problem, is distinctly more dangerous than that investigated in Sect. 14.3. The

TDBZ problem not only causes the potential damage to the actuator, but also leads to the crash of system with the magnitude of system states tending to infinity.

To handle the TDBZ problem, the ZG method is used. Based on the ZG method, an EF is defined as $\epsilon_{\rm t} = \phi_{\rm t}^2/2$ with $\phi_{\rm t} = 2x_2^2 + \alpha_3(\alpha_1 + u) - \ddot{y}_{\rm d} + \lambda(2x_1x_2 + x_2 - \dot{y}_{\rm d}) + \lambda z_2$. Then, the GD design formula is adopted and the ZG controller is obtained as follows:

$$\dot{u} = -\gamma \frac{\partial \epsilon_{\rm t}}{\partial u} = -\gamma \alpha_3 \phi_{\rm t}. \tag{14.10}$$

This controller is a z2g1 controller.

14.4.3 Simulation, Verification and Comparison

In the simulations, the parameters and initial values for z2g0 controller (14.9) and z2g1 controller (14.10) are set as $\lambda = 10$, $\gamma = 10^3$, $x_1(0) = 1$, $x_2(0) = 0$ and $u(0) = 0$ illustratively. Without loss of generality, the desired trajectory $y_{\rm d} = 0.5\sin(3t) + 0.25$ is considered.

As seen from Fig. 14.3, the simulation of system (14.5) equipped with z2g0 controller (14.9) stops at the time instant $t_{\rm s} = \pi/2$ when the value of α_3 (i.e., the denominator of the z2g0 controller) becomes zero. This is consistent with Fig. 14.4a, which shows that the magnitude of x_2 becomes extremely large at the time instant $t_{\rm s}$. Note that, before time instant $t_{\rm s}$, the actual output y of the system tracks well the desired output $y_{\rm d}$. Besides, as shown in Fig. 14.4e, the absolute tracking error $|e|$ keeps decreasing until the system encounters the DBZ point (i.e., $\alpha_3 = 0$). From Fig. 14.5a, it can be observed that the magnitude of control input u is extremely

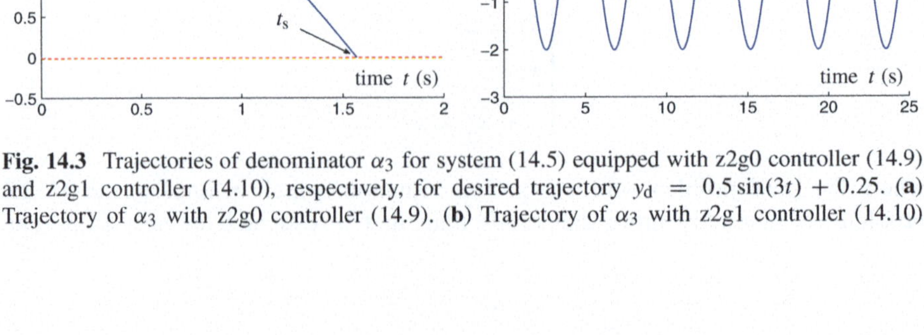

Fig. 14.3 Trajectories of denominator α_3 for system (14.5) equipped with z2g0 controller (14.9) and z2g1 controller (14.10), respectively, for desired trajectory $y_{\rm d} = 0.5\sin(3t) + 0.25$. (**a**) Trajectory of α_3 with z2g0 controller (14.9). (**b**) Trajectory of α_3 with z2g1 controller (14.10)

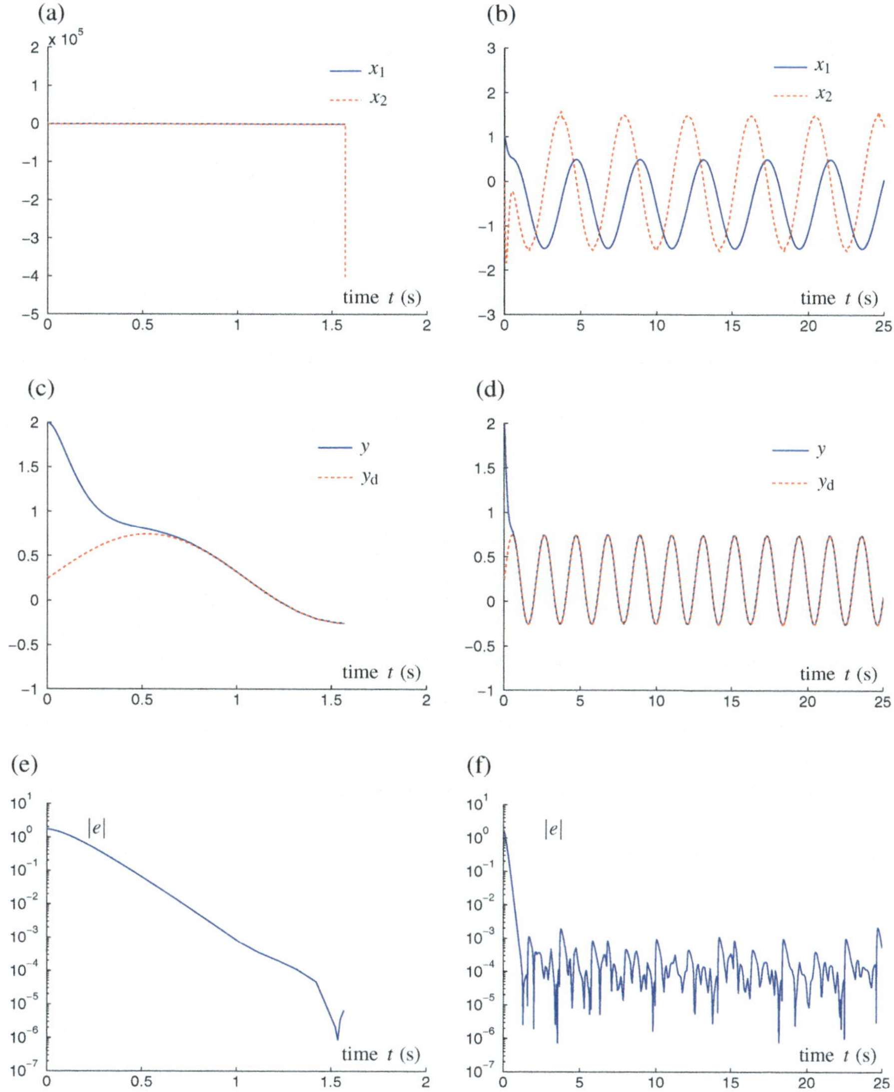

Fig. 14.4 Tracking performance of system (14.5) equipped with z2g0 controller (14.9) and z2g1 controller (14.10), respectively, for desired trajectory $y_d = 0.5\sin(3t) + 0.25$. (**a**) System states with z2g0 controller (14.9). (**b**) System states with z2g1 controller (14.10). (**c**) Output trajectory with z2g0 controller (14.9) and desired trajectory. (**d**) Output trajectory with z2g1 controller (14.10) and desired trajectory. (**e**) Absolute tracking error with z2g0 controller (14.9). (**f**) Absolute tracking error with z2g1 controller (14.10)

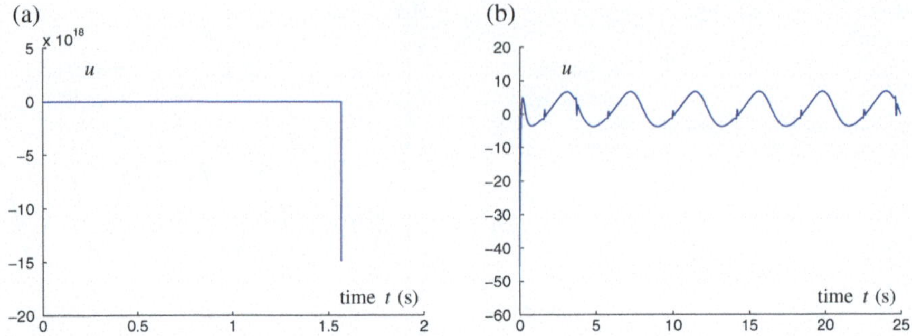

Fig. 14.5 Control inputs of system (14.5) equipped with z2g0 controller (14.9) and z2g1 controller (14.10), respectively, for desired trajectory $y_d = 0.5\sin(3t) + 0.25$. (**a**) Control input with z2g0 controller (14.9). (**b**) Control input with z2g1 controller (14.10)

large at the DBZ time instant, which is of order 10^{19}. Evidently, compared with the aforementioned PDBZ situation, these simulation results reveal that the TDBZ situation is more dangerous to the system.

Besides, in view of Figs. 14.3, 14.4, and 14.5, the efficacy of the ZG controller in conquering the TDBZ problem is substantiated. Specifically, the simulation of the system equipped with z2g1 controller (14.10) does not stop during the simulation execution. As seen from Fig. 14.3b, the system always gets through the DBZ points (i.e., $\alpha_3 = 0$) of z2g0 controller (14.9). The curves of states x_1 and x_2 in Fig. 14.4b are smooth and bounded. Besides, Fig. 14.4d, f show that the output trajectory y of the system tracks the desired trajectory y_d with the steady-state tracking error being small (specifically, of order 10^{-3}). In view of these results as well as the curve of control input u shown in Fig. 14.5b, the ZG method is an effective solution to the TDBZ problem.

14.5 Application to Two-Wheeled Mobile Robot

In this section, the ZG method is applied to the tracking control of a two-wheeled mobile robot. The schematic diagram of the mobile robot is shown in Fig. 14.6. Based on the usual assumption that the wheels of the mobile robot do not skid sideways, the motion of the mobile robot on a floor surface can be described as follows [27]:

$$\begin{cases} \dot{x}_1 = u_1 \cos x_3, \\ \dot{x}_2 = u_1 \sin x_3, \\ \dot{x}_3 = u_2, \end{cases} \tag{14.11}$$

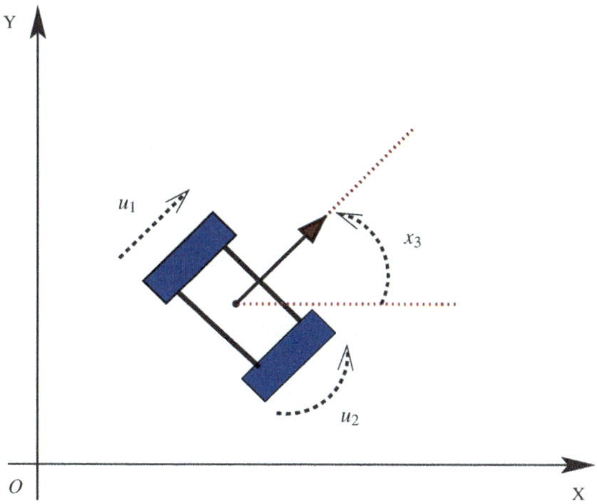

Fig. 14.6 Schematic diagram of two-wheeled mobile robot

where state variables $x_1 \in \mathbb{R}$ and $x_2 \in \mathbb{R}$ denote the X-axis and Y-axis positions of the mobile robot, respectively; x_1 and x_2 are also the outputs of the mobile robot system; state variable $x_3 \in \mathbb{R}$ denotes the orientation of the mobile robot; control inputs $u_1 \in \mathbb{R}$ and $u_2 \in \mathbb{R}$ denote the translational velocity and angular velocity of the mobile robot, respectively. To lay a basis for further discussion, the smooth desired X-axis position and Y-axis position of the mobile robot are denoted by $x_{1d} \in \mathbb{R}$ and $x_{2d} \in \mathbb{R}$, respectively. The absolute tracking errors are defined as $|e_1| = |x_1 - x_{1d}|$ and $|e_2| = |x_2 - x_{2d}|$. Based on the ZG method, the design procedure of the controller group for mobile robot system (14.11) is presented as follows.

In terms of output x_1, a ZF is defined as $z_1 = x_1 - x_{1d}$. Adopting the ZD design formula with (14.11) considered, we have $u_1 \cos x_3 - \dot{x}_{1d} + \lambda(x_1 - x_{1d}) = 0$. Thus, $u_1 = (\dot{x}_{1d} - \lambda x_1 + \lambda x_{1d})/\cos x_3$ can be obtained, which, however, leads to the TDBZ problem according to Proposition 14.2. By defining an EF $\epsilon_1 = (u_1 \cos x_3 - \dot{x}_{1d} + \lambda x_1 - \lambda x_{1d})^2/2$ and adopting the GD design formula, the following z1g1 controller is obtained:

$$\dot{u}_1 = -\gamma \cos x_3 (u_1 \cos x_3 - \dot{x}_{1d} + \lambda x_1 - \lambda x_{1d}). \tag{14.12}$$

Besides, in terms of output x_2, we define a ZF $z_2 = x_2 - x_{2d}$ for x_2. Adopting the ZD design formula with (14.11) considered, we have $u_1 \sin x_3 - \dot{x}_{2d} + \lambda(x_2 - x_{2d}) = 0$. As u_1 is used to control x_1, another ZF is defined as $z_3 = u_1 \sin x_3 - \dot{x}_{2d} + \lambda(x_2 - x_{2d})$. Utilizing the ZD design formula again with (14.11) and (14.12) considered, we have

$$-\gamma \sin x_3 \cos x_3 (u_1 \cos x_3 - \dot{x}_{1d} + \lambda x_1 - \lambda x_{1d}) + u_1 u_2 \cos x_3 - \ddot{x}_{2d} \\ +2\lambda(u_1 \sin x_3 - \dot{x}_{2d}) + \lambda^2 (x_2 - x_{2d}) = 0. \tag{14.13}$$

Evidently, a z2g0 controller can be directly obtained from (14.13), i.e.,

$$\begin{aligned}u_2 =&\big(\gamma \sin x_3 \cos x_3 (u_1 \cos x_3 - \dot{x}_{1\mathrm{d}} + \lambda x_1 - \lambda x_{1\mathrm{d}}) + \ddot{x}_{2\mathrm{d}} \\ &- 2\lambda(u_1 \sin x_3 - \dot{x}_{2\mathrm{d}}) - \lambda^2(x_2 - x_{2\mathrm{d}})\big)/(u_1 \cos x_3),\end{aligned}$$

which leads to the TDBZ problem in view of Proposition 14.2. Therefore, according to the ZG method, we further define an EF $\epsilon = \phi_2^2/2$ with

$$\begin{aligned}\phi_2 =& -\gamma \sin x_3 \cos x_3 (u_1 \cos x_3 - \dot{x}_{1\mathrm{d}} + \lambda x_1 - \lambda x_{1\mathrm{d}}) + u_1 u_2 \cos x_3 - \ddot{x}_{2\mathrm{d}} \\ &+ 2\lambda(u_1 \sin x_3 - \dot{x}_{2\mathrm{d}}) + \lambda^2(x_2 - x_{2\mathrm{d}}).\end{aligned}$$

Then, using the GD design formula, we obtain the following z2g1 controller:

$$\dot{u}_2 = -\gamma u_1 \cos x_3 \phi_2. \tag{14.14}$$

In view of z1g1 controller (14.12) and z2g1 controller (14.14), the corresponding ZG controller group for the tracking control of two-wheeled mobile robot (14.11) is obtained as follows:

$$\begin{cases}\dot{u}_1 = -\gamma \cos x_3 (u_1 \cos x_3 - \dot{x}_{1\mathrm{d}} + \lambda x_1 - \lambda x_{1\mathrm{d}}), \\ \dot{u}_2 = -\gamma u_1 \cos x_3 \phi_2.\end{cases} \tag{14.15}$$

In the application example, the circular trajectory with the radius being 1 m is considered, and the corresponding desired X-axis and Y-axis positions are

$$\begin{cases}x_{1\mathrm{d}} = \cos(2\pi \sin^2(\pi t/(2T))) + 2, \\ x_{2\mathrm{d}} = \sin(2\pi \sin^2(\pi t/(2T))) + 5,\end{cases} \tag{14.16}$$

where $T = 20$ s is the task duration. The initial positions of the mobile robot are set as $x_1(0) = 3.02$ m and $x_2(0) = 5$ m with the initial orientation $x_3(0) = \pi/2$ rad, and the initial translational velocity $u_1(0)$ and angular velocity $u_2(0)$ are set as 0.2 m/s and 0.1 rad/s, respectively. Besides, the design parameters of ZG controller group (14.15) are set as $\lambda = 1.2$ and $\gamma = 10^6$. The tracking performance of two-wheeled mobile robot (14.11) equipped with ZG controller group (14.15) for desired circular trajectory (14.16) is shown in Fig. 14.7. Specifically, as seen from Fig. 14.7a, b, the mobile robot quickly and successfully tracks the desired trajectory with the maximal steady-state tracking error being of order 10^{-4} m. Besides, Fig. 14.7c shows the trajectories of the state variables of the mobile robot system during the tracking-control process. In addition, Fig. 14.7d shows the trajectories of the control inputs (i.e., translational velocity u_1 and angular velocity u_2 of the mobile robot) with the maximal magnitude of the translational velocity u_1 being smaller than 0.5 m/s and that of the angular velocity u_2 being smaller than

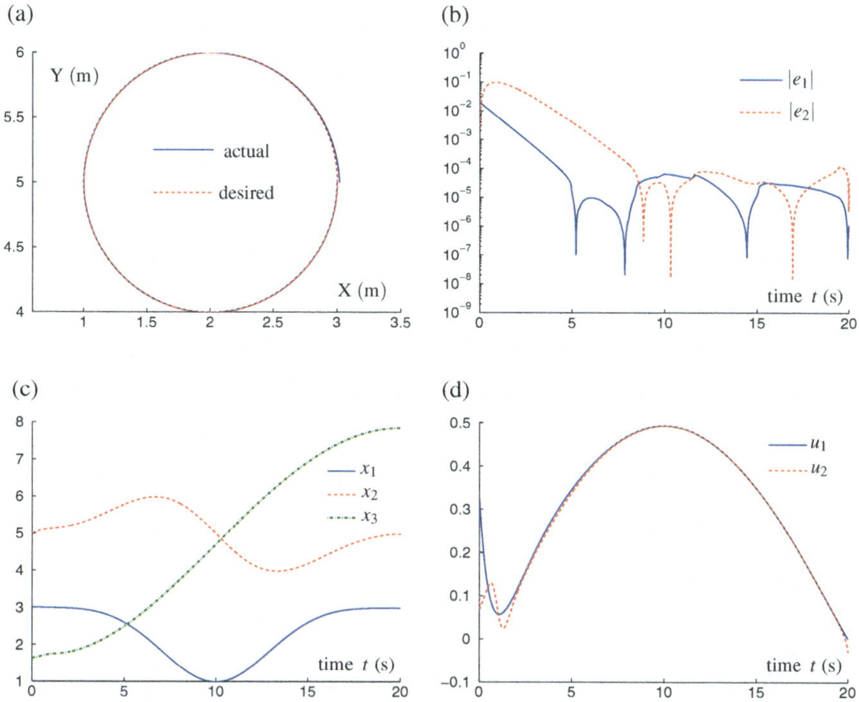

Fig. 14.7 Tracking performance of two-wheeled mobile robot (14.11) equipped with ZG controller group (14.15) for desired circular trajectory (14.16). (**a**) Actual trajectory and desired trajectory. (**b**) Absolute tracking errors. (**c**) State variables. (**d**) Translational velocity and angular velocity

0.5 rad/s. These results further substantiate the efficacy of the ZG method for solving the tracking-control problem of nonlinear systems with physical meanings while handling the TDBZ problem.

14.6 Chapter Summary

In this chapter, the DBZ problem in the control field has been discussed, classified and solved. The concepts of PDBZ and TDBZ problems have been presented, providing a new perspective for dealing with the DBZ problem. Besides, the specific examples of PDBZ and TDBZ problems have been shown and solved, respectively. The corresponding simulation results have indicated that the PDBZ problem may not affect the stability of the system states, while the TDBZ problem does affect the stability, indicating that the latter is more dangerous for the tracking control of systems. Besides, the superiority of the ZG controllers in conquering DBZ problems has been substantiated well. Particularly, the application of the ZG method to the two-wheeled mobile robot has substantiated the efficacy of the ZG method for the tracking control of nonlinear system with physical meaning in conquering the TDBZ problem.

References

1. Chang DE, Levine J, Jo J, Choi K-H (2013) Control of roll-to-roll web systems via differential flatness and dynamic feedback linearization. IEEE Trans Control Syst Technol 21(4):1309–1317
2. Na J, Ren X, Zheng D (2013) Adaptive control for nonlinear pure-feedback systems with high-order sliding mode observer. IEEE Trans Neural Netw Learn Syst 24(3):128–135
3. Niu B, Zhao J (2013) Tracking control for output-constrained nonlinear switched systems with a barrier Lyapunov function. Int J Syst Sci 44(5):978–985
4. Zhang Z, Shen H, Li Z, Zhang S (2015) Zero-error tracking control of uncertain nonlinear systems in the presence of actuator hysteresis. Int J Syst Sci 46(15):2853–2864
5. Ladiod S, Boucherit MS, Guerra, TM (2005) Adaptive fuzzy control of a class of MIMO nonlinear systems. Fuzzy Set Syst 151(1):59–77
6. Zhang T, Ge SS, Hang CC (2000) Stable adaptive control for a class of nonlinear systems using a modified Lyapunov function. IEEE Trans Autom Control 45(1):129–132
7. Zhang Y, Luo F, Yin Y, Liu J, Yu X (2013) Singularity-conquering ZG controller for output tracking of a class of nonlinear systems. In: Proceedings of Chinese control conference, pp 477–482
8. Zhang Y, Yu X, Yin Y, Peng C, Fan Z (2014) Singularity-conquering ZG controllers of z2g1 type for tracking control of the IPC system. Int J Control 87(9):1729–1746
9. Li T-S, Wang D, Feng G, Tong S-C (2010) A DSC approach to robust adaptive NN tracking control for strict-feedback nonlinear systems. IEEE Trans Syst Man Cybern B Cybern 40(3):915–927
10. Nam K, Lee S, Won S (1994) A local stabilizing control scheme using an approximate feedback linearization. IEEE Trans Autom Control 39(11):2311–2314
11. Aracil J, Gordillo F, Ponce E (2005) Stabilization of oscillations through backstepping in high-dimensional systems. IEEE Trans Autom Control 50(5):705–710
12. Hua C, Guan X, Shi P (2005) Robust backstepping control for a class of time delayed systems. IEEE Trans Autom Control 50(6):894–899
13. Kulkarni A, Perwar S (2009) Wavelet based adaptive backstepping controller for a class of nonregular systems with input constraints. Expert Syst Appl 36(3):6686–6696
14. Wan Y, Zhao J (2013) Extended backstepping method for single-machine infinite-bus power systems with SMES. IEEE Trans Control Syst Technol 21(3):915–923
15. Castillo NL, Juarez AL, Chairez I (2004) Active disturbance rejection robust control for uncertain systems with ill-defined relative degree. In: Proceedings of European control conference, pp 987–992
16. Huang J, Hung TV, Tseng M (2015) Smooth switching robust adaptive control for omnidirectional mobile robots. IEEE Trans Control Syst Technol 23(5):1986–1993
17. Hwang Y, Chen M, Wu T (2003) Division controllers for homogeneous dyadic bilinear systems. IEEE Trans Autom Control 48(4):701–705
18. Yang Y, Feng G, Ren J (2004) A combined backstepping and small-gain approach to robust adaptive fuzzy control for strict-feedback nonlinear systems. IEEE Trans Syst Man Cybern A Syst Hum 34(3):406–420

19. Arefi MM, Jahed-Motlagh, MR (2013) Observer-based adaptive neural control of uncertain MIMO nonlinear systems with unknown control direction. Int J Adapt Control Signal Process 27(9):741–754
20. Xiao L, Zhang Y (2012) Two new types of Zhang neural networks solving systems of time-varying nonlinear inequalities. IEEE Trans Circuits Syst Regul Pap 59(10):2363–2373
21. Zhang Z, Zhang Y (2013) Design and experimentation of acceleration-level driftfree scheme aided by two recurrent neural networks. IET Control Theory Appl 7(1):25–42
22. Zhang Y, Mu B, Zheng, H (2013) Link between and comparison and combination of Zhang neural network and quasi-Newton BFGS method for time-varying quadratic minimization. IEEE Trans Cybern 43(2):490–503
23. Liao B, Zhang Y (2014) Different complex ZFs leading to different complex ZNN models for time-varying complex generalized inverse matrices. IEEE Trans Neural Netw Learn Syst 25(9):1621–1631
24. Zhang Y, Yi C (2011) Zhang neural networks and neural-dynamic method. Nova Science Publishers, New York
25. Zhang Y, Yin Y, Wu H, Guo D (2012) Zhang dynamics and gradient dynamics with tracking-control application. In: Proceedings of the 5th international symposium on computational intelligence and design, pp 235–238
26. Zhang Y, Zhang Y, Chen D, Xiao Z, Yan X (2017) Division by zero, pseudo-division by zero, Zhang dynamics method and Zhang-gradient method about control singularity conquering. Int J Syst Sci 48(1):1–12
27. Urakubo T (2015) Feedback stabilization of a nonholonomic system with potential fields: application to a two-wheeled mobile robot among obstacles. Nonlinear Dyn 81(3):1475–1487

Part V
Time-Varying Systems Using ZG Control

Chapter 15
ZG Output Tracking of TVL System with DBZ Handled

Abstract In this chapter, the output tracking of TVL system is investigated. For solving such an output-tracking problem, three different types of controllers are presented, i.e., the conventional controller, ZD controller and ZG controller. Simulation results with two illustrative examples show that such three types of controllers are feasible and effective for the output-tracking problem solving. Especially, the presented ZG controller is capable of conquering the DBZ problem of TVL system.

15.1 Introduction

TVL system is extensively encountered in many fields, which has attracted much attention of researchers [1–3]. Most of the control theory, including the output tracking control, is devoted to the research of time-invariant linear systems, and the main reason is that the time-invariant linear systems are simpler. However, it is known that almost nothing is time-invariant in reality. As a matter of fact, the general TVL systems are normally too difficult to analyze and study due to the difficulties (e.g., the knotty DBZ problem) existing in TVL systems. Hence, the output tracking of TVL systems is an interesting and important topic that is worthy of research.

Recent studies have shown that ZD and GD are two types of powerful methods for time-varying problem solving [1, 4–8]. In this chapter [1], the output tracking of TVL system is considered and the ZD and ZG controllers are designed by exploiting the ZD and GD methods.

15.2 Problem Description

A general TVL system can be described as

$$\begin{cases} \dot{\mathbf{x}} = \mathbf{f}(\mathbf{x}, \mathbf{u}) = A\mathbf{x} + B\mathbf{u}, \\ \mathbf{y} = C\mathbf{x} + D\mathbf{u}, \end{cases} \tag{15.1}$$

Y. Zhang et al., *Zhang-Gradient Control*,
https://doi.org/10.1007/978-981-15-8257-8_15

where $\mathbf{x} \in \mathbb{R}^n$ is the system state, $\mathbf{u} \in \mathbb{R}^m$ is the control input, $\mathbf{y} \in \mathbb{R}^l$ is the system output, and $A \in \mathbb{R}^{n\times n}$, $B \in \mathbb{R}^{n\times m}$, $C \in \mathbb{R}^{l\times n}$ and $D \in \mathbb{R}^{l\times m}$ are the time-varying system matrices. For tracking the desired trajectory $\mathbf{y}_\mathrm{d} \in \mathbb{R}^l$, the output-tracking error of system (15.1) is defined as

$$\mathbf{e} = \mathbf{y} - \mathbf{y}_\mathrm{d} \in \mathbb{R}^l.$$

Note that the conventional output-tracking problem [2, 3] can be described as follows: given the desired trajectory $\mathbf{y}_\mathrm{d}$, a control input in the general form of $\mathbf{u}$ (or termed, $\mathbf{u}$-form) is to be designed such that it drives the output trajectory $\mathbf{y}$ to track the desired trajectory $\mathbf{y}_\mathrm{d}$ as close as possible (i.e., $\mathbf{e}$ asymptotically approaches zero). Note that the output-tracking error $\mathbf{e}$ is actually nonzero due to various kinds of errors (e.g., the truncation and round-off errors) existing in the computer simulation.

15.3 Design of Controllers

In this section, three different types of ZG controllers (i.e., conventional, ZD and ZG controllers) are designed for the output tracking of TVL system (15.1).

15.3.1 Design of Conventional Controller

From TVL system (15.1), the output-tracking error $\mathbf{e}$ and the basic idea of output tracking, it follows that

$$\mathbf{e} = \mathbf{y} - \mathbf{y}_\mathrm{d} = C\mathbf{x} + D\mathbf{u} - \mathbf{y}_\mathrm{d} = 0,$$

which can be rewritten as

$$D\mathbf{u} = \mathbf{y}_\mathrm{d} - C\mathbf{x}.$$

Then, the conventional controller in the $\mathbf{u}$-form is designed as

$$\mathbf{u} = D^{+}\left(\mathbf{y}_\mathrm{d} - C\mathbf{x}\right), \tag{15.2}$$

where D^{+} denotes the pseudoinverse of D [9]. Note that, if $D \equiv 0$, such a conventional controller becomes inapplicable in this case.

15.3.2 Design of ZD Controller

For more complex and general situations to be handled, the following ZF is first constructed:

$$\mathbf{z} = \mathbf{e} = \mathbf{y} - \mathbf{y}_\mathrm{d} = C\mathbf{x} + D\mathbf{u} - \mathbf{y}_\mathrm{d}. \quad (15.3)$$

Via the ZD method [4–7], the following ZD design formula is then utilized:

$$\dot{\mathbf{z}} = -\lambda \mathbf{z}, \quad (15.4)$$

where the design parameter $\lambda \in \mathbb{R}^+$ is used to scale the exponential convergence rate of the ZD solution.

It follows from (15.3) and (15.4) that

$$\dot{C}\mathbf{x} + C\dot{\mathbf{x}} + \dot{D}\mathbf{u} + D\dot{\mathbf{u}} - \dot{\mathbf{y}}_\mathrm{d} = -\lambda(C\mathbf{x} + D\mathbf{u} - \mathbf{y}_\mathrm{d}),$$

which can be further written as

$$\dot{C}\mathbf{x} + CA\mathbf{x} + CB\mathbf{u} + \dot{D}\mathbf{u} + D\dot{\mathbf{u}} - \dot{\mathbf{y}}_\mathrm{d} + \lambda(C\mathbf{x} + D\mathbf{u} - \mathbf{y}_\mathrm{d}) = 0. \quad (15.5)$$

Then, we have

$$D\dot{\mathbf{u}} = -(\dot{C}\mathbf{x} + CA\mathbf{x} + CB\mathbf{u} + \dot{D}\mathbf{u} - \dot{\mathbf{y}}_\mathrm{d} + \lambda(C\mathbf{x} + D\mathbf{u} - \mathbf{y}_\mathrm{d})).$$

Thus, the ZD controller in the $\dot{\mathbf{u}}$-form, which is different from the conventional $\mathbf{u}$-form, is obtained as

$$\dot{\mathbf{u}} = -D^+(\dot{C}\mathbf{x} + CA\mathbf{x} + CB\mathbf{u} + \dot{D}\mathbf{u} - \dot{\mathbf{y}}_\mathrm{d} + \lambda(C\mathbf{x} + D\mathbf{u} - \mathbf{y}_\mathrm{d})). \quad (15.6)$$

15.3.3 Design of ZG Controller

From (15.5), we can further define

$$\boldsymbol{\phi} = \dot{C}\mathbf{x} + CA\mathbf{x} + CB\mathbf{u} + \dot{D}\mathbf{u} + D\dot{\mathbf{u}} - \dot{\mathbf{y}}_\mathrm{d} + \lambda(C\mathbf{x} + D\mathbf{u} - \mathbf{y}_\mathrm{d}), \quad (15.7)$$

which should be zero theoretically. Based on the GD method [8], a norm-based scalar-valued nonnegative EF is defined as $\varepsilon = \|\boldsymbol{\phi}\|_2^2/2$.

Then, according to the GD method [8], we can adopt the GD design formula $\dot{\mathbf{v}} = -\gamma\partial\varepsilon/\partial\mathbf{v}$ with $\mathbf{v} = \dot{\mathbf{u}}$, where the design parameter $\gamma \in \mathbb{R}^+$ is used to scale the convergence rate of the GD solution.

Finally, substituting (15.7) into the GD design formula yields the following ZG controller in the $\ddot{\mathbf{u}}$-form:

$$\ddot{\mathbf{u}} = -\gamma D^{\mathrm{T}}(\dot{C}\mathbf{x} + CA\mathbf{x} + CB\mathbf{u} + \dot{D}\mathbf{u} + D\dot{\mathbf{u}} - \dot{\mathbf{y}}_{\mathrm{d}} + \lambda(C\mathbf{x} + D\mathbf{u} - \mathbf{y}_{\mathrm{d}})). \quad (15.8)$$

15.4 Simulation, Verification and Comparison

In order to verify the efficacy of the presented three controllers, the tracking performance of these controllers for two TVL systems are compared in this section. Such two systems are different, as the DBZ problem does not exist in the first system while existing in the second one.

Example 15.1 In this example, the following TVL system is considered:

$$\begin{cases} \dot{\mathbf{x}} = \begin{bmatrix} -2 & 1 & 0 \\ 1 & -1-t & 0 \\ 0 & 0 & -0.1t^2 \end{bmatrix} \mathbf{x} + \begin{bmatrix} 1 & t & 1 \\ \cos(t) & 3 & 2 \\ t & t & 0 \end{bmatrix} \mathbf{u}, \\ y = \begin{bmatrix} \sin(t) & t^2 & t \end{bmatrix} \mathbf{x} + \begin{bmatrix} t+1 & \sin(t) & \cos(t) \end{bmatrix} \mathbf{u}, \end{cases} \quad (15.9)$$

where the DBZ problem does not exist for the coefficient before **u**. In the simulation, the design parameters are set as $\lambda = 1$ and $\gamma = 1000$ for the ZD and ZG controllers. Besides, the desired trajectory and the running time are selected as $y_{\mathrm{d}} = \sin(0.5t)$ and 30 s, respectively, and the initial values $x_1(0)$, $x_2(0)$ and $x_3(0)$ are set as 0.5. Corresponding to the three controllers, the simulation results on the tracking control of TVL system (15.9) are shown in Figs. 15.1 and 15.3, respectively.

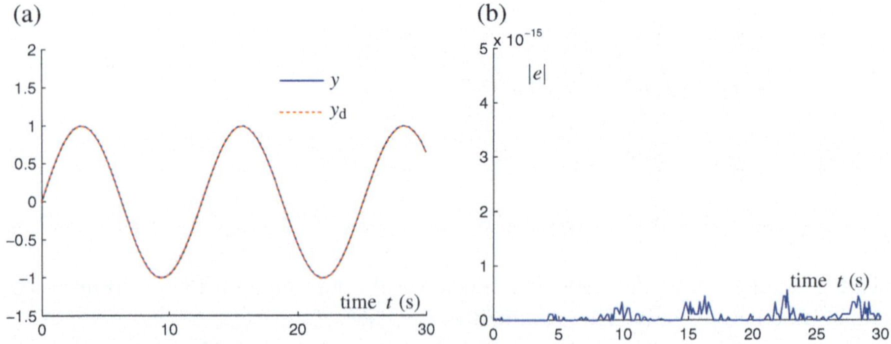

Fig. 15.1 Tracking performance of TVL system (15.9) equipped with conventional controller (15.2) for desired trajectory $y_{\mathrm{d}} = \sin(0.5t)$. (**a**) Output trajectory and desired trajectory. (**b**) Absolute tracking error

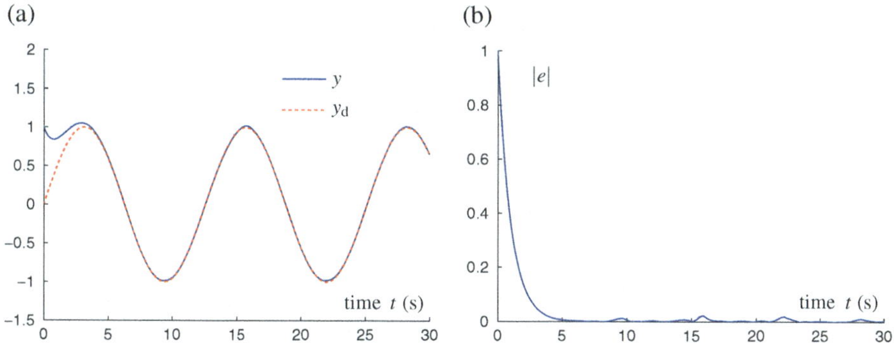

Fig. 15.2 Tracking performance of TVL system (15.9) equipped with ZD controller (15.6) for desired trajectory $y_d = \sin(0.5t)$. (**a**) Output trajectory and desired trajectory. (**b**) Absolute tracking error

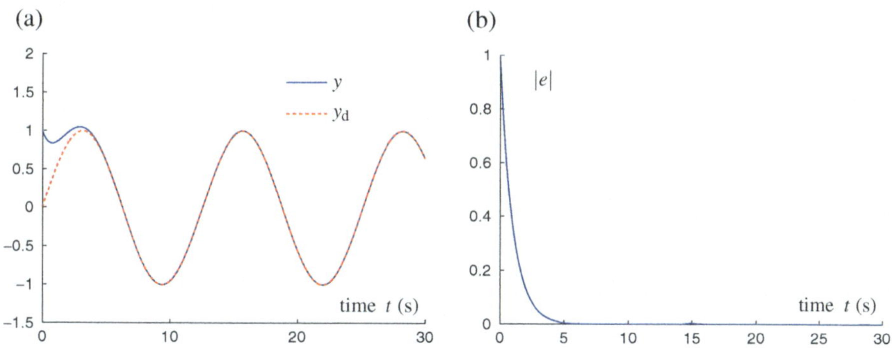

Fig. 15.3 Tracking performance of TVL system (15.9) equipped with ZG controller (15.8) for desired trajectory $y_d = \sin(0.5t)$. (**a**) Output trajectory and desired trajectory. (**b**) Absolute tracking error

Specifically, Figs. 15.1a, 15.2a, and 15.3a display that the outputs of TVL system (15.9) equipped with the three controllers can track well the desired trajectory $y_d = \sin(0.5t)$. Besides, compared with the absolute tracking errors in Figs. 15.1b and 15.3b, the absolute tracking error in Fig. 15.2b is larger and decreases more slowly. This illustrates that the tracking performance of TVL system (15.9) equipped with the conventional or ZG controller is better than that of TVL system (15.9) equipped with the ZD controller.

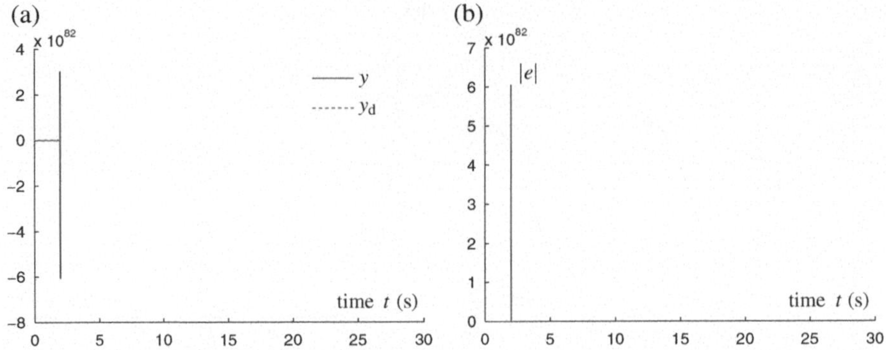

Fig. 15.4 Tracking performance of TVL system (15.10) equipped with conventional controller (15.2) for desired trajectory $y_d = 10\sin(2t) + 0.5t$. (**a**) Output trajectory and desired trajectory. (**b**) Absolute tracking error

Example 15.2 In this example, another TVL system is considered as below:

$$\begin{cases} \dot{\mathbf{x}} = \begin{bmatrix} -(t-10)^2 & 1 \\ -1 & -(t-20)^4 \end{bmatrix} \mathbf{x} + \begin{bmatrix} 1 \\ 2 \end{bmatrix} u, \\ y = \begin{bmatrix} 1 & 5 \end{bmatrix} \mathbf{x} + \begin{bmatrix} t-2 \end{bmatrix} u. \end{cases} \tag{15.10}$$

Similarly, the design parameters are set as $\lambda = 1$ and $\gamma = 1000$ for the ZD and ZG controllers, and the running time is selected as 30 s. In addition, the initial values $x_1(0)$ and $x_2(0)$ are set as 0.5. For further verification, the desired trajectory $y_d = 10\sin(2t) + 0.5t$ is intentionally selected in the simulation. Note that, in terms of the conventional and ZD controllers designed for the TVL system, there exists the DBZ problem when time t approaches 2 s, which makes such two controllers crash in this example. The corresponding simulation results are shown in Figs. 15.4, 15.5, and 15.6.

Specifically, Fig. 15.4 shows that the output-tracking process of TVL system (15.10) with the conventional controller stops when time t is near 2 s. Specifically, as time t approaches 2 s, the output magnitude of TVL system (15.10) equipped with the conventional controller becomes extremely large, which leads to system crash. Besides, as observed from Fig. 15.5, the output-tracking process of TVL system (15.10) with the ZD controller also stops at around $t = 2$ s. The reason is that, when time t approaches 2 s, the magnitude of the term D^+ of the conventional or ZD controller tends to be extremely large, which makes the value of control input become too large to implement, and consequently the TVL system crashes with its output being out of control.

By contrast, as seen from Fig. 15.6, the output of the TVL system with the ZG controller is capable of tracking the desired trajectory $y_d = 10\sin(2t) + 0.5t$ successfully. More specifically, as time t approaches 2 s, the control input becomes

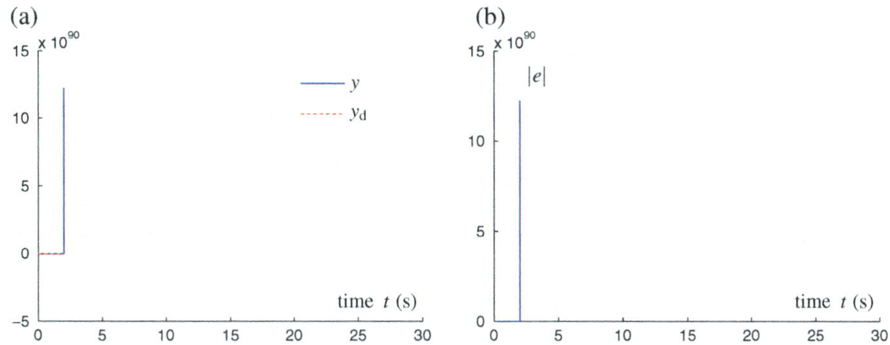

Fig. 15.5 Tracking performance of TVL system (15.10) equipped with ZD controller (15.6) for desired trajectory $y_d = 10\sin(2t) + 0.5t$. (**a**) Output trajectory and desired trajectory. (**b**) Absolute tracking error

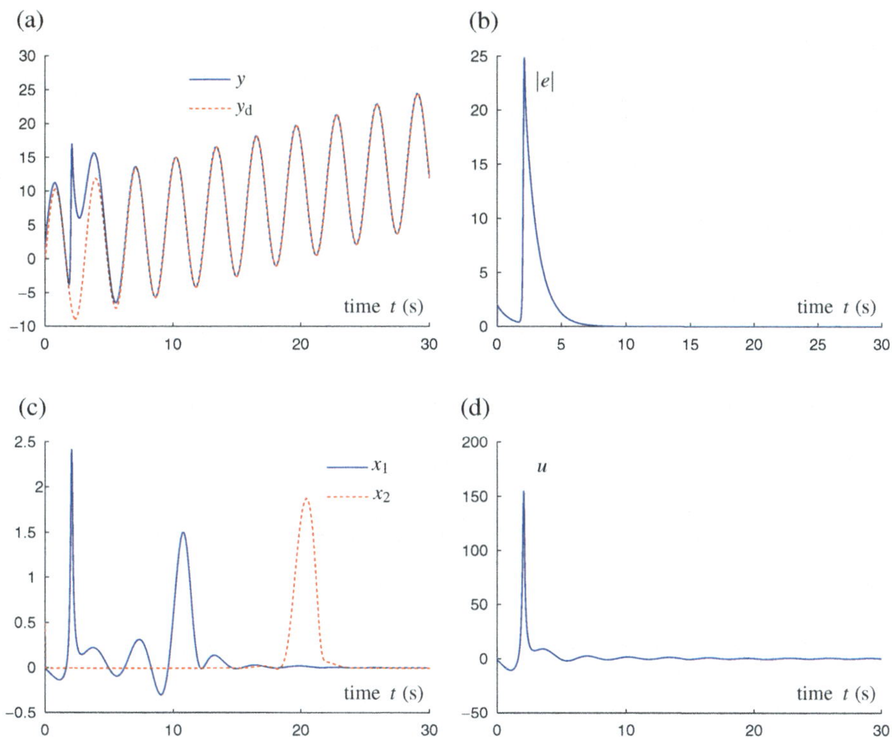

Fig. 15.6 Tracking performance of TVL system (15.10) equipped with ZG controller (15.8) for desired output trajectory $y_d = 10\sin(2t) + 0.5t$. (**a**) Output trajectory and desired trajectory. (**b**) Absolute tracking error. (**c**) System states. (**d**) Control input

large and the system states start to fluctuate, which drives the output to deviate the trajectory; however, after 2 s (i.e., $t > 2$ s), the TVL system can adjust itself and the output can track the desired trajectory automatically again. This illustrates that the ZG controller conquers the DBZ problem successfully while the other two controllers (i.e., conventional and ZD controllers) fail.

In summary, the simulation results of the above two examples substantiate that the conventional, ZD and ZG controllers are all effective for the output tracking of the TVL system. More importantly, the ZG controller shows its superiority for conquering the DBZ problem of the TVL system.

15.5 Chapter Summary

In this chapter, the output tracking of TVL system has been investigated. Based on the ZG method, three types of controllers (i.e., conventional, ZD and ZG controllers) have been designed for solving such a tracking-control problem. The illustrative and comparative simulation results have been presented to show the efficacy of these three controllers for the output tracking of TVL system. More importantly, the ZG controller has conquered the DBZ problem successfully.

References

1. Zhang Y, Liu J, Yin Y, Luo F, Deng J (2013) Zhang-gradient controllers of Z0G0, Z1G0 and Z1G1 types for output tracking of time-varying linear systems with control-singularity conquered finally. In: Proceedings of the 10th international symposium on neural networks, pp 533–540
2. Chen M-S (1998) A tracking controller for linear time-varying systems. J Dyn Syst Meas Control 120(1):111–116
3. Zheng D (2002) Linear system theory. Tsinghua University Press, Beijing
4. Zhang Y, Ma W, Cai B (2009) From Zhang neural network to Newton iteration for matrix inversion. IEEE Trans Circuits Syst Regul Pap 56(7):1405–1415
5. Zhang Y, Yi C, Guo D, Zheng J (2011) Comparison on Zhang neural dynamics and gradient-based neural dynamics for online solution of nonlinear time-varying equation. Neural Comput Appl 20(1):1–7
6. Zhang Y, Yi C, Ma W (2009) Simulation and verification of Zhang neural network for online time-varying matrix inversion. Simul Model Pract Theory 17(10):1603–1617
7. Zhang Y, Li Z (2009) Zhang neural network for online solution of time-varying convex quadratic program subject to time-varying linear-equality constraints. Phys Lett A 373(18):1639–1643
8. Zhang Y, Ke Z, Xu P, Yi C (2010) Time-varying square roots finding via Zhang dynamics versus gradient dynamics and the former's link and new explanation to Newton-Raphson iteration. Inform Process Lett 110(24):1103–1109
9. Zhang Y, Yang Y, Tang N, Cai B (2011) Zhang neural network solving for time-varying full-rank matrix Moore-Penrose inverse. Comput 92(2):97–121

Chapter 16
ZG Stabilization of TVL System with PDBZ Shown

16.1 Introduction

In modern engineering applications, more and more structures themselves possess time-varying characteristics [1, 2], such as the rotation of solar panels and the movement of robotic arms. Some processes often need to take into account the time-varying characteristics of systems. Thereinto, TVL system is an important part of research on linear system theory [1]. Strictly speaking, the actual physical systems cannot be linear systems. However, through approximately processing and reasonably simplifying, a large number of physical systems can be idealized or simplified as linear systems in a sufficiently accurate sense and within a specified tolerance range, in view of the fact that linear systems are easier to handle in comparison with nonlinear systems on many occasions.

The stabilization problem of TVL system is a basic problem in control theory, and has wide theoretical significance and application value. In addition, stabilization is the primary condition for control system to function properly, and the analysis of stabilization for TVL system is quite complex and difficult [1]. For instance, in [3], the ZD method was used to stabilize a 4th-order hyper-chaotic Lu system to the zero equilibrium point steadily. Besides, in [4], a ZG stabilization controller was presented to realize the stabilization control of a bilinear system. Note that the aforementioned systems are time-invariant.

In this chapter [2], we focus on the stabilization of TVL system with PDBZ phenomenon shown. As we know, the PDBZ problem is a special case extensively studied in DBZ problem [2, 5]. In order to overcome the DBZ problem, various kinds of methods are put forward and carried out. For example, in [6], the ZG method was used for the tracking control of nonlinear systems with the DBZ problem conquered. Besides, in [7], a dead-zone was overlaid onto the controller to avert the DBZ problem. Moreover, in [8], a smooth switching adaptive robust controller was presented to overcome the DBZ problem.

Y. Zhang et al., *Zhang-Gradient Control*,
https://doi.org/10.1007/978-981-15-8257-8_16

The ZG method is an efficient solution to conquer the DBZ problem [5, 9], which is made up of the ZD and GD methods [10–12]. For example, the ZG method is successfully used to solve the tracking control of the ship course with the DBZ (specifically, PDBZ) problem conquered [13]. In this chapter [2], we also employ the ZG method to solve the PDBZ problem. In the meantime, the ZD method is used to show the PDBZ problem. Although these two methods are able to achieve the stabilization of TVL system through computer simulations, there are some differences between the two methods. For example, the ZD controller itself contains DBZ points, while the ZG controller itself does not contain the DBZ points. Furthermore, we can more intuitively find the difference between the ZD and ZG methods when simulated to the time instant of DBZ for ZD controller.

16.2 Problem Description

In this section, we mainly present the system equation of TVL system, which is investigated in this chapter [2] as below:

$$\begin{cases} \dot{x}_1 = f_1(x_1, x_2, t) = \big(\sin(10t) + 2 + \sin(0.1t)\big)x_1 + x_2, \\ \dot{x}_2 = f_2(x_1, u, t) = 2\cos(3t)x_1 + 50\cos(wt)u, \end{cases} \tag{16.1}$$

where w is an unknown constant parameter. The objective of designing stabilization controller is to make system (16.1) satisfy the following conditions:

$$\begin{cases} x_1 \to 0, \\ x_2 \to 0. \end{cases}$$

Note that, based on the ensuing controller design processes as well as computer simulation results, it can be found that both the ZD and ZG methods can achieve the stabilization control of TVL system (16.1).

16.3 Design of ZD Controller

In this section, we employ the ZD method to design ZD stabilization controller for solving the stabilization control problem with the PDBZ phenomenon shown. For obtaining the ZD stabilization controller, we define the first ZF [5, 10, 13]:

$$z_1 = x_1. \tag{16.2}$$

Based on the ZD method, the first ZD design formula [11] is used as follows:

$$\dot{z}_1 = -\lambda_1 z_1, \tag{16.3}$$

where the design parameter $\lambda_1 \in \mathbb{R}^+$ is used to scale the convergence rate of the ZD solution [3, 5]. Substituting (16.2) into (16.3), we obtain

$$\dot{x}_1 = -\lambda_1 x_1. \quad (16.4)$$

Then, substituting the expression of $\dot{x}_1$ in (16.1) into (16.4) yields

$$(\sin(10t) + 2 + \sin(0.1t))x_1 + x_2 + \lambda_1 x_1 = 0. \quad (16.5)$$

Besides, we define $\alpha_1 = \sin(10t) + 2 + \sin(0.1t)$, and then Eq. (16.5) is further written as

$$\alpha_1 x_1 + x_2 + \lambda_1 x_1 = 0.$$

As seen from the above equation, the control input u is not explicitly included. Thus, the second ZF is defined as below:

$$z_2 = \alpha_1 x_1 + x_2 + \lambda_1 x_1. \quad (16.6)$$

The second ZD design formula is employed as follows:

$$\dot{z}_2 = -\lambda_2 z_2, \quad (16.7)$$

where the design parameter $\lambda_2 \in \mathbb{R}^+$ is used to scale the convergence rate of the ZD solution as well. Substituting (16.6) into (16.7), we obtain

$$\dot{\alpha}_1 x_1 + \alpha_1 \dot{x}_1 + \dot{x}_2 + \lambda_1 \dot{x}_1 + \lambda_2 z_2 = 0.$$

In view of $\dot{x}_2 = 2\cos(3t)x_1 + 50\cos(wt)u$, the above equation is further presented as

$$\dot{\alpha}_1 x_1 + \alpha_1 \dot{x}_1 + 2\cos(3t)x_1 + 50\cos(wt)u + \lambda_1 \dot{x}_1 + \lambda_2 z_2 = 0. \quad (16.8)$$

Apparently, Eq. (16.8) contains the control input u. Therefore, the ZD stabilization controller is obtained as below:

$$u = -\frac{\dot{\alpha}_1 x_1 + \alpha_1 \dot{x}_1 + 2\cos(3t)x_1 + \lambda_1 \dot{x}_1 + \lambda_2 z_2}{50\cos(wt)}. \quad (16.9)$$

In particular, ZD stabilization controller (16.9) is acquired by adopting the ZD method twice but without adopting the GD method. In consideration of the fact just mentioned, controller (16.9) is also termed z2g0 controller. To further simplify the expression of ZD stabilization controller (16.9), we obtain

$$u = \frac{\alpha_2}{50\cos(wt)}, \tag{16.10}$$

where $\alpha_2 = -(\dot{\alpha}_1 x_1 + \alpha_1 \dot{x}_1 + 2\cos(3t)x_1 + \lambda_1 \dot{x}_1 + \lambda_2 z_2)$. Distinctly, the denominator $50\cos(wt)$ is contained in the expression of ZD stabilization controller (16.10). As we know, when $\cos(wt) = 0$, the value of u tends to infinity, which is called DBZ problem [6]. However, substituting (16.10) into system equation (16.1), we further obtain

$$\begin{cases} \dot{x}_1 = (\sin(10t) + 2 + \sin(0.1t))x_1 + x_2, \\ \dot{x}_2 = 2\cos(3t)x_1 + \alpha_2. \end{cases} \tag{16.11}$$

Evidently, Eq. (16.11) does not contain the DBZ problem. In fact, this is a PDBZ problem [2]. The presented controller (16.10) achieves the stabilization control of TVL system (16.1), despite of the controller itself containing DBZ points. From the above, for TVL system (16.1), the ZD method is certainly a straightforward and efficient method to deal with the stabilization problem.

16.4 Design of ZG Controller

In this section, we apply the ZG method to design ZG stabilization controller for solving the stabilization control problem as well as handling the PDBZ phenomenon shown in ZD stabilization controller (16.10). As mentioned earlier, the ZG method is made up of the ZD and GD methods [5, 6]. Before adopting the GD method, we need to define the third ZF as follows:

$$z_3 = x_2. \tag{16.12}$$

Besides, the third ZD design formula is adopted as below:

$$\dot{z}_3 = -\lambda_3 z_3, \tag{16.13}$$

where $\lambda_3 \in \mathbb{R}^+$ is the design parameter. Substituting (16.12) into (16.13) yields

$$\dot{x}_2 = -\lambda_3 x_2.$$

In view of $\dot{x}_2 = 2\cos(3t)x_1 + 50\cos(wt)u$, the above equation is further written as

$$2\cos(3t)x_1 + 50\cos(wt)u + \lambda_3 x_2 = 0. \tag{16.14}$$

As seen from Eq. (16.14), it contains the control input u. Furthermore, from Eqs. (16.8) and (16.14), we define $\phi_1 = \dot{\alpha}_1 x_1 + \alpha_1 \dot{x}_1 + 2\cos(3t)x_1 + 50\cos(wt)u + \lambda_1 \dot{x}_1 + \lambda_2 z_2$ and $\phi_2 = 2\cos(3t)x_1 + 50\cos(wt)u + \lambda_3 x_2$, which should theoretically be zero. Then, we utilize the GD method. Particularly, an EF is defined as $\epsilon = \phi_1^2/2 + \phi_2^2/2$. Next, the GD design formula is adopted as follows:

$$\dot{u} = -\gamma \frac{\partial \epsilon}{\partial u}, \tag{16.15}$$

where the design parameter $\gamma \in \mathbb{R}^+$ is used to scale the convergence rate of the GD solution [4, 5]. Formula (16.15) can be rewritten as

$$\dot{u} = -\gamma \left(\frac{\partial \epsilon}{\partial \phi_1} \frac{\partial \phi_1}{\partial u} + \frac{\partial \epsilon}{\partial \phi_2} \frac{\partial \phi_2}{\partial u} \right). \tag{16.16}$$

Thus, ZG stabilization controller (16.16) is obtained by adopting the ZD method for three times and the GD method for one time. In view of the fact just mentioned, controller (16.16) is also termed z3g1 controller. The above-mentioned ZG stabilization controller (16.16) for system (16.1) is further written as

$$\dot{u} = -50\gamma(\phi_1 + \phi_2)\cos(wt). \tag{16.17}$$

We can observe that ZG stabilization controller (16.17) does not contain the division operation. Therefore, the PDBZ problem existing in ZD stabilization controller (16.10) can be effectively solved by the ZG method.

16.5 Simulation, Verification and Comparison

In this section, we conduct the corresponding computer simulations. For ZD stabilization controller (16.10), the initial states $x_1(0)$ and $x_2(0)$ are relatively arbitrarily set as 2 and 3, respectively. For ZG stabilization controller (16.17), $x_1(0)$ and $x_2(0)$ are also set to be 2 and 3, respectively, and $u(0)$ is set as 1. The design parameters λ_1, λ_2, λ_3, and γ are set as 10, 10, 10, and 1000, respectively. In addition, the task duration is set as 20 s. In order to give the comparative analysis of simulation results, without loss of generality, w is set as 0.2, 2, and 20 rad/s, respectively.

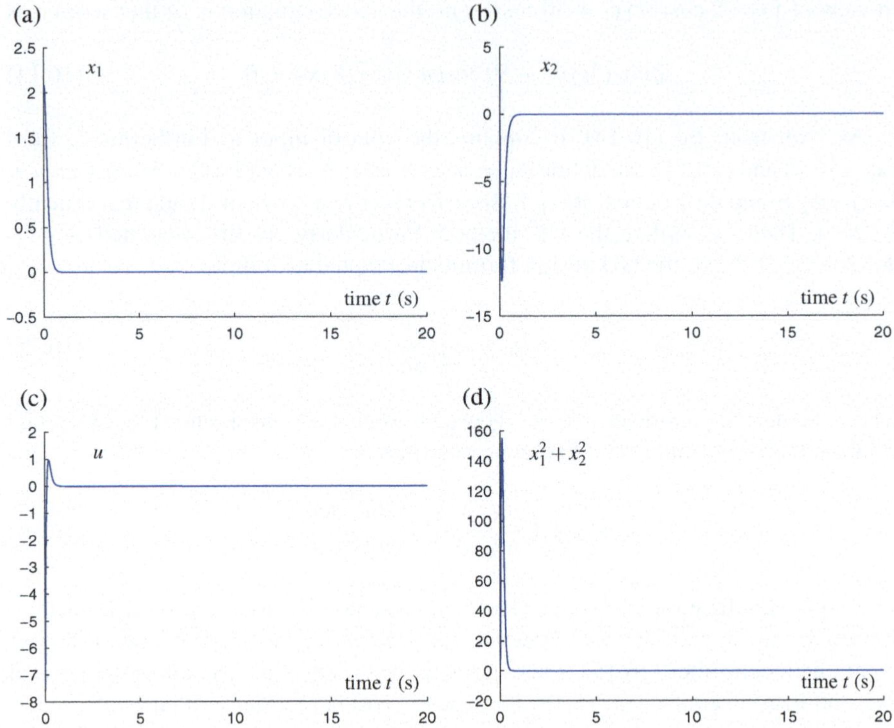

Fig. 16.1 Stabilization performance of TVL system (16.1) equipped with ZD controller (16.10) using $w = 0.2$ rad/s. (**a**) System state x_1. (**b**) System state x_2. (**c**) Control input. (**d**) Square sum of x_1 and x_2

First, when w is set as 0.2 rad/s, the corresponding simulation results of ZD stabilization controller (16.10) and ZG stabilization controller (16.17) are shown in Figs. 16.1 and 16.2, respectively. Specifically, Fig. 16.1a describes the trajectory of system state x_1, and Fig. 16.1b describes the trajectory of system state x_2. In addition, Fig. 16.1c describes the trajectory of control input u, and Fig. 16.1d describes the trajectory of square sum of x_1 and x_2. As time t evolves, the control input u converges to zero quickly. Accordingly, system states x_1 and x_2 also converge to zero quickly, thereby implying that the stabilization control of system (16.1) is realized. Calculations show that, in theory, ZD stabilization controller (16.10) encounters the first DBZ point at about 7.853 s. Nevertheless, the control input u has converged to zero before this time instant. From Fig. 16.2, we can also observe that the control input u converges to zero quickly, with the system states x_1 and x_2 also converging to zero quickly.

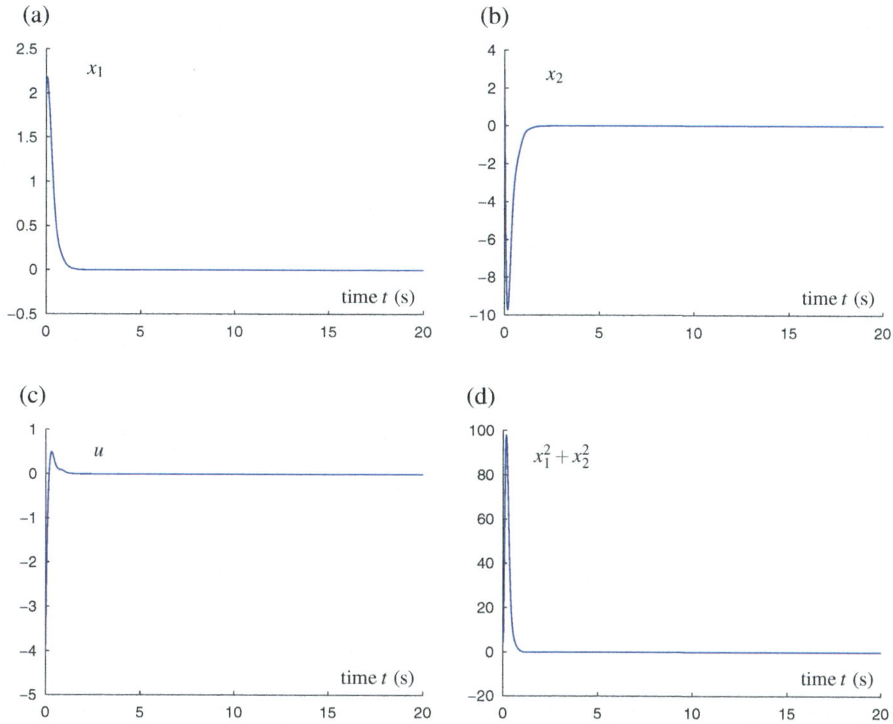

Fig. 16.2 Stabilization performance of TVL system (16.1) equipped with ZG controller (16.17) using $w = 0.2$ rad/s. (**a**) System state x_1. (**b**) System state x_2. (**c**) Control input. (**d**) Square sum of x_1 and x_2

Next, when w is set as 2 rad/s, the corresponding simulation results of ZD stabilization controller (16.10) and ZG stabilization controller (16.17) are shown in Figs. 16.3 and 16.4, respectively. From Fig. 16.3, we can observe that the control input u contains a DBZ point. By calculations, we conclude that the first DBZ point of ZD stabilization controller (16.10) emerges at about 0.7853 s, which is in accordance with simulation results. After the first DBZ point emerges, the control input u converges to zero soon afterwards. Moreover, the system states x_1 and x_2 converge to zero quickly, and the square sum of x_1 and x_2 also converges to zero ultimately. From Fig. 16.4, the system states x_1 and x_2 also converge to zero, followed by the control input u converging to zero. However, the value of control input u does not go to infinity at about 0.7853 s, which is discussed in detail later.

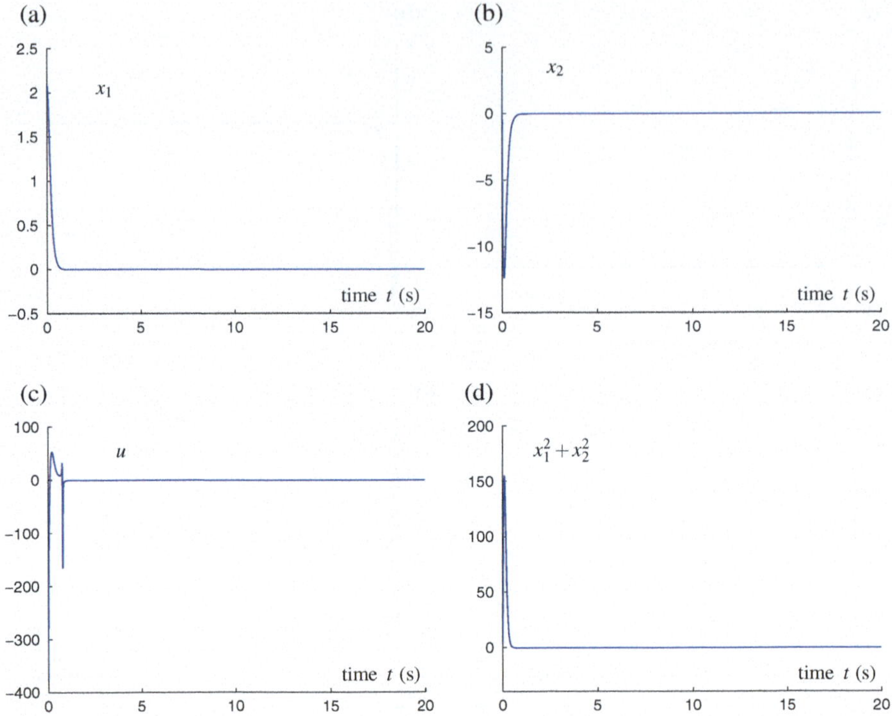

Fig. 16.3 Stabilization performance of TVL system (16.1) equipped with ZD controller (16.10) using $w = 2$ rad/s. (**a**) System state x_1. (**b**) System state x_2. (**c**) Control input. (**d**) Square sum of x_1 and x_2

Then, when w is set as 20 rad/s, the corresponding simulation results of ZD stabilization controller (16.10) and ZG stabilization controller (16.17) are shown in Figs. 16.5 and 16.6, respectively. Apparently, in comparison with Fig. 16.3c, there are more DBZ points in Fig. 16.4c, which emerge around every 0.157 s. However, the control input u converges to zero soon afterwards, similar to that in Fig. 16.3. The system states x_1 and x_2, as well as the square sum of x_1 and x_2, also converge to zero quickly as shown in Fig. 16.5. From Fig. 16.6, we can observe that the system states x_1 and x_2 converge to zero quickly. Nevertheless, the control input u does not contain DBZ points like that in Fig. 16.4, which is different from ZD stabilization controller (16.10).

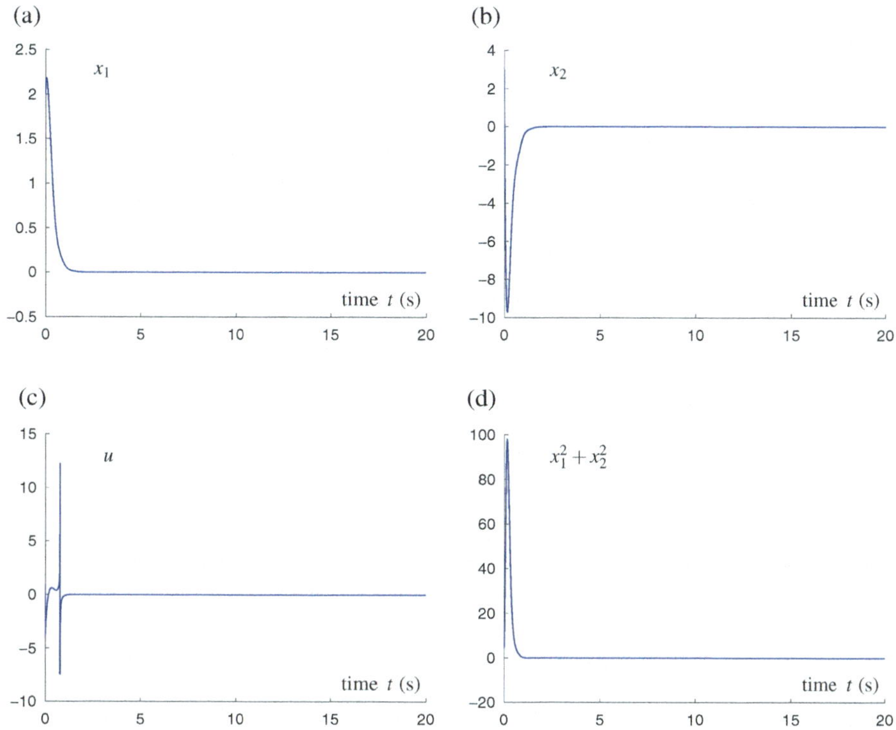

Fig. 16.4 Stabilization performance of TVL system (16.1) equipped with ZG controller (16.17) using $w = 2$ rad/s. (**a**) System state x_1. (**b**) System state x_2. (**c**) Control input. (**d**) Square sum of x_1 and x_2

In order to further explain the distinctions between ZD stabilization controller (16.10) and ZG stabilization controller (16.17) in detail, we specially present the simulation results at about 0.7853 s with w set as 2 rad/s. From Figs. 16.7 and 16.8, we can more intuitively observe that the control input of ZD stabilization controller (16.10) goes to infinity when simulated to the DBZ time instant. However, the control input of ZG stabilization controller (16.17) has a finite value when simulated to the time instant of ZD stabilization controller (16.10) encountering the DBZ point. Through the comparison, we discover that: (1) ZD stabilization controller (16.10) itself contains DBZ points; (2) ZD stabilization controller (16.10) and ZG stabilization controller (16.17) both can achieve the stabilization of TVL system (16.1); (3) ZG stabilization controller (16.17) can conquer DBZ points contained in ZD stabilization controller (16.10). In addition, through conducting computer simulations, we further discover that, when w is relatively small, before ZD stabilization controller (16.10) encounters DBZ points, it has converged to zero; when w is relatively large, ZD stabilization controller (16.10) may encounters many DBZ points.

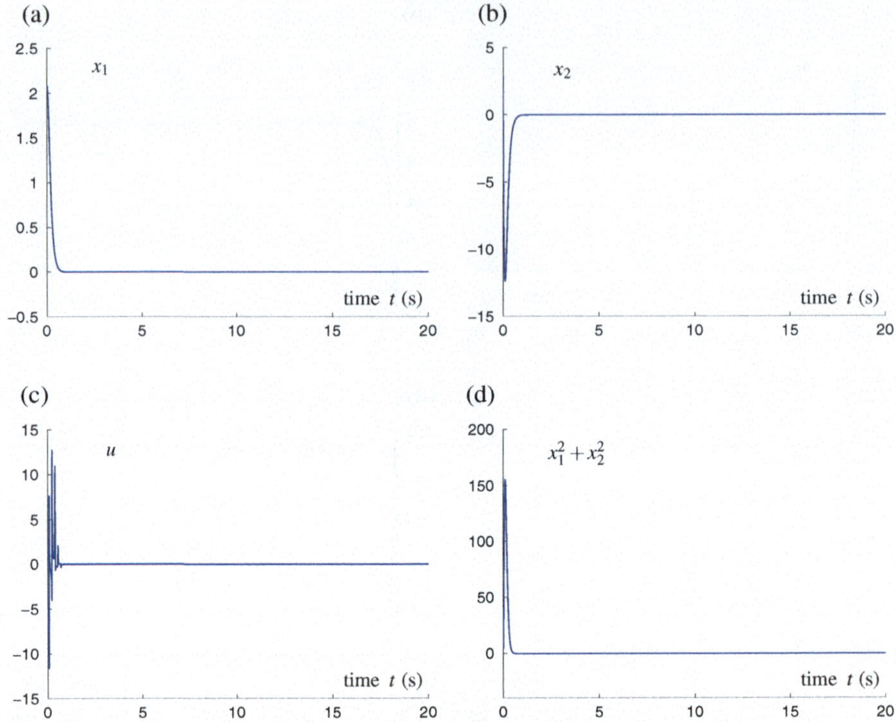

Fig. 16.5 Stabilization performance of TVL system (16.1) equipped with ZD controller (16.10) using $w = 20$ rad/s. (**a**) System state x_1. (**b**) System state x_2. (**c**) Control input. (**d**) Square sum of x_1 and x_2

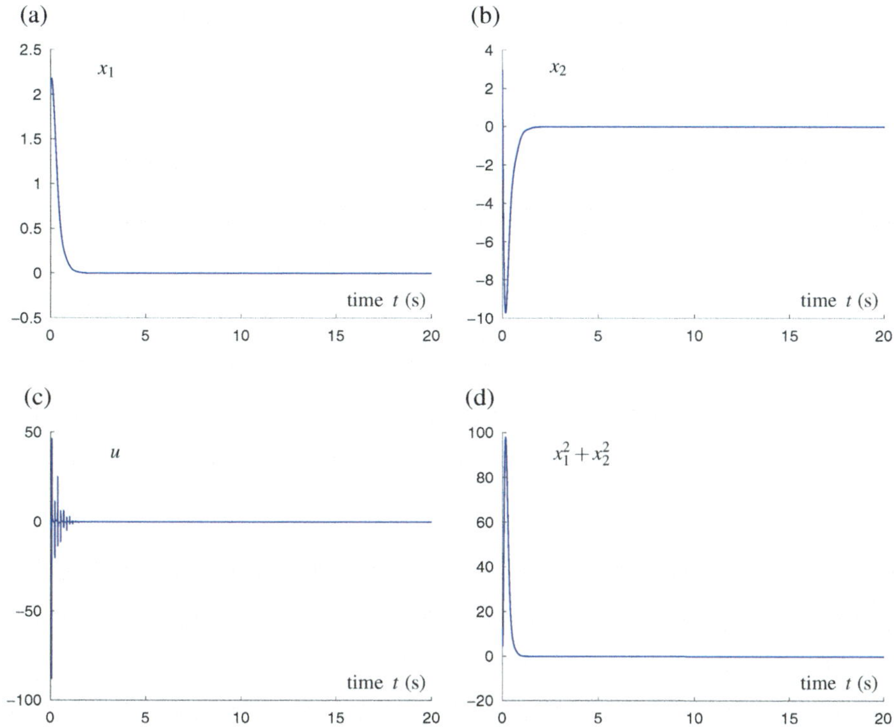

Fig. 16.6 Stabilization performance of TVL system (16.1) equipped with ZG controller (16.17) using $w = 20$ rad/s. (**a**) System state x_1. (**b**) System state x_2. (**c**) Control input. (**d**) Square sum of x_1 and x_2

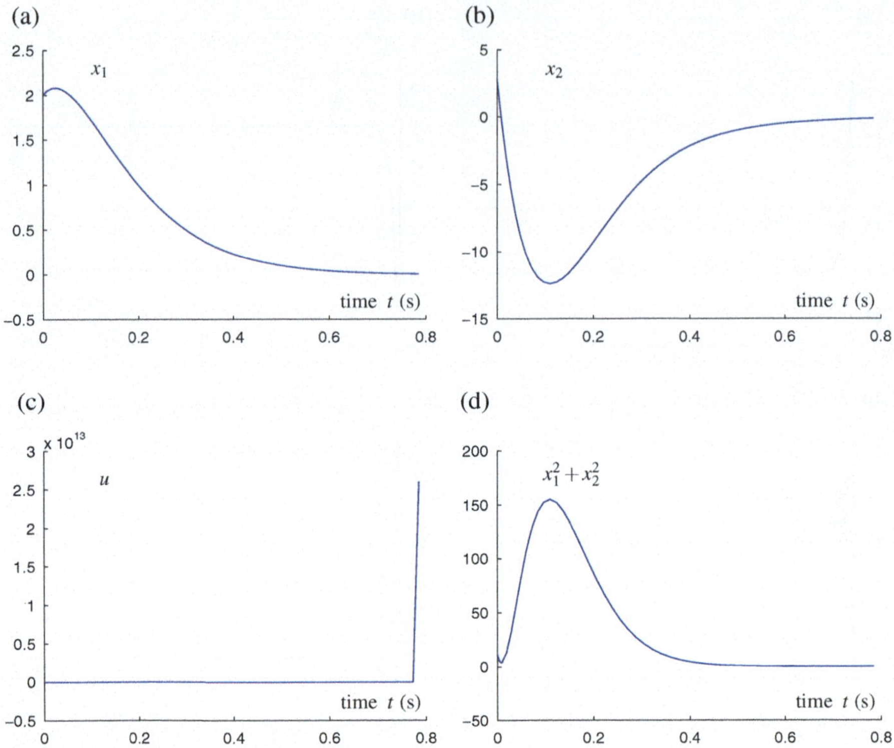

Fig. 16.7 Stabilization performance of TVL system (16.1) equipped with ZD controller (16.10) using $w = 2$ rad/s when specially simulated to DBZ time instant. (**a**) System state x_1. (**b**) System state x_2. (**c**) Control input. (**d**) Square sum of x_1 and x_2

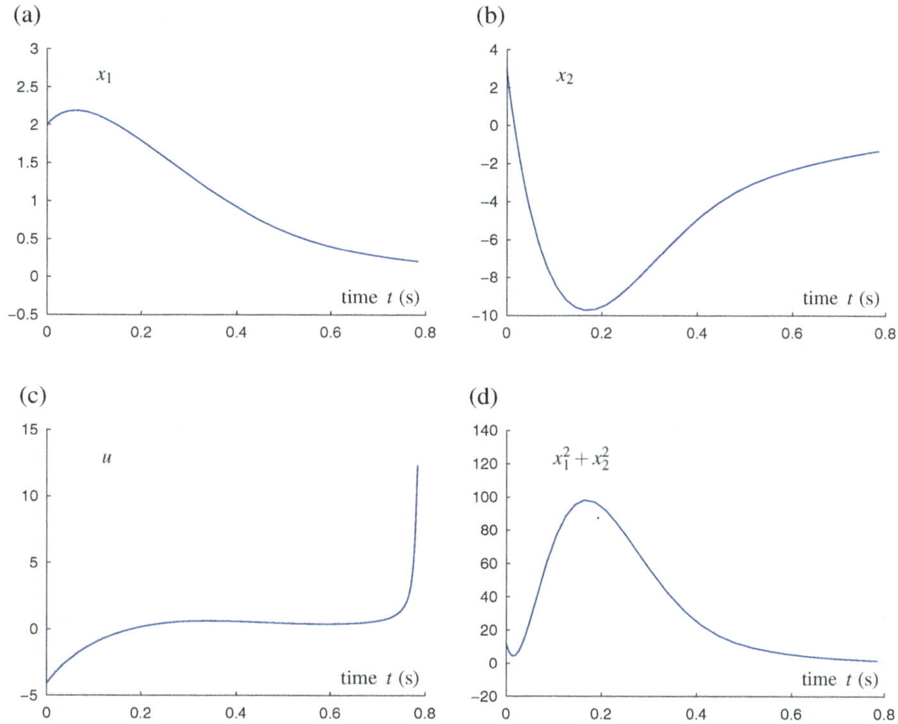

Fig. 16.8 Stabilization performance of TVL system (16.1) equipped with ZG controller (16.17) using $w = 2$ rad/s when specially simulated to DBZ time instant. (**a**) System state x_1. (**b**) System state x_2. (**c**) Control input. (**d**) Square sum of x_1 and x_2

16.6 Chapter Summary

In this chapter, the stabilization control of the TVL system has been investigated by applying the ZG method with PDBZ phenomenon shown and solved. By conducting the computer simulations, it can be concluded that the presented ZD controller can realize the stabilization of the TVL system in spite of the controller itself containing DBZ points, and that the presented ZG controller not only realizes the stabilization of the TVL system, but also solves the PDBZ problem contained in the ZD stabilization controller. Through the contrastive simulation analysis, we can discover that the ZD and ZG methods have similarities; namely, such two methods are simple and effective for achieving the stabilization control of the TVL system.

References

1. DaCunha JJ (2005) Stability for time varying linear dynamic systems on time scales. J Comput Appl Math 176(2):381–410
2. Zhang Y, Guo J, Qiu B, Li J, Yang Z (2017) Stabilization of time-varying linear system using ZD and ZG methods respectively with pseudo division-by-zero phenomenon shown. In: Proceedings of Chinese automation congress, pp 141–146
3. Zhang Y, Qiao T, Zhang D, Tan H, Liang D (2016) Simple effective Zhang-dynamics stabilization control of the 4th-order hyper-chaotic Lu system with one input. In: Proceedings of the 12th international conference on natural computation, fuzzy systems and knowledge discovery, pp 325–330
4. Zhang Y, Zhang Y, Yan X, Qiu B, Tan H (2015) ZG stabilization and tracking control for bilinear system of u-integration type. In: Proceedings of the 27th Chinese control and decision conference, pp 1262–1267
5. Zhang Y, Luo F, Yin Y, Liu J, Yu X (2013) Singularity-conquering ZG controller for output tracking of a class of nonlinear systems. In: Proceedings of the 32nd Chinese control conference, pp 477–482
6. Zhang Y, Zhang Y, Chen D, Xiao Z, Yan X (2017) Division by zero, pseudo-division by zero, Zhang dynamics method and Zhang-gradient method about control singularity conquering. Int J Syst Sci 48(1):1–12
7. Hwang Y, Chen M, Wu T (2003) Division controllers for homogeneous dyadic bilinear systems. IEEE Trans Autom Control 48(4):701–705
8. Huang J, Huang T, Tseng M (2015) Smooth switching robust adaptive control for omnidirectional mobile robots. IEEE Trans Control Syst Technol 23(5):1986–1993
9. Li J, Mao M, Zhang Y, Chen D, Yin Y (2017) ZD, ZG and IOL controllers and comparisons for nonlinear system output tracking with DBZ problem conquered in different relative-degree cases. Asian J Control 19(4):1–14
10. Li J, Mao M, Zhang Y (2017) Simpler ZD-achieving controller for chaotic systems synchronization with parameter perturbation, model uncertainty and external disturbance as compared with other controllers. Optik 131:364–373
11. Zhang Y, Qiu B, Liao B, Yang Z (2017) Control of pendulum tracking (including swinging up) of IPC system using zeroing-gradient method. Nonlinear Dyn 89(1):1–25
12. Zhang Y, Wang J, Chen D, Qiu B, Qiao T (2016) ZD controller for synchronization of Lu chaotic systems with one input. In: Proceedings of the 5th international conference on computer science and network technology, pp 816–819
13. Yin Y, Xie Q, Wang Y, Chen D, Zhang Y (2013) ZG control for ship course tracking with singularity considered and solved. In: Proceedings of the 11th international conference on dependable, autonomic and secure computing, pp 352–357

Chapter 17
ZG Output Tracking of TVL and TVN Systems

Abstract In this chapter, the ZG method is utilized to design ZD and ZG controllers for the output tracking of TVL and TVN systems. Particularly, the investigated TVL and TVN systems may both have PDBZ phenomena. From the simulation results, although the presented ZD and ZG controllers fulfill well the output tracking of TVL and TVN systems, the infinite value of the former and the finite value of the latter at DBZ time instants indicate that the ZG controller is more effective in dealing with the PDBZ problem.

17.1 Introduction

During the last several decades, much progress has been made in the research of output-tracking control problems, which is important for practical engineering applications [1–3]. In the previous studies, many approaches were focused on time-invariant systems, which are relatively simple [4]. However, most of practical systems are time-varying, meaning that we need to make an intensive study of time-varying systems. For the purpose of deeper research, many methods have been presented for solving the output-tracking control problem of time-varying system [5], such as the linear quadratic regulator [6] and the input-output linearization approach [7]. Nevertheless, many researchers do not handle the DBZ problem, or even pay no attention to the existence of this problem. In this chapter [8], we employ the ZD method for solving the output-tracking problems of time-varying systems, and we also adopt the GD method that can effectively conquer the DBZ problem [9, 10]. Then, the ZG method [11–15] can be applied to the output tracking of time-varying systems, especially TVN systems, which is capable of solving the DBZ problem [16–18]. In order to investigate the output-tracking control problem more comprehensively, we consider the TVL and TVN systems simultaneously, both of which are with PDBZ phenomena.

Y. Zhang et al., *Zhang-Gradient Control*,
https://doi.org/10.1007/978-981-15-8257-8_17

17.2 Design of Controllers for TVL System

In this section, based on the general description of TVL system [7], we put forward a specific TVL system as a case study, which is described as below:

$$\begin{cases} \dot{x} = f_1(x, u, t) = \sin(t)x + 10\cos(0.3t)u, \\ y = x \to y_d, \end{cases} \tag{17.1}$$

where x denotes the state variable; u denotes the control input; y denotes the actual output of the system; and y_d denotes the desired output.

17.2.1 Design of ZD Controller for TVL System

According to the ZD method, the output-tracking controller can be derived via the following procedure. We first define ZF $z_1 = y - y_d = x - y_d$ as the output-tracking error. To make z_1 converge to zero, the ZD design formula [9] is applied as below:

$$\dot{z}_1 = -\lambda z_1, \tag{17.2}$$

where the design parameter $\lambda \in \mathbb{R}^+$ is used to scale the convergence rate of the ZD solution. Substituting (17.1) into (17.2) yields $\sin(t)x + 10\cos(0.3t)u - \dot{y}_d = -\lambda(x - y_d)$, which can be rewritten as

$$\sin(t)x + 10\cos(0.3t)u - \dot{y}_d + \lambda(x - y_d) = 0. \tag{17.3}$$

Then, we have

$$u = \frac{-\lambda(x - y_d) - \sin(t)x + \dot{y}_d}{10\cos(0.3t)}. \tag{17.4}$$

Hence, we obtain ZD controller (17.4) of TVL system (17.1).

Particularly, the denominator of (17.4) equals zero when $t = 5(2k + 1)\pi/3\,\mathrm{s}$ (with $k = 0, 1, 2, \cdots$). Therefore, the values of ZD controller (17.4) are infinite at these DBZ time instants, which is termed DBZ problem. Notably, the denominator of (17.4) is the sole controller coefficient in (17.1), which means that the denominator will eventually be eliminated by substituting (17.4) into (17.1). This is a PDBZ situation; that is, ZD controller (17.4) may make the actual output of system (17.1) converge to the desired output, although (17.4) itself has DBZ time instants.

17.2.2 Design of ZG Controller for TVL System

According to Eq. (17.3) obtained by the ZD method, we can define

$$\phi_1 = \sin(t)x + 10\cos(0.3t)u - \dot{y}_\mathrm{d} + \lambda\big(x - y_\mathrm{d}\big),$$

which should theoretically be zero. Based on the GD method [10], an EF is defined as below:

$$\epsilon_1 = \frac{1}{2}\phi_1^2.$$

Then, we apply the GD design formula as follows:

$$\dot{u} = -\gamma\frac{\partial\epsilon_1}{\partial u},$$

where the design parameter $\gamma \in \mathbb{R}^+$ is used to scale the convergence rate of the GD solution. The ZG controller for system (17.1) is thus obtained as

$$\dot{u} = -10\gamma\cos(0.3t)\phi_1. \tag{17.5}$$

Compared with ZD controller (17.4), ZG controller (17.5) does not contain a division operation, which means that the ZG method is capable of effectively solving the DBZ problem (more specifically, the PDBZ problem).

17.3 Design of Controllers for TVN System

In the previous section, the derivations of ZD and ZG controllers are presented for the output tracking of TVL system (17.1). In this section, in order to present the tracking control problem more comprehensively and comparatively, we consider a specific TVN system as follows:

$$\begin{cases} \dot{x} = f_1(x, u, t) = \sin(t)x^2 + 10\cos(0.3t)u, \\ y = x \to y_\mathrm{d}. \end{cases} \tag{17.6}$$

17.3.1 Design of ZD Controller for TVN System

Similar to the previous section, based on the ZD method, ZF $z_2 = y - y_d = x - y_d$ is constructed. Then, adopting the ZD design formula, we obtain

$$\dot{z}_2 = -\lambda z_2,$$

which can be rewritten as

$$\sin(t)x^2 + 10\cos(0.3t)u - \dot{y}_d + \lambda(x - y_d) = 0. \tag{17.7}$$

Finally, the ZD controller designed for TVN system (17.6) is expressed as

$$u = \frac{-\lambda(x - y_d) - \sin(t)x^2 + \dot{y}_d}{10\cos(0.3t)}. \tag{17.8}$$

Note that the denominator of expression (17.8) equals zero at particular instants, also termed DBZ time instants, such as $t = 5(2k + 1)\pi/3\,\text{s}$ (with $k = 0, 1, 2, \cdots$), which will make the control input of ZD controller (17.8) infinite. Nevertheless, it is actually a PDBZ problem like ZD controller (17.4) designed for TVL system (17.1) in the previous section. The denominator will be eliminated by substituting (17.8) into (17.6). However, in terms of controller (17.8) itself, the value of u becomes infinite at these DBZ time instants.

17.3.2 Design of ZG Controller for TVN System

On the basis of Eq. (17.7) obtained by the ZD method, we define

$$\phi_2 = \sin(t)x^2 + 10\cos(0.3t)u - \dot{y}_d + \lambda\big(x - y_d\big),$$

which should theoretically be zero as well. Based on the GD method, an EF is defined as below:

$$\epsilon_2 = \frac{1}{2}\phi_2^2.$$

Applying the GD design formula, we have

$$\dot{u} = -\gamma\frac{\partial \epsilon_2}{\partial u}.$$

The expression of ZG controller for system (17.6) is obtained as

$$\dot{u} = -10\gamma \cos(0.3t)\phi_2. \tag{17.9}$$

Similar to ZG controller (17.5) for TVL system (17.1), ZG controller (17.9) for TVN system (17.6) does not contain a division operation. This implies that the ZG method can also solve the DBZ problem (more specifically, PDBZ problem) in the case of TVN system.

17.4 Simulation, Verification and Comparison

In this section, with different desired output trajectories for TVL system (17.1) and TVN system (17.6) considered, the computer simulations of ZD and ZG controllers designed for these two systems are conducted. The simulation results are shown in Figs. 17.1, 17.2, 17.3, and 17.4. Thereinto, the PDBZ phenomena can be observed from subfigures (c) and (d) of Figs. 17.1 and 17.3. Remarkably, through simulation and verification for the specific DBZ time instants, such as $t = 5\pi/3 \approx 5.236\,\text{s}$ (for $k = 0$), the comparisons between the ZD and ZG controllers are presented. It is substantiated that the ZG method has better capability of solving the PDBZ problem. Besides, in the simulations, the value of λ is set as 10, and the value of γ is set as 10^5. In what follows, $|e| = |y - y_{\mathrm{d}}| = |x - y_{\mathrm{d}}|$ represents the absolute tracking error.

Example 17.1 In this example, the simulation results synthesized by the ZD and ZG controllers for TVL system (17.1) are presented, with the desired output trajectory selected as $y_{\mathrm{d}} = \cos(t) + \sin(2t) + 2$. The corresponding simulation results are shown in Figs. 17.1 and 17.2, respectively. As shown in Figs. 17.1a and 17.2a, the actual output trajectory for system (17.1) coincides well with the desired output trajectory. These validate that ZD controller (17.4) and ZG controller (17.5) can both fulfill the tracking control task for TVL system (17.1), with the PDBZ problem exists in the former. Furthermore, Figs. 17.1b and 17.2b show the absolute errors $|e|$ of output tracking, respectively, by using the ZD and ZG controllers. Besides, the control inputs of the ZD and ZG controllers are shown in Figs. 17.1c and 17.2c, respectively. Moreover, in Fig. 17.1d, the corresponding value of ZD controller (17.4) at the DBZ time instant is of order 10^{15}. By contrast, at the same time instant, the corresponding value of ZG controller (17.5) shown in Fig. 17.2d is within the interval $(35, 40)$, which is acceptable in practice.

Example 17.2 In this example, we show the simulation results synthesized by the ZD and ZG controllers for TVN system (17.6), with the desired output trajectory $y_{\mathrm{d}} = \exp(\sin(t))$ considered. The corresponding simulation results of ZD controller (17.8) and ZG controller (17.9) are shown in Figs. 17.3 and 17.4, respectively. As illustrated in Figs. 17.3a and 17.4a, the actual output trajectory of TVN system (17.6) coincides well with the desired output trajectory. These

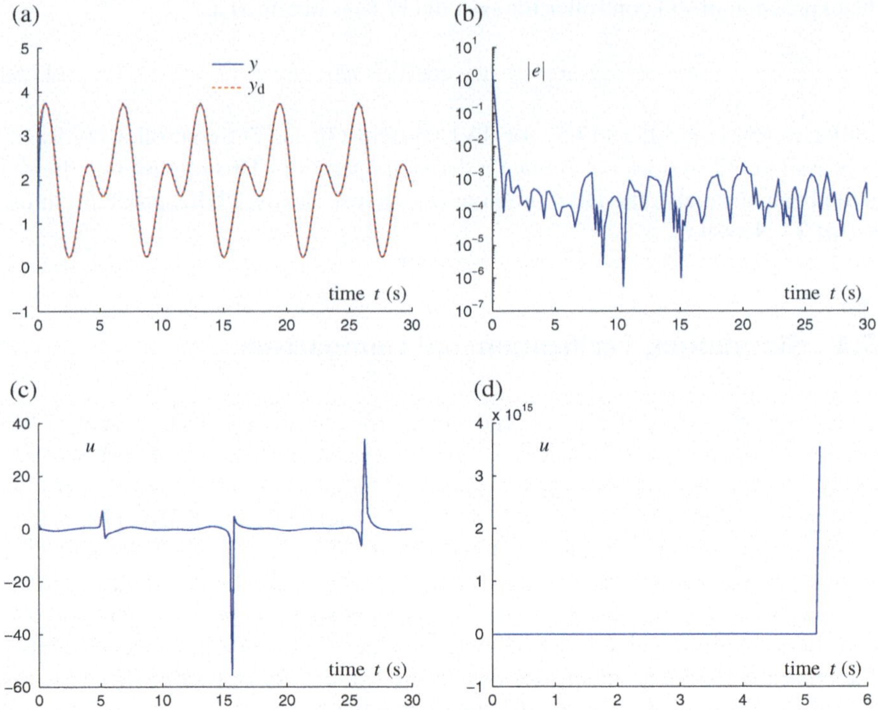

Fig. 17.1 Tracking performance of TVL system (17.1) equipped with ZD controller (17.4) for desired trajectory $y_{\mathrm{d}} = \cos(t) + \sin(2t) + 2$. (**a**) Output trajectory and desired trajectory. (**b**) Absolute tracking error. (**c**) Control input. (**d**) Control input from beginning to DBZ time instant $t \approx 5.236$ s

substantiate that ZD controller (17.8) and ZG controller (17.9) can both fulfill the tracking control task for TVN system (17.6). Moreover, the absolute errors $|e|$ of ZD and ZG controllers for output tracking are shown in Figs. 17.3b and 17.4b, respectively. The control inputs of ZD and ZG controllers are shown in Figs. 17.3c and 17.4c, respectively. As seen from Fig. 17.3d, the corresponding value of ZD controller (17.8) at the DBZ time instant is of order 10^{14}. By contrast, at the same time instant, the corresponding value of ZG controller (17.9) shown in Fig. 17.4d is within the interval (10, 12), which is practically acceptable.

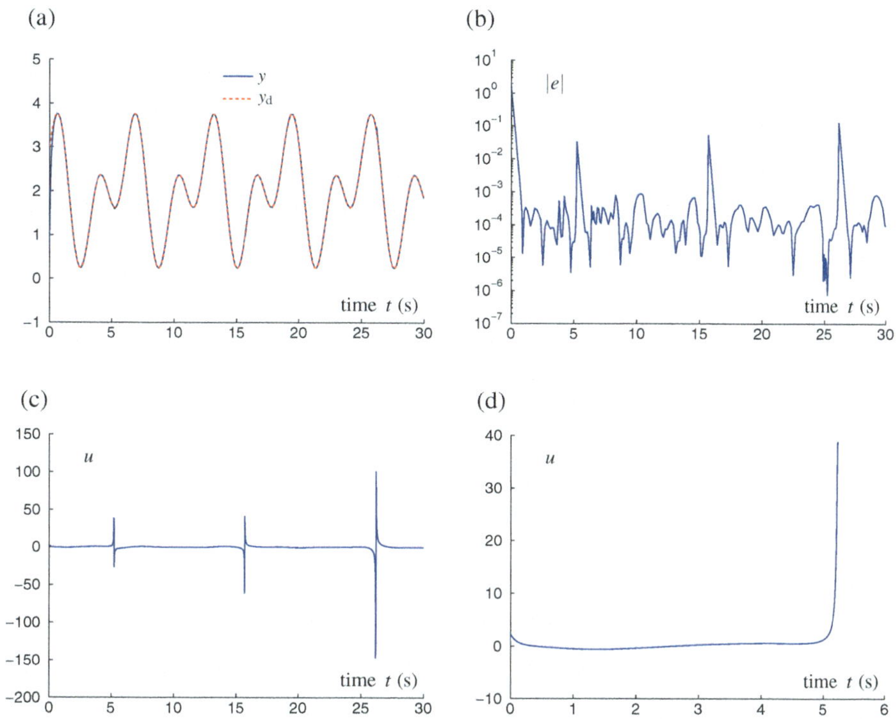

Fig. 17.2 Tracking performance of TVL system (17.1) equipped with ZG controller (17.5) for desired trajectory $y_d = \cos(t) + \sin(2t) + 2$. (**a**) Output trajectory and desired trajectory. (**b**) Absolute tracking error. (**c**) Control input. (**d**) Control input from beginning to DBZ time instant $t \approx 5.236$ s

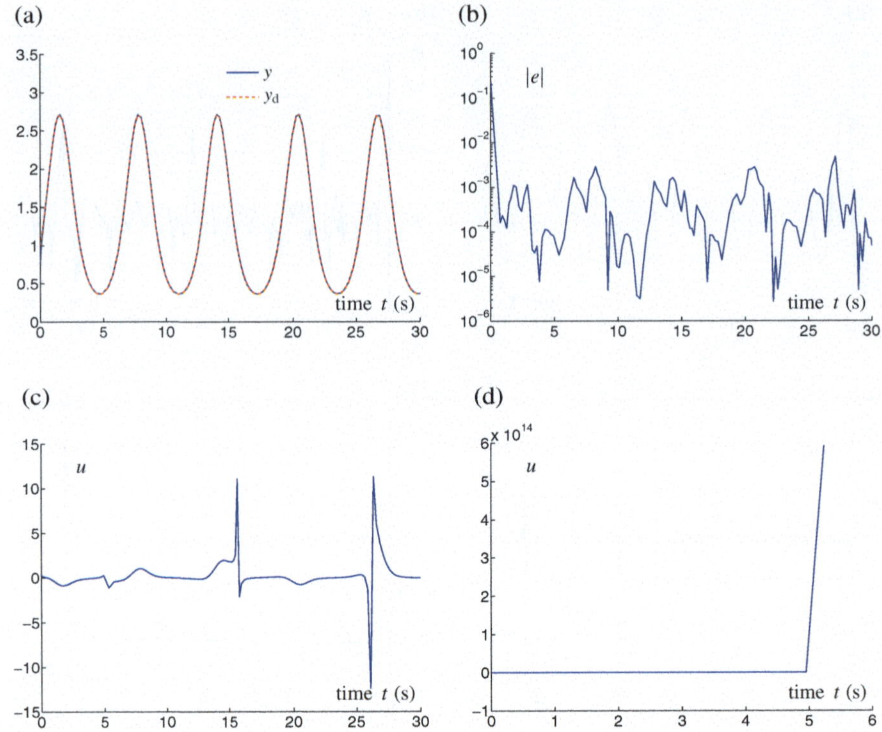

Fig. 17.3 Tracking performance of TVN system (17.6) equipped with ZD controller (17.8) for desired trajectory $y_d = \exp(\sin(t))$. (**a**) Output trajectory and desired trajectory. (**b**) Absolute tracking error. (**c**) Control input. (**d**) Control input from beginning to DBZ time instant $t \approx 5.236$ s

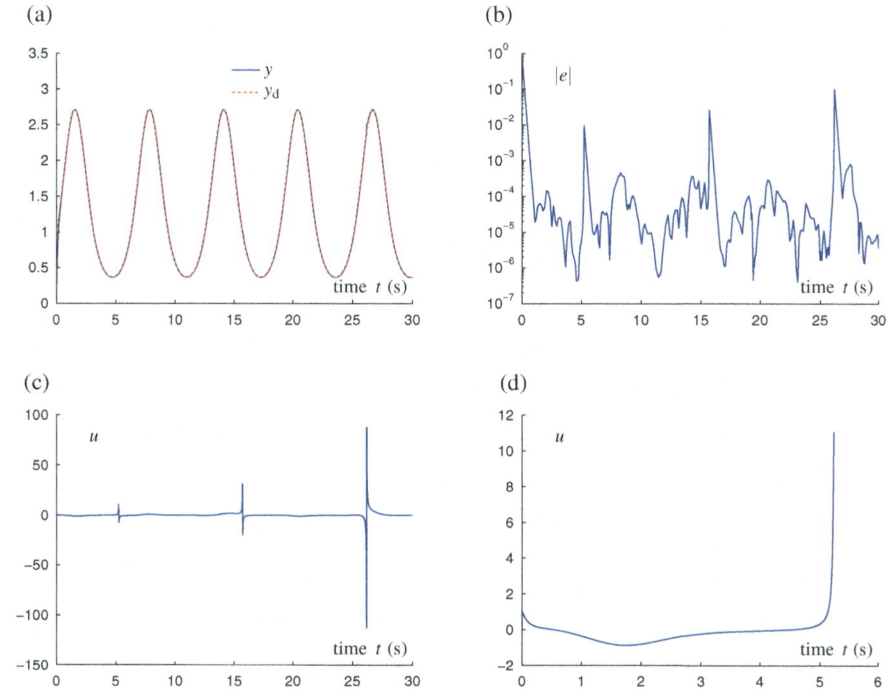

Fig. 17.4 Tracking performance of TVN system (17.6) equipped with ZG controller (17.9) for desired trajectory $y_d = \exp(\sin(t))$. (**a**) Output trajectory and desired trajectory. (**b**) Absolute tracking error. (**c**) Control input. (**d**) Control input from beginning to DBZ time instant $t \approx 5.236$ s

17.5 Chapter Summary

In this chapter, we have designed and investigated the ZD and ZG controllers for the output tracking of the TVL and TVN systems simultaneously and comparatively. The simulation results have substantiated that both the ZD and ZG controllers can solve the tracking control problem. The difference between the ZD and ZG controllers has been illustrated by the simulation at the DBZ time instant. The values of ZD controllers at the DBZ time instants are infinite. In contrast, the values of ZG controllers at the same time instants are finite, which has verified that the ZG controllers have better capability for dealing with the PDBZ problem.

References

1. Gao H, Chen T (2008) Network-based H_∞ output tracking control. IEEE Trans Autom Control 53(3):655–667
2. Yin Y, Xie Q, Wang Y, Chen D, Zhang Y (2013) ZG control for ship course tracking with singularity considered and solved. In: Proceedings of the 11th IEEE international conference on dependable, autonomic and secure computing, pp 352–357
3. Ye D, Yang G (2006) Adaptive fault-tolerant tracking control against actuator faults with application to flight control. IEEE Trans Control Syst Technol 14(6):1088–1096
4. Zhang Y, Zhang D, Qiu B, Wang J, Li J (2016) Sigmoid function aided Zhang dynamics control for output tracking of time-varying linear system with bounded input. In: Proceedings of the 35th Chinese control conference, pp 5901–5905
5. Jiang Z-P, Lin Y, Wang Y (2008) Stabilization of time-varying nonlinear systems: a control Lyapunov function approach. In: Proceedings of IEEE international conference on control and automation, pp 404–409
6. Zheng D (2002) Linear system theory. Tsinghua University Press, Beijing
7. Jo NH, Seo JH (2000) Input output linearization approach to state observer design for nonlinear systems. IEEE Trans Autom Control 45(12):2388–2393
8. Guo J, Yang X, Zhang Y, Li J, Yang M (2018) Output tracking of time-varying linear and nonlinear systems using ZN and ZG controllers with pseudo division-by-zero phenomena shown. In: Proceedings of the 30th Chinese control and decision conference, pp 4505–4510
9. Zhang Y, Yi C, Guo D, Zheng J (2011) Comparison on Zhang neural dynamics and gradient-based neural dynamics for online solution of nonlinear time-varying equation. Neural Comput Appl 20(1):1–7
10. Zhang Y, Yu X, Yin Y, Xiao L, Fan Z (2013) Using GD to conquer the singularity problem of conventional controller for output tracking of nonlinear system of a class. Phys Lett A 377(25–27):1611–1614
11. Zhang Y, Luo F, Yin Y, Liu J, Yu X (2013) Singularity-conquering ZG controller for output tracking of a class of nonlinear systems. In: Proceedings of the 32nd Chinese control conference, pp 477–482
12. Zhang Y, Liu M, Jin L, Zhang Y, Tan H (2015) Synchronization of two chaotic systems with three or two inputs via ZG method. In: Proceedings of the 34th Chinese control conference, pp 563–568
13. Zhang Y, Zhai K, Chen D, Jin L, Hu C (2016) Challenging simulation practice (failure and success) on implicit tracking control of double-integrator system via Zhang-gradient method. Math Comput Simul 120:104–119
14. Zhang Y, Zhai K, Wang Y, Chen D, Peng C (2014) Design and illustration of ZG controllers for linear and nonlinear tracking control of double-integrator system. In: Proceedings of the 33rd Chinese control conference, pp 3462–3467
15. Zhang Y, Yin Y, Wu H, Guo D (2013) Zhang dynamics and gradient dynamics with tracking-control application. In: Proceedings of the 5th international symposium on computational intelligence and design, pp 235–238
16. Zhang Y, Liu J, Yin Y, Luo F, Deng J (2013) Zhang-gradient controllers of Z0G0, Z1G0 and Z1G1 types for output tracking of time-varying linear system with control-singularity conquered finally. In: Proceedings of the 10th international symposium on neural networks, pp 533–540
17. Zhang Y, Peng C, Yu X, Yin Y, Ling Y (2013) ZD and ZG controllers for explicit and implicit tracking of pendulum with singularity finally conquered. In: Proceedings of the 12th international conference on machine learning and cybernetics, pp 777–782
18. Zhang Y, Yu X, Yin Y, Peng C, Fan Z (2014) Singularity-conquering ZG controllers of z2g1 Type for tracking control of the IPC system. Int J Control 87(9):1729–1746

Index

Y. Zhang et al., *Zhang-Gradient Control*,
https://doi.org/10.1007/978-981-15-8257-8

The manufacturer's authorised representative in the EU is Springer Nature Customer Service Centre GmbH, Europaplatz 3, 69115 Heidelberg, Germany. If you have any concerns regarding our products, please contact ProductSafety@springernature.com

Printed and bound by CPI Group (UK) Ltd, Croydon, CR0 4YY
27/07/2026
02177376-0003